T0207146

The Art of Reinforcement Learning

Michael Hu

The Art of Reinforcement Learning

Fundamentals, Mathematics, and Implementations with Python

Apress®

Michael Hu
Shanghai, Shanghai, China

ISBN-13 (pbk): 978-1-4842-9605-9   ISBN-13 (electronic): 978-1-4842-9606-6
https://doi.org/10.1007/978-1-4842-9606-6

Managing Director, Apress Media LLC: Welmoed Spahr
Acquisitions Editor: Celestin Suresh John
Development Editor: Laura Berendson
Editorial Assistant: Gryffin Winkler

Cover designed by eStudioCalamar

Cover image designed by Freepik (www.freepik.com)

Distributed to the book trade worldwide by Springer Science+Business Media New York, 1 New York Plaza, Suite 4600, New York, NY 10004-1562, USA. Phone 1-800-SPRINGER, fax (201) 348-4505, e-mail orders-ny@springer-sbm.com, or visit www.springeronline.com. Apress Media, LLC is a California LLC and the sole member (owner) is Springer Science + Business Media Finance Inc (SSBM Finance Inc). SSBM Finance Inc is a **Delaware** corporation.

For information on translations, please e-mail booktranslations@springernature.com; for reprint, paperback, or audio rights, please e-mail bookpermissions@springernature.com.

Apress titles may be purchased in bulk for academic, corporate, or promotional use. eBook versions and licenses are also available for most titles. For more information, reference our Print and eBook Bulk Sales web page at http://www.apress.com/bulk-sales.

Any source code or other supplementary material referenced by the author in this book is available to readers on GitHub (https://github.com/Apress). For more detailed information, please visit https://www.apress.com/gp/services/source-code.

Paper in this product is recyclable

To my beloved family,

This book is dedicated to each of you, who have been a constant source of love and support throughout my writing journey.

To my hardworking parents, whose tireless efforts in raising us have been truly remarkable. Thank you for nurturing my dreams and instilling in me a love for knowledge. Your unwavering dedication has played a pivotal role in my accomplishments.

To my sisters and their children, your presence and love have brought immense joy and inspiration to my life. I am grateful for the laughter and shared moments that have sparked my creativity.

And to my loving wife, your consistent support and understanding have been my guiding light. Thank you for standing by me through the highs and lows, and for being my biggest cheerleader.

—Michael Hu

Contents

About the Author

Michael Hu is an exceptional software engineer with a wealth of expertise spanning over a decade, specializing in the design and implementation of enterprise-level applications. His current focus revolves around leveraging the power of machine learning (ML) and artificial intelligence (AI) to revolutionize operational systems within enterprises. A true coding enthusiast, Michael finds solace in the realms of mathematics and continuously explores cutting-edge technologies, particularly machine learning and deep learning. His unwavering passion lies in the realm of deep reinforcement learning, where he constantly seeks to push the boundaries of knowledge. Demonstrating his commitment to the field, he has built various numerous open source projects on GitHub that closely emulate state-of-the-art reinforcement learning algorithms pioneered by DeepMind, including notable examples like AlphaZero, MuZero, and Agent57. Through these projects, Michael demonstrates his commitment to advancing the field and sharing his knowledge with fellow enthusiasts. He currently resides in the city of Shanghai, China.

About the Technical Reviewer

Shovon Sengupta has over 14 years of expertise and a deepened understanding of advanced predictive analytics, machine learning, deep learning, and reinforcement learning. He has established a place for himself by creating innovative financial solutions that have won numerous awards. He is currently working for one of the leading multinational financial services corporations in the United States as the Principal Data Scientist at the AI Center of Excellence. His job entails leading innovative initiatives that rely on artificial intelligence to address challenging business problems. He has a US patent (United States Patent: Sengupta et al.: Automated Predictive Call Routing Using Reinforcement Learning [US 10,356,244 B1]) to his credit. He is also a Ph.D. scholar at BITS Pilani. He has reviewed quite a few popular titles from leading publishers like Packt and Apress and has also authored a few courses for Packt and CodeRed (EC-Council) in the realm of machine learning. Apart from that, he has presented at various international conferences on machine learning, time series forecasting, and building trustworthy AI. His primary research is concentrated on deep reinforcement learning, deep learning, natural language processing (NLP), knowledge graph, causality analysis, and time series analysis. For more details about Shovon's work, please check out his LinkedIn page: www.linkedin.com/in/shovon-sengupta-272aa917.

Preface

Reinforcement learning (RL) is a highly promising yet challenging subfield of artificial intelligence (AI) that plays a crucial role in shaping the future of intelligent systems. From robotics and autonomous agents to recommendation systems and strategic decision-making, RL enables machines to learn and adapt through interactions with their environment. Its remarkable success stories include RL agents achieving human-level performance in video games and even surpassing world champions in strategic board games like Go. These achievements highlight the immense potential of RL in solving complex problems and pushing the boundaries of AI.

What sets RL apart from other AI subfields is its fundamental approach: agents learn by interacting with the environment, mirroring how humans acquire knowledge. However, RL poses challenges that distinguish it from other AI disciplines. Unlike methods that rely on precollected training data, RL agents generate their own training samples. These agents are not explicitly instructed on how to achieve a goal; instead, they receive state representations of the environment and a reward signal, forcing them to explore and discover optimal strategies on their own. Moreover, RL involves complex mathematics that underpin the formulation and solution of RL problems.

While numerous books on RL exist, they typically fall into two categories. The first category emphasizes the fundamentals and mathematics of RL, serving as reference material for researchers and university students. However, these books often lack implementation details. The second category focuses on practical hands-on coding of RL algorithms, neglecting the underlying theory and mathematics. This apparent gap between theory and implementation prompted us to create this book, aiming to strike a balance by equally emphasizing fundamentals, mathematics, and the implementation of successful RL algorithms.

This book is designed to be accessible and informative for a diverse audience. It is targeted toward researchers, university students, and practitioners seeking a comprehensive understanding of RL. By following a structured approach, the book equips readers with the necessary knowledge and tools to apply RL techniques effectively in various domains.

The book is divided into four parts, each building upon the previous one. Part I focuses on the fundamentals and mathematics of RL, which form the foundation for almost all discussed algorithms. We begin by solving simple RL problems using tabular methods. Chapter 2, the cornerstone of this part, explores Markov decision processes (MDPs) and the associated value functions, which are recurring concepts throughout the book. Chapters 3 to 5 delve deeper into these fundamental concepts by discussing how to use dynamic programming (DP), Monte Carlo methods, and temporal difference (TD) learning methods to solve small MDPs.

Part II tackles the challenge of solving large-scale RL problems that render tabular methods infeasible due to their complexity (e.g., large or infinite state spaces). Here, we shift our focus to value function approximation, with particular emphasis on leveraging (deep) neural networks. Chapter 6 provides a brief introduction to linear value function approximation, while Chap. 7 delves into the renowned Deep Q-Network (DQN) algorithm. In Chap. 8, we discuss enhancements to the DQN algorithm.

Part III explores policy-based methods as an alternative approach to solving RL problems. While Parts I and II primarily focus on value-based methods (learning the value function), Part III concentrates on learning the policy directly. We delve into the theory behind policy gradient methods and the REINFORCE algorithm in Chap. 9. Additionally, we explore Actor-Critic algorithms, which combine policy-based and value-based approaches, in Chap. 10. Furthermore, Chap. 11 covers advanced policy-based algorithms, including surrogate objective functions and the renowned Proximal Policy Optimization (PPO) algorithm.

The final part of the book addresses advanced RL topics. Chapter 12 discusses how distributed RL can enhance agent performance, while Chap. 13 explores the challenges of hard-to-explore RL problems and presents curiosity-driven exploration as a potential solution. In the concluding chapter, Chap. 14, we delve into model-based RL by providing a comprehensive examination of the famous AlphaZero algorithm.

Unlike a typical hands-on coding handbook, this book does not primarily focus on coding exercises. Instead, we dedicate our resources and time to explaining the fundamentals and core ideas behind each algorithm. Nevertheless, we provide complete source code for all examples and algorithms discussed in the book. Our code implementations are done from scratch, without relying on third-party RL libraries, except for essential tools like Python, OpenAI Gym, Numpy, and the PyTorch deep learning framework. While third-party RL libraries expedite the implementation process in real-world scenarios, we believe coding each algorithm independently is the best approach for learning RL fundamentals and mastering the various RL algorithms.

Throughout the book, we employ mathematical notations and equations, which some readers may perceive as heavy. However, we prioritize intuition over rigorous proofs, making the material accessible to a broader audience. A foundational understanding of calculus at a basic college level, minimal familiarity with linear algebra, and elementary knowledge of probability and statistics are sufficient to embark on this journey. We strive to ensure that interested readers from diverse backgrounds can benefit from the book's content.

We assume that readers have programming experience in Python since all the source code is written in this language. While we briefly cover the basics of deep learning in Chap. 7, including neural networks and their workings, we recommend some prior familiarity with machine learning, specifically deep learning concepts such as training a deep neural network. However, beyond the introductory coverage, readers can explore additional resources and materials to expand their knowledge of deep learning.

This book draws inspiration from Reinforcement Learning: An Introduction by Richard S. Sutton and Andrew G. Barto, a renowned RL publication. Additionally, it is influenced by prestigious university RL courses, particularly the mathematical style and notation derived from Professor Emma Brunskill's RL course at Stanford University. Although our approach may differ slightly from Sutton and Barto's work, we strive to provide simpler explanations. Additionally, we have derived some examples from Professor David Silver's RL course at University College London, which offers a comprehensive resource for understanding the fundamentals presented in Part I. We would like to express our gratitude to Professor Dimitri P. Bertsekas for his invaluable guidance and inspiration in the field of optimal control and reinforcement learning. Furthermore, the content of this book incorporates valuable insights from research papers published by various organizations and individual researchers.

In conclusion, this book aims to bridge the gap between the fundamental concepts, mathematics, and practical implementation of RL algorithms. By striking a balance between theory and implementation, we provide readers with a comprehensive understanding of RL, empowering them to apply these techniques in various domains. We present the necessary mathematics and offer complete source code for implementation to help readers gain a deep understanding of RL principles. We hope this book serves as a valuable resource for readers seeking to explore the fundamentals, mathematics, and practical aspects of RL algorithms. We must acknowledge that despite careful editing from our editors and multiple round of reviews, we cannot guarantee the book's content is error free. Your feedback and corrections are invaluable to us. Please do not hesitate to contact us with any concerns or suggestions for improvement.

Source Code

You can download the source code used in this book from github.com/apress/art-of-reinforcement-lear ning.

Michael Hu

Introduction

1

Artificial intelligence has made impressive progress in recent years, with breakthroughs achieved in areas such as image recognition, natural language processing, and playing games. In particular, reinforcement learning, a type of machine learning that focuses on learning by interacting with an environment, has led to remarkable achievements in the field.

In this book, we focus on the combination of reinforcement learning and deep neural networks, which have become central to the success of agents that can master complex games such as board game Go and Atari video games.

This first chapter provides an overview of reinforcement learning, including key concepts such as states, rewards, policies, and the common terms used in reinforcement learning, like the difference between episodic and continuing reinforcement learning problems, model-free vs. model-based methods.

Despite the impressive progress in the field, reinforcement learning still faces significant challenges. For example, it can be difficult to learn from sparse rewards, and the methods can suffer from instability. Additionally, scaling to large state and action spaces can be a challenge.

Throughout this book, we will explore these concepts in greater detail and discuss state-of-the-art techniques used to address these challenges. By the end of this book, you will have a comprehensive understanding of the principles of reinforcement learning and how they can be applied to real-world problems.

We hope this introduction has sparked your curiosity about the potential of reinforcement learning, and we invite you to join us on this journey of discovery.

1.1 AI Breakthrough in Games

Atari

The Atari 2600 is a home video game console developed by Atari Interactive, Inc. in the 1970s. It features a collection of iconic video games. These games, such as Pong, Breakout, Space Invaders, and Pac-Man, have become classic examples of early video gaming culture. In this platform, players can interact with these classic games using a joystick controller.

Fig. 1.1 A DQN agent learning to play Atari's Breakout. The goal of the game is to use a paddle to bounce a ball up and break through a wall of bricks. The agent only takes in the raw pixels from the screen, and it has to figure out what's the right action to take in order to maximize the score. Idea adapted from Mnih et al. [1]. Game owned by Atari Interactive, Inc.

The breakthrough in Atari games came in 2015 when Mnih et al. [1] from DeepMind developed an AI agent called DQN to play a list of Atari video games, some even better than humans.

What makes the DQN agent so influential is how it was trained to play the game. Similar to a human player, the agent was only given the raw pixel image of the screen as inputs, as illustrated in Fig. 1.1, and it has to figure out the rules of the game all by itself and decide what to do during the game to maximize the score. No human expert knowledge, such as predefined rules or sample games of human play, was given to the agent.

The DQN agent is a type of reinforcement learning agent that learns by interacting with an environment and receiving a reward signal. In the case of Atari games, the DQN agent receives a score for each action it takes.

Mnih et al. [1] trained and tested their DQN agents on 57 Atari video games. They trained one DQN agent for one Atari game, with each agent playing only the game it was trained on; the training was over millions of frames. The DQN agent can play half of the games (30 of 57 games) at or better than a human player, as shown by Mnih et al. [1]. This means that the agent was able to learn and develop strategies that were better than what a human player could come up with.

Since then, various organizations and researchers have made improvements to the DQN agent, incorporating several new techniques. The Atari video games have become one of the most used test beds for evaluating the performance of reinforcement learning agents and algorithms. The Arcade Learning Environment (ALE) [2], which provides an interface to hundreds of Atari 2600 game environments, is commonly used by researchers for training and testing reinforcement learning agents.

In summary, the Atari video games have become a classic example of early video gaming culture, and the Atari 2600 platform provides a rich environment for training agents in the field of reinforcement learning. The breakthrough of DeepMind's DQN agent, trained and tested on 57 Atari video games, demonstrated the capability of an AI agent to learn and make decisions through trial-and-error interactions with classic games. This breakthrough has spurred many improvements and advancements in the field of reinforcement learning, and the Atari games have become a popular test bed for evaluating the performance of reinforcement learning algorithms.

Go

Go is an ancient Chinese strategy board game played by two players, who take turns laying pieces of stones on a 19x19 board with the goal of surrounding more territory than the opponent. Each player

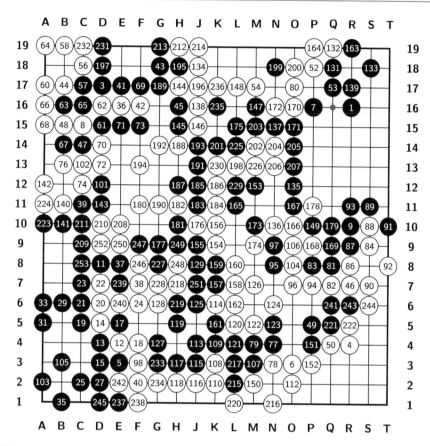

Fig. 1.2 Yoda Norimoto (black) vs. Kiyonari Tetsuya (white), Go game from the 66th NHK Cup, 2018. White won by 0.5 points. Game record from CWI [4]

has a set of black or white stones, and the game begins with an empty board. Players alternate placing stones on the board, with the black player going first.

The stones are placed on the intersections of the lines on the board, rather than in the squares. Once a stone is placed on the board, it cannot be moved, but it can be captured by the opponent if it is completely surrounded by their stones. Stones that are surrounded and captured are removed from the board.

The game continues until both players pass, at which point the territory on the board is counted. A player's territory is the set of empty intersections that are completely surrounded by their stones, plus any captured stones. The player with the larger territory wins the game. In the case of the final board position shown in Fig. 1.2, the white won by 0.5 points.

Although the rules of the game are relatively simple, the game is extremely complex. For instance, the number of legal board positions in Go is enormously large compared to Chess. According to research by Tromp and Farnebäck [3], the number of legal board positions in Go is approximately 2.1×10^{170}, which is vastly greater than the number of atoms in the universe.

This complexity presents a significant challenge for artificial intelligence (AI) agents that attempt to play Go. In March 2016, an AI agent called AlphaGo developed by Silver et al. [5] from DeepMind made history by beating the legendary Korean player Lee Sedol with a score of 4-1 in Go. Lee Sedol is a winner of 18 world titles and is considered one of the greatest Go player of the past decade.

AlphaGo's victory was remarkable because it used a combination of deep neural networks and tree search algorithms, as well as the technique of reinforcement learning.

AlphaGo was trained using a combination of supervised learning from human expert games and reinforcement learning from games of self-play. This training enabled the agent to develop creative and innovative moves that surprised both Lee Sedol and the Go community.

The success of AlphaGo has sparked renewed interest in the field of reinforcement learning and has demonstrated the potential for AI to solve complex problems that were once thought to be the exclusive domain of human intelligence. One year later, Silver et al. [6] from DeepMind introduced a new and more powerful agent, AlphaGo Zero. AlphaGo Zero was trained using pure self-play, without any human expert moves in its training, achieving a higher level of play than the previous AlphaGo agent. They also made other improvements like simplifying the training processes.

To evaluate the performance of the new agent, they set it to play games against the exact same AlphaGo agent that beat the world champion Lee Sedol in 2016, and this time the new AlphaGo Zero beats AlphaGo with score 100-0.

In the following year, Schrittwieser et al. [7] from DeepMind generalized the AlphaGo Zero agent to play not only Go but also other board games like Chess and Shogi (Japanese chess), and they called this generalized agent AlphaZero. AlphaZero is a more general reinforcement learning algorithm that can be applied to a variety of board games, not just Go, Chess, and Shogi.

Reinforcement learning is a type of machine learning in which an agent learns to make decisions based on the feedback it receives from its environment. Both DQN and AlphaGo (and its successor) agents use this technique, and their achievements are very impressive. Although these agents are designed to play games, this does not mean that reinforcement learning is only capable of playing games. In fact, there are many more challenging problems in the real world, such as navigating a robot, driving an autonomous car, and automating web advertising. Games are relatively easy to simulate and implement compared to these other real-world problems, but reinforcement learning has the potential to be applied to a wide range of complex challenges beyond game playing.

1.2   What Is Reinforcement Learning

In computer science, reinforcement learning is a subfield of machine learning that focuses on learning how to act in a world or an environment. The goal of reinforcement learning is for an agent to learn from interacting with the environment in order to make a sequence of decisions that maximize accumulated reward in the long run. This process is known as *goal-directed learning*.

Unlike other machine learning approaches like supervised learning, reinforcement learning does not rely on labeled data to learn from. Instead, the agent must learn through trial and error, without being directly told the rules of the environment or what action to take at any given moment. This makes reinforcement learning a powerful tool for modeling and solving real-world problems where the rules and optimal actions may not be known or easily determined.

Reinforcement learning is not limited to computer science, however. Similar ideas are studied in other fields under different names, such as operations research and optimal control in engineering. While the specific methods and details may vary, the underlying principles of goal-directed learning and decision-making are the same.

Examples of reinforcement learning in the real world are all around us. Human beings, for example, are naturally good at learning from interacting with the world around us. From learning to walk as a baby to learning to speak our native language to learning to drive a car, we learn through trial and error and by receiving feedback from the environment. Similarly, animals can also be trained to perform a variety of tasks through a process similar to reinforcement learning. For instance, service dogs can be trained to assist individuals in wheelchairs, while police dogs can be trained to help search for missing people.

One vivid example that illustrates the idea of reinforcement learning is a video of a dog with a big stick in its mouth trying to cross a narrow bridge.[1] The video shows the dog attempting to pass the bridge, but failing multiple times. However, after some trial and error, the dog eventually discovers that by tilting its head, it can pass the bridge with its favorite stick. This simple example demonstrates the power of reinforcement learning in solving complex problems by learning from the environment through trial and error.

1.3 Agent-Environment in Reinforcement Learning

Reinforcement learning is a type of machine learning that focuses on how an agent can learn to make optimal decisions by interacting with an environment. The agent-environment loop is the core of reinforcement learning, as shown in Fig. 1.3. In this loop, the agent observes the state of the environment and a reward signal, takes an action, and receives a new state and reward signal from the environment. This process continues iteratively, with the agent learning from the rewards it receives and adjusting its actions to maximize future rewards.

Environment

The environment is the world in which the agent operates. It can be a physical system, such as a robot navigating a maze, or a virtual environment, such as a game or a simulation. The environment

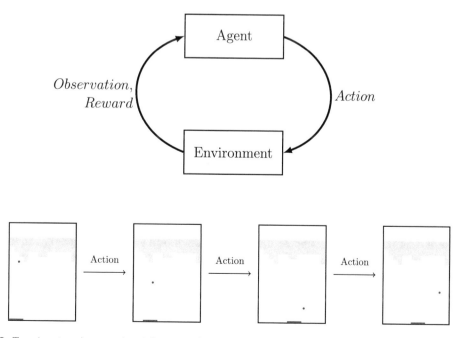

Fig. 1.3 Top: Agent-environment in reinforcement learning in a loop. Bottom: The loop unrolled by time

---

[1] Dog Thinks Through A Problem: www.youtube.com/watch?v=m_CrIu01SnM.

provides the agent with two pieces of information: the state of the environment and a reward signal. The state describes the relevant information about the environment that the agent needs to make a decision, such as the position of the robot or the cards in a poker game. The reward signal is a scalar value that indicates how well the agent is doing in its task. The agent's objective is to maximize its cumulative reward over time.

The environment has its own set of rules, which determine how the state and reward signal change based on the agent's actions. These rules are often called the dynamics of the environment. In many cases, the agent does not have access to the underlying dynamics of the environment and must learn them through trial and error. This is similar to how we humans interact with the physical world every day, normally we have a pretty good sense of what's going on around us, but it's difficult to fully understand the dynamics of the universe.

Game environments are a popular choice for reinforcement learning because they provide a clear objective and well-defined rules. For example, a reinforcement learning agent could learn to play the game of Pong by observing the screen and receiving a reward signal based on whether it wins or loses the game.

In a robotic environment, the agent is a robot that must learn to navigate a physical space or perform a task. For example, a reinforcement learning agent could learn to navigate a maze by using sensors to detect its surroundings and receiving a reward signal based on how quickly it reaches the end of the maze.

State

In reinforcement learning, an environment state or simply state is the statistical data provided by the environment to represent the current state of the environment. The state can be discrete or continuous. For instance, when driving a stick shift car, the speed of the car is a continuous variable, while the current gear is a discrete variable.

Ideally, the environment state should contain all relevant information that's necessary for the agent to make decisions. For example, in a single-player video game like Breakout, the pixels of frames of the game contain all the information necessary for the agent to make a decision. Similarly, in an autonomous driving scenario, the sensor data from the car's cameras, lidar, and other sensors provide relevant information about the surrounding environment.

However, in practice, the available information may depend on the task and domain. In a two-player board game like Go, for instance, although we have perfect information about the board position, we don't have perfect knowledge about the opponent player, such as what they are thinking in their head or what their next move will be. This makes the state representation more challenging in such scenarios.

Furthermore, the environment state might also include noisy data. For example, a reinforcement learning agent driving an autonomous car might use multiple cameras at different angles to capture images of the surrounding area. Suppose the car is driving near a park on a windy day. In that case, the onboard cameras could also capture images of some trees in the park that are swaying in the wind. Since the movement of these trees should not affect the agent's ability to drive, because the trees are inside the park and not on the road or near the road, we can consider these movements of the trees as noise to the self-driving agent. However, it can be challenging to ignore them from the captured images. To tackle this problem, researchers might use various techniques such as filtering and smoothing to eliminate the noisy data and obtain a cleaner representation of the environment state.

Reward

In reinforcement learning, the reward signal is a numerical value that the environment provides to the agent after the agent takes some action. The reward can be any numerical value, positive, negative, or zero. However, in practice, the reward function often varies from task to task, and we need to carefully design a reward function that is specific to our reinforcement learning problem.

Designing an appropriate reward function is crucial for the success of the agent. The reward function should be designed to encourage the agent to take actions that will ultimately lead to achieving our desired goal. For example, in the game of Go, the reward is 0 at every step before the game is over, and +1 or −1 if the agent wins or loses the game, respectively. This design incentivizes the agent to win the game, without explicitly telling it how to win.

Similarly, in the game of Breakout, the reward can be a positive number if the agent destroys some bricks negative number if the agent failed to catch the ball, and zero reward otherwise. This design incentivizes the agent to destroy as many bricks as possible while avoiding losing the ball, without explicitly telling it how to achieve a high score.

The reward function plays a crucial role in the reinforcement learning process. The goal of the agent is to maximize the accumulated rewards over time. By optimizing the reward function, we can guide the agent to learn a policy that will achieve our desired goal. Without the reward signal, the agent would not know what the goal is and would not be able to learn effectively.

In summary, the reward signal is a key component of reinforcement learning that incentivizes the agent to take actions that ultimately lead to achieving the desired goal. By carefully designing the reward function, we can guide the agent to learn an optimal policy.

Agent

In reinforcement learning, an agent is an entity that interacts with an environment by making decisions based on the received state and reward signal from the environment. The agent's goal is to maximize its cumulative reward in the long run. The agent must learn to make the best decisions by trial and error, which involves exploring different actions and observing the resulting rewards.

In addition to the external interactions with the environment, the agent may also has its internal state represents its knowledge about the world. This internal state can include things like memory of past experiences and learned strategies.

It's important to distinguish the agent's internal state from the environment state. The environment state represents the current state of the world that the agent is trying to influence through its actions. The agent, however, has no direct control over the environment state. It can only affect the environment state by taking actions and observing the resulting changes in the environment. For example, if the agent is playing a game, the environment state might include the current positions of game pieces, while the agent's internal state might include the memory of past moves and the strategies it has learned.

In this book, we will typically use the term "state" to refer to the environment state. However, it's important to keep in mind the distinction between the agent's internal state and the environment state. By understanding the role of the agent and its interactions with the environment, we can better understand the principles behind reinforcement learning algorithms. It is worth noting that the terms "agent" and "algorithm" are frequently used interchangeably in this book, particularly in later chapters.

Action

In reinforcement learning, the agent interacts with an environment by selecting actions that affect the state of the environment. Actions are chosen from a predefined set of possibilities, which are specific to each problem. For example, in the game of Breakout, the agent can choose to move the paddle to the left or right or take no action. It cannot perform actions like jumping or rolling over. In contrast, in the game of Pong, the agent can choose to move the paddle up or down but not left or right.

The chosen action affects the future state of the environment. The agent's current action may have long-term consequences, meaning that it will affect the environment's states and rewards for many future time steps, not just the next immediate stage of the process.

Actions can be either discrete or continuous. In problems with discrete actions, the set of possible actions is finite and well defined. Examples of such problems include Atari and Go board games. In contrast, problems with continuous actions have an infinite set of possible actions, often within a continuous range of values. An example of a problem with continuous actions is robotic control, where the degree of angle movement of a robot arm is often a continuous action.

Reinforcement learning problems with discrete actions are generally easier to solve than those with continuous actions. Therefore, this book will focus on solving reinforcement learning problems with discrete actions. However, many of the concepts and techniques discussed in this book can be applied to problems with continuous actions as well.

Policy

A policy is a key concept in reinforcement learning that defines the behavior of an agent. In particular, it maps each possible state in the environment to the probabilities of chose different actions. By specifying how the agent should behave, a policy guides the agent to interact with its environment and maximize its cumulative reward. We will delve into the details of policies and how they interact with the MDP framework in Chap. 2.

For example, suppose an agent is navigating a grid-world environment. A simple policy might dictate that the agent should always move to the right until it reaches the goal location. Alternatively, a more sophisticated policy could specify that the agent should choose its actions based on its current position and the probabilities of moving to different neighboring states.

Model

In reinforcement learning, a model refers to a mathematical description of the dynamics function and reward function of the environment. The dynamics function describes how the environment evolves from one state to another, while the reward function specifies the reward that the agent receives for taking certain actions in certain states.

In many cases, the agent does not have access to a perfect model of the environment. This makes learning a good policy challenging, since the agent must learn from experience how to interact with the environment to maximize its reward. However, there are some cases where a perfect model is available. For example, if the agent is playing a game with fixed rules and known outcomes, the agent can use this knowledge to select its actions strategically. We will explore this scenario in detail in Chap. 2.

In reinforcement learning, the agent-environment boundary can be ambiguous. Despite a house cleaning robot appearing to be a single agent, the agent's direct control typically defines its boundary, while the remaining components comprise the environment. In this case, the robot's wheels and other hardwares are considered to be part of the environment since they aren't directly controlled by the agent. We can think of the robot as a complex system composed of several parts, such as hardware, software, and the reinforcement learning agent, which can control the robot's movement by signaling the software interface, which then communicates with microchips to manage the wheel movement.

1.4   Examples of Reinforcement Learning

Reinforcement learning is a versatile technique that can be applied to a variety of real-world problems. While its success in playing games is well known, there are many other areas where it can be used as an effective solution. In this section, we explore a few examples of how reinforcement learning can be applied to real-world problems.

Autonomous Driving

Reinforcement learning can be used to train autonomous vehicles to navigate complex and unpredictable environments. The goal for the agent is to safely and efficiently drive the vehicle to a desired location while adhering to traffic rules and regulations. The reward signal could be a positive number for successful arrival at the destination within a specified time frame and a negative number for any accidents or violations of traffic rules. The environment state could contain information about the vehicle's location, velocity, and orientation, as well as sensory data such as camera feeds and radar readings. Additionally, the state could include the current traffic conditions and weather, which would help the agent to make better decisions while driving.

Navigating Robots in a Factory Floor

One practical application of reinforcement learning is to train robots to navigate a factory floor. The goal for the agent is to safely and efficiently transport goods from one point to another without disrupting the work of human employees or other robots. In this case, the reward signal could be a positive number for successful delivery within a specified time frame and a negative number for any accidents or damages caused. The environment state could contain information about the robot's location, the weight and size of the goods being transported, the location of other robots, and sensory data such as camera feeds and battery level. Additionally, the state could include information about the production schedule, which would help the agent to prioritize its tasks.

Automating Web Advertising

Another application of reinforcement learning is to automate web advertising. The goal for the agent is to select the most effective type of ad to display to a user, based on their browsing history and profile. The reward signal could be a positive number for when the user clicks on the ad, and zero otherwise. The environment state could contain information such as the user's search history, demographics, and current trends on the Internet. Additionally, the state could include information about the context of the web page, which would help the agent to choose the most relevant ad.

Video Compression

Reinforcement learning can also be used to improve video compression. DeepMind's MuZero agent has been adapted to optimize video compression for some YouTube videos. In this case, the goal for the agent is to compress the video as much as possible without compromising the quality. The reward signal could be a positive number for high-quality compression and a negative number for low-quality compression. The environment state could contain information such as the video's resolution, bit rate, frame rate, and the complexity of the scenes. Additionally, the state could include information about the viewing device, which would help the agent to optimize the compression for the specific device.

Overall, reinforcement learning has enormous potential for solving real-world problems in various industries. The key to successful implementation is to carefully design the reward signal and the environment state to reflect the specific goals and constraints of the problem. Additionally, it is important to continually monitor and evaluate the performance of the agent to ensure that it is making the best decisions.

1.5   Common Terms in Reinforcement Learning

Episodic vs. Continuing Tasks

In reinforcement learning, the type of problem or task is categorized as episodic or continuing depending on whether it has a natural ending. Natural ending refers to a point in a task or problem where it is reasonable to consider the task or problem is completed.

Episodic problems have a natural termination point or terminal state, at which point the task is over, and a new episode starts. A new episode is independent of previous episodes. Examples of episodic problems include playing an Atari video game, where the game is over when the agent loses all lives or won the game, and a new episode always starts when we reset the environment, regardless of whether the agent won or lost the previous game. Other examples of episodic problems include Tic-Tac-Toe, chess, or Go games, where each game is independent of the previous game.

On the other hand, continuing problems do not have a natural endpoint, and the process could go on indefinitely. Examples of continuing problems include personalized advertising or recommendation systems, where the agent's goal is to maximize a user's satisfaction or click-through rate over an indefinite period. Another example of a continuing problem is automated stock trading, where the agent wants to maximize their profits in the stock market by buying and selling stocks. In this scenario, the agent's actions, such as the stocks they buy and the timing of their trades, can influence the future prices and thus affect their future profits. The agent's goal is to maximize their long-term profits by making trades continuously, and the stock prices will continue to fluctuate in the future. Thus, the agent's past trades will affect their future decisions, and there is no natural termination point for the problem.

It is possible to design some continuing reinforcement learning problems as episodic by using a time-constrained approach. For example, the episode could be over when the market is closed. However, in this book, we only consider natural episodic problems that is, the problems with natural termination.

Understanding the differences between episodic and continuing problems is crucial for designing effective reinforcement learning algorithms for various applications. For example, episodic problems may require a different algorithmic approach than continuing problems due to the differences in their termination conditions. Furthermore, in real-world scenarios, distinguishing between episodic and continuing problems can help identify the most appropriate reinforcement learning approach to use for a particular task or problem.

Deterministic vs. Stochastic Tasks

In reinforcement learning, it is important to distinguish between deterministic and stochastic problems. A problem is deterministic if the outcome is always the same when the agent takes the same action in the same environment state. For example, Atari video game or a game of Go is a deterministic problem. In these games, the rules of the game are fixed; when the agent repeatedly takes the same action under the same environment condition (state), the outcome (reward signal and next state) is always the same.

The reason that these games are considered deterministic is that the environment's dynamics and reward functions do not change over time. The rules of the game are fixed, and the environment always behaves in the same way for a given set of actions and states. This allows the agent to learn a policy that maximizes the expected reward by simply observing the outcomes of its actions.

On the other hand, a problem is stochastic if the outcome is not always the same when the agent takes the same action in the same environment state. One example of a stochastic environment is playing poker. The outcome of a particular hand is not entirely determined by the actions of the player. Other players at the table can also take actions that influence the outcome of the hand. Additionally, the cards dealt to each player are determined by a shuffled deck, which introduces an element of chance into the game.

For example, let's say a player is dealt a pair of aces in a game of Texas hold'em. The player might decide to raise the bet, hoping to win a large pot. However, the other players at the table also have their own sets of cards and can make their own decisions based on the cards they hold and the actions of the other players.

If another player has a pair of kings, they might also decide to raise the bet, hoping to win the pot. If a third player has a pair of twos, they might decide to fold, as their hand is unlikely to win. The outcome of the hand depends not only on the actions of the player with the pair of aces but also on the actions of the other players at the table, as well as the cards dealt to them.

This uncertainty and complexity make poker a stochastic problem. While it is possible to use various strategies to improve one's chances of winning in poker, the outcome of any given hand is never certain, and a skilled player must be able to adjust their strategy based on the actions of the other players and the cards dealt.

Another example of stochastic environment is the stock market. The stock market is a stochastic environment because the outcome of an investment is not always the same when the same action is taken in the same environment state. There are many factors that can influence the price of a stock, such as company performance, economic conditions, geopolitical events, and investor sentiment. These factors are constantly changing and can be difficult to predict, making it impossible to know with certainty what the outcome of an investment will be.

For example, let's say you decide to invest in a particular stock because you believe that the company is undervalued and has strong growth prospects. You buy 100 shares at a price of $145.0 per share. However, the next day, the company announces that it has lost a major customer and its revenue projections for the next quarter are lower than expected. The stock price drops to $135.0 per share, and you have lost $1000 on your investment. It's most likely the stochastic nature of the environment led to the loss outcome other than the action (buying 100 shares).

While it is possible to use statistical analysis and other tools to try to predict stock price movements, there is always a level of uncertainty and risk involved in investing in the stock market. This uncertainty and risk are what make the stock market a stochastic environment, and why it is important to use appropriate risk management techniques when making investment decisions.

In this book, we focus on deterministic reinforcement learning problems. By understanding the fundamentals of deterministic reinforcement learning, readers will be well equipped to tackle more complex and challenging problems in the future.

Model-Free vs. Model-Based Reinforcement Learning

In reinforcement learning, an environment is a system in which an agent interacts with in order to achieve a goal. A model is a mathematical representation of the environment's dynamics and reward functions. Model-free reinforcement learning means the agent does not use the model of the environment to help it make decisions. This may occur because either the agent lacks access to the accurate model of the environment or the model is too complex to use during decision-making.

In model-free reinforcement learning, the agent learns to take actions based on its experiences of the environment without explicitly simulating future outcomes. Examples of model-free reinforcement learning methods include Q-learning, SARSA (State-Action-Reward-State-Action), and deep reinforcement learning algorithms such as DQN, which we'll introduce later in the book.

On the other hand, in model-based reinforcement learning, the agent uses a model of the environment to simulate future outcomes and plan its actions accordingly. This may involve constructing a complete model of the environment or using a simplified model that captures only the most essential aspects of the environment's dynamics. Model-based reinforcement learning can be more sample-efficient than model-free methods in certain scenarios, especially when the environment is relatively simple and the model is accurate. Examples of model-based reinforcement learning methods include dynamic programming algorithms, such as value iteration and policy iteration, and probabilistic planning methods, such as Monte Carlo Tree Search in AlphaZero agent, which we'll introduce later in the book.

In summary, model-free and model-based reinforcement learning are two different approaches to solving the same problem of maximizing rewards in an environment. The choice between these approaches depends on the properties of the environment, the available data, and the computational resources.

1.6   Why Study Reinforcement Learning

Machine learning is a vast and rapidly evolving field, with many different approaches and techniques. As such, it can be challenging for practitioners to know which type of machine learning to use for a given problem. By discussing the strengths and limitations of different branches of machine learning, we can better understand which approach might be best suited to a particular task. This can help us make more informed decisions when developing machine learning solutions and ultimately lead to more effective and efficient systems.

There are three branches of machine learning. One of the most popular and widely adopted in the real world is supervised learning, which is used in domains like image recognition, speech recognition, and text classification. The idea of supervised learning is very simple: given a set of training data and the corresponding labels, the objective is for the system to generalize and predict the label for data that's not present in the training dataset. These training labels are typically provided by some supervisors (e.g., humans). Hence, we've got the name supervised learning.

Another branch of machine learning is unsupervised learning. In unsupervised learning, the objective is to discover the hidden structures or features of the training data without being provided with any labels. This can be useful in domains such as image clustering, where we want the system to group similar images together without knowing ahead of time which images belong to which group. Another application of unsupervised learning is in dimensionality reduction, where we want to represent high-dimensional data in a lower-dimensional space while preserving as much information as possible.

Reinforcement learning is a type of machine learning in which an agent learns to take actions in an environment in order to maximize a reward signal. It's particularly useful in domains where there is no clear notion of "correct" output, such as in robotics or game playing. Reinforcement learning has potential applications in areas like robotics, healthcare, and finance.

Supervised learning has already been widely used in computer vision and natural language processing. For example, the ImageNet classification challenge is an annual computer vision competition where deep convolutional neural networks (CNNs) dominate. The challenge provides a training dataset with labels for 1.2 million images across 1000 categories, and the goal is to predict the labels for a separate evaluation dataset of about 100,000 images. In 2012, Krizhevsky et al. [8] developed AlexNet, the first deep CNN system used in this challenge. AlexNet achieved an 18% improvement in accuracy compared to previous state-of-the-art methods, which marked a major breakthrough in computer vision.

Since the advent of AlexNet, almost all leading solutions to the ImageNet challenge have been based on deep CNNs. Another breakthrough came in 2015 when researchers He et al. [9] from Microsoft developed ResNet, a new architecture designed to improve the training of very deep CNNs with hundreds of layers. Training deep CNNs is challenging due to vanishing gradients, which makes it difficult to propagate the gradients backward through the network during backpropagation. ResNet addressed this challenge by introducing skip connections, which allowed the network to bypass one or more layers during forward propagation, thereby reducing the depth of the network that the gradients have to propagate through.

While supervised learning is capable of discovering hidden patterns and features from data, it is limited in that it merely mimics what it is told to do during training and cannot interact with the world and learn from its own experience. One limitation of supervised learning is the need to label every possible stage of the process. For example, if we want to use supervised learning to train an agent to play Go, then we would need to collect the labels for every possible board position, which is impossible due to the enormous number of possible combinations. Similarly, in Atari video games, a single pixel change would require relabeling, making supervised learning inapplicable in these cases. However, supervised learning has been successful in many other applications, such as language translation and image classification.

Unsupervised learning tries to discover hidden patterns or features without labels, but its objective is completely different from that of RL, which is to maximize accumulated reward signals. Humans and animals learn by interacting with their environment, and this is where reinforcement learning (RL) comes in. In RL, the agent is not told which action is good or bad, but rather it must discover that for itself through trial and error. This trial-and-error search process is unique to RL. However, there are other challenges that are unique to RL, such as dealing with delayed consequences and balancing exploration and exploitation.

While RL is a distinct branch of machine learning, it shares some commonalities with other branches, such as supervised and unsupervised learning. For example, improvements in supervised learning and deep convolutional neural networks (CNNs) have been adapted to DeepMind's DQN, AlphaGo, and other RL agents. Similarly, unsupervised learning can be used to pretrain the weights of RL agents to improve their performance. Furthermore, many of the mathematical concepts used in RL, such as optimization and how to train a neural network, are shared with other branches of machine learning. Therefore, while RL has unique challenges and applications, it also benefits from and contributes to the development of other branches of machine learning.

1.7   The Challenges in Reinforcement Learning

Reinforcement learning (RL) is a type of machine learning in which an agent learns to interact with an environment to maximize some notion of cumulative reward. While RL has shown great promise in a variety of applications, it also comes with several common challenges, as discussed in the following sections:

Exploration vs. Exploitation Dilemma

The exploration-exploitation dilemma refers to the fundamental challenge in reinforcement learning of balancing the need to explore the environment to learn more about it with the need to exploit previous knowledge to maximize cumulative reward. The agent must continually search for new actions that may yield greater rewards while also taking advantage of actions that have already proven successful.

In the initial exploration phase, the agent is uncertain about the environment and must try out a variety of actions to gather information about how the environment responds. This is similar to how humans might learn to play a new video game by trying different button combinations to see what happens. However, as the agent learns more about the environment, it becomes increasingly important to focus on exploiting the knowledge it has gained in order to maximize cumulative reward. This is similar to how a human might learn to play a game more effectively by focusing on the actions that have already produced high scores.

There are many strategies for addressing the exploration-exploitation trade-off, ranging from simple heuristics to more sophisticated algorithms. One common approach is to use an ϵ-greedy policy, in which the agent selects the action with the highest estimated value with probability $1 - \epsilon$ and selects a random action with probability ϵ in order to encourage further exploration. Another approach is to use a Thompson sampling algorithm, which balances exploration and exploitation by selecting actions based on a probabilistic estimate of their expected value.

It is important to note that the exploration-exploitation trade-off is not a simple problem to solve, and the optimal balance between exploration and exploitation will depend on many factors, including the complexity of the environment and the agent's prior knowledge. As a result, there is ongoing research in the field of reinforcement learning aimed at developing more effective strategies for addressing this challenge.

Credit Assignment Problem

In reinforcement learning (RL), the credit assignment problem refers to the challenge of determining which actions an agent took that led to a particular reward. This is a fundamental problem in RL because the agent must learn from its own experiences in order to improve its performance.

To illustrate this challenge, let's consider the game of Tic-Tac-Toe, where two players take turns placing Xs and Os on a 3×3 grid until one player gets three in a row. Suppose the agent is trying to learn to play Tic-Tac-Toe using RL, and the reward is +1 for a win, −1 for a loss, and 0 for a draw. The agent's goal is to learn a policy that maximizes its cumulative reward.

Now, suppose the agent wins a game of Tic-Tac-Toe. How can the agent assign credit to the actions that led to the win? This can be a difficult problem to solve, especially if the agent is playing against another RL agent that is also learning and adapting its strategies.

To tackle the credit assignment problem in RL, there are various techniques that can be used, such as Monte Carlo methods or temporal difference learning. These methods use statistical analysis to estimate the value of each action taken by the agent, based on the rewards received and the states visited. By using these methods, the agent can gradually learn to assign credit to the actions that contribute to its success and adjust its policy accordingly.

In summary, credit assignment is a key challenge in reinforcement learning, and it is essential to develop effective techniques for solving this problem in order to achieve optimal performance.

Reward Engineering Problem

The reward engineering problem refers to the process of designing a good reward function that encourages the desired behavior in a reinforcement learning (RL) agent. The reward function determines what the agent is trying to optimize, so it is crucial to make sure it reflects the desired goal we want the the agent to achieve.

An example of good reward engineering is in the game of Atari Breakout, where the goal of the agent is to clear all the bricks at the top of the screen by bouncing a ball off a paddle. One way to design a reward function for this game is to give the agent a positive reward for each brick it clears and a negative reward for each time the ball passes the paddle and goes out of bounds. However, this reward function alone may not lead to optimal behavior, as the agent may learn to exploit a loophole by simply bouncing the ball back and forth on the same side of the screen without actually clearing any bricks.

To address this challenge, the reward function can be designed to encourage more desirable behavior. For example, the reward function can be modified to give the agent a larger positive reward for clearing multiple bricks in a row or for clearing the bricks on the edges of the screen first. This can encourage the agent to take more strategic shots and aim for areas of the screen that will clear more bricks at once.

An example of bad reward engineering is the CoastRunners video game in Atari; it's a very simple boat racing game. The goal of the game is to finish the boat race as quickly as possible. But there's one small issue with the game, the player can earn higher scores by hitting some targets laid out along the route. There's a video that shows that a reinforcement learning agent plays the game by repeatedly hitting the targets instead of finishing the race.[2] This example should not be viewed as the failure of the reinforcement learning agent, but rather humans failed to design and use the correct reward function.

Overall, reward engineering is a crucial part of designing an effective RL agent. A well-designed reward function can encourage the desired behavior and lead to optimal performance, while a poorly designed reward function can lead to suboptimal behavior and may even encourage undesired behavior.

---

[2] Reinforcement learning agent playing the CoastRunners game: www.youtube.com/watch?v=tlOIHko8ySg.

Generalization Problem

In reinforcement learning (RL), the generalization problem refers to the ability of an agent to apply what it has learned to new and previously unseen situations. To understand this concept, consider the example of a self-driving car. Suppose the agent is trying to learn to navigate a particular intersection, with a traffic light and crosswalk. The agent receives rewards for reaching its destination quickly and safely, but it must also follow traffic laws and avoid collisions with other vehicles and pedestrians.

During training, the agent is exposed to a variety of situations at the intersection, such as different traffic patterns and weather conditions. It learns to associate certain actions with higher rewards, such as slowing down at the yellow light and stopping at the red light. Over time, the agent becomes more adept at navigating the intersection and earns higher cumulative rewards.

However, when the agent is faced with a new intersection, with different traffic patterns and weather conditions, it may struggle to apply what it has learned. This is where generalization comes in. If the agent has successfully generalized its knowledge, it will be able to navigate the new intersection based on its past experiences, even though it has not seen this exact intersection before. For example, it may slow down at a yellow light, even if the timing is slightly different than what it has seen before, or it may recognize a pedestrian crossing and come to a stop, even if the appearance of the crosswalk is slightly different.

If the agent has not successfully generalized its knowledge, it may struggle to navigate the new intersection and may make mistakes that lead to lower cumulative rewards. For example, it may miss a red light or fail to recognize a pedestrian crossing, because it has only learned to recognize these situations in a particular context.

Therefore, generalization is a crucial aspect of RL, as it allows the agent to apply its past experiences to new and previously unseen situations, which can improve its overall performance and make it more robust to changes in the environment.

Sample Efficiency Problem

The sample efficiency problem in reinforcement learning refers to the ability of an RL agent to learn an optimal policy with a limited number of interactions with the environment. This can be challenging, especially in complex environments where the agent may need to explore a large state space or take a large number of actions to learn the optimal policy.

To better understand sample efficiency, let's consider an example of an RL agent playing a game of Super Mario Bros. In this game, the agent must navigate Mario through a series of levels while avoiding enemies and obstacles, collecting coins, and reaching the flag at the end of each level.

To learn how to play Super Mario Bros., the agent must interact with the environment, taking actions such as moving left or right, jumping, and shooting fireballs. Each action leads to a new state of the environment, and the agent receives a reward based on its actions and the resulting state.

For example, the agent may receive a reward for collecting a coin or reaching the flag and a penalty for colliding with an enemy or falling into a pit. By learning from these rewards, the agent can update its policy to choose actions that lead to higher cumulative rewards over time.

However, learning the optimal policy in Super Mario Bros. can be challenging due to the large state space and the high dimensionality of the input data, which includes the position of Mario, the enemies, and the obstacles on the screen.

To address the challenge of sample efficiency, the agent may use a variety of techniques to learn from a limited number of interactions with the environment. For example, the agent may use function approximation to estimate the value or policy function based on a small set of training examples. The agent may also use off-policy learning, which involves learning from data collected by a different policy than the one being optimized.

Overall, sample efficiency is an important challenge in reinforcement learning, especially in complex environments. Techniques such as function approximation, off-policy learning, can help address this challenge and enable RL agents to learn optimal policies with a limited number of interactions with the environment.

1.8 Summary

In the first chapter of the book, readers were introduced to the concept of reinforcement learning (RL) and its applications. The chapter began by discussing the breakthroughs in AI in games, showcasing the success of RL in complex games such as Go. The chapter then provided an overview of the agent-environment interaction that forms the basis of RL, including key concepts such as environment, agent, reward, state, action, and policy. Several examples of RL were presented, including Atari video game playing, board game Go, and robot control tasks.

Additionally, the chapter introduced common terms used in RL, including episodic vs. continuing tasks, deterministic vs. stochastic tasks, and model-free vs. model-based reinforcement learning. The importance of studying RL was then discussed, including its potential to solve complex problems and its relevance to real-world applications. The challenges faced in RL, such as the exploration-exploitation dilemma, the credit assignment problem, and the generalization problem, were also explored.

The next chapter of the book will focus on Markov decision processes (MDPs), which is a formal framework used to model RL problems.

References

[1] Volodymyr Mnih, Koray Kavukcuoglu, David Silver, Andrei A. Rusu, Joel Veness, Marc G. Bellemare, Alex Graves, Martin Riedmiller, Andreas K. Fidjeland, Georg Ostrovski, Stig Petersen, Charles Beattie, Amir Sadik, Ioannis Antonoglou, Helen King, Dharshan Kumaran, Daan Wierstra, Shane Legg, and Demis Hassabis. Human-level control through deep reinforcement learning. *Nature*, 518(7540):529–533, Feb 2015.

[2] M. G. Bellemare, Y. Naddaf, J. Veness, and M. Bowling. The arcade learning environment: An evaluation platform for general agents. *Journal of Artificial Intelligence Research*, 47:253–279, Jun 2013.

[3] John Tromp and Gunnar Farnebäck. Combinatorics of go. In H. Jaap van den Herik, Paolo Ciancarini, and H. H. L. M. (Jeroen) Donkers, editors, *Computers and Games*, pages 84–99, Berlin, Heidelberg, 2007. Springer Berlin Heidelberg.

[4] CWI. 66th NHK Cup. https://homepages.cwi.nl/~aeb/go/games/games/NHK/66/index.html, 2018.

[5] David Silver, Aja Huang, Chris J. Maddison, Arthur Guez, Laurent Sifre, George van den Driessche, Julian Schrittwieser, Ioannis Antonoglou, Veda Panneershelvam, Marc Lanctot, Sander Dieleman, Dominik Grewe, John Nham, Nal Kalchbrenner, Ilya Sutskever, Timothy Lillicrap, Madeleine Leach, Koray Kavukcuoglu, Thore Graepel, and Demis Hassabis. Mastering the game of go with deep neural networks and tree search. *Nature*, 529(7587):484–489, Jan 2016.

[6] David Silver, Julian Schrittwieser, Karen Simonyan, Ioannis Antonoglou, Aja Huang, Arthur Guez, Thomas Hubert, Lucas Baker, Matthew Lai, Adrian Bolton, Yutian Chen, Timothy Lillicrap, Fan Hui, Laurent Sifre, George van den Driessche, Thore Graepel, and Demis Hassabis. Mastering the game of go without human knowledge. *Nature*, 550(7676):354–359, Oct 2017.

[7] David Silver, Thomas Hubert, Julian Schrittwieser, Ioannis Antonoglou, Matthew Lai, Arthur Guez, Marc Lanctot, Laurent Sifre, Dharshan Kumaran, Thore Graepel, Timothy Lillicrap, Karen Simonyan, and Demis Hassabis. Mastering chess and shogi by self-play with a general reinforcement learning algorithm, 2017.

[8] Alex Krizhevsky, Ilya Sutskever, and Geoffrey E Hinton. Imagenet classification with deep convolutional neural networks. In F. Pereira, C.J. Burges, L. Bottou, and K.Q. Weinberger, editors, *Advances in Neural Information Processing Systems*, volume 25. Curran Associates, Inc., 2012.

[9] Kaiming He, Xiangyu Zhang, Shaoqing Ren, and Jian Sun. Deep residual learning for image recognition, 2015.

Markov Decision Processes

2

Markov decision processes (MDPs) offer a powerful framework for tackling sequential decision-making problems in reinforcement learning. Their applications span various domains, including robotics, finance, and optimal control.

In this chapter, we provide an overview of the key components of Markov decision processes (MDPs) and demonstrate the formulation of a basic reinforcement learning problem using the MDP framework. We delve into the concepts of policy and value functions, examine the Bellman equations, and illustrate their utilization in updating values for states or state-action pairs. Our focus lies exclusively on finite MDPs, wherein the state and action spaces are finite. While we primarily assume a deterministic problem setting, we also address the mathematical aspects applicable to stochastic problems.

If we were to choose the single most significant chapter in the entire book, this particular chapter would undeniably be at the forefront of our list. Its concepts and mathematical equations hold such importance that they are consistently referenced and utilized throughout the book.

2.1 Overview of MDP

At a high level, a Markov decision process (MDP) is a mathematical framework for modeling sequential decision-making problems under uncertainty. The main idea is to represent the problem in terms of states, actions, a transition model, and a reward function and then use this representation to find an optimal policy that maximizes the expected sum of rewards over time.

To be more specific, an MDP consists of the following components:

- **States** ($\mathcal{S}$): The set of all possible configurations or observations of the environment that the agent can be in. For example, in a game of chess, the state might be the current board configuration, while in a financial portfolio management problem, the state might be the current prices of various stocks. Other examples of states include the position and velocity of a robot, the location and orientation of a vehicle, or the amount of inventory in a supply chain.
- **Actions** ($\mathcal{A}$): The set of all possible actions that the agent can take. In a game of chess, this might include moving a piece, while in a financial portfolio management problem, this might include buying or selling a particular stock. Other examples of actions include accelerating or decelerating a robot, turning a vehicle, or restocking inventory in a supply chain.

M. Hu, *The Art of Reinforcement Learning*,
https://doi.org/10.1007/978-1-4842-9606-6_2

- **Transition model or dynamics function** ($\mathcal{P}$): A function that defines the probability of transitioning to a new state s' given the current state s and the action taken a. In other words, it models how the environment responds to the agent's actions. For example, in a game of chess, the transition model might be determined by the rules of the game and the player's move. In a finance problem, the transition model could be the result of a stock price fluctuation. Other examples of transition models include the physics of a robot's motion, the dynamics of a vehicle's movement, or the demand and supply dynamics of a supply chain.
- **Reward function** ($\mathcal{R}$): A function that specifies the reward or cost associated with taking an action in a given state. In other words, it models the goal of the agent's task. For example, in a game of chess, the reward function might assign a positive value to winning the game and a negative value to losing, while in a finance problem, the reward function might be based on maximizing profit or minimizing risk. Other examples of reward functions include the energy efficiency of a robot's motion, the fuel efficiency of a vehicle's movement, or the profit margins of a supply chain.

Why Are MDPs Useful?

MDPs provide a powerful framework for modeling decision-making problems because they allow us to use mathematical concepts to model real-world problems. For example, MDPs can be used to

- Model robot navigation problems, where the robot must decide which actions to take in order to reach a particular goal while avoiding obstacles. For example, the state might include its current position and the obstacles in its environment, and the actions might include moving in different directions. The transition model could be determined by the physics of the robot's motion, and the reward function could be based on reaching the goal as quickly as possible while avoiding collisions with obstacles.
- Optimize portfolio management strategies, where the agent must decide which stocks to buy or sell in order to maximize profits while minimizing risk. For example, the state might include the current prices of different stocks and the agent's portfolio holdings, and the actions might include buying or selling stocks. The transition model could be the result of stock price fluctuations, and the reward function could be based on the agent's profits or risk-adjusted returns.
- Design personalized recommendation systems, where the agent must decide which items to recommend to a particular user based on their past behavior. For example, the state might include the user's past purchases and the agent's current recommendations, and the actions might include recommending different items. The transition model could be the user's response to the recommendations, and the reward function could be based on how much the user likes the recommended items or makes a purchase.
- Solve many other decision-making problems in various domains, such as traffic control, resource allocation, and game playing. In each case, the MDP framework provides a way to model the problem in terms of states, actions, transition probabilities, and rewards and then solve it by finding a policy that maximizes the expected sum of rewards over time.

Our goal of modeling a problem using MDP is to eventually solve the MDP problem. To solve an MDP, we must find a policy π that maps states to actions in a way that maximizes the expected sum of rewards over time. In other words, the policy tells the agent what action to take in each state to achieve its goal. One way to find the optimal policy is to use the value iteration algorithm or the policy iteration algorithm, which iteratively updates the values of the state (or state-action pair) based on the Bellman equations, until the optimal policy is found. These fundamental concepts will be explored in this chapter and the subsequent chapters.

In summary, MDPs provide a flexible and powerful way to model decision-making problems in various domains. By formulating a problem as an MDP, we can use mathematical concepts to analyze the problem and find an optimal policy that maximizes the expected sum of rewards over time. The key components of an MDP are the states, actions, dynamics function, and reward function, which can be customized to fit the specific problem at hand.

2.2 Model Reinforcement Learning Problem Using MDP

A Markov decision process (MDP) models a sequence of states, actions, and rewards in a way that is useful for studying reinforcement learning. For example, a robot navigating a maze can be modeled as an MDP, where the states represent the robot's location in the maze, the actions represent the robot's movement choices, and the rewards represent the robot's progress toward the goal.

In this context, we use a subscript t to index the different stages of the process (or so-called time step of the sequence), where t could be any discrete value like $t = 0, 1, 2, \cdots$. Note that the time step is not a regular time interval like seconds or minutes but refers to the different stages of the process. For example, in this book S_t, A_t, and R_t often mean the state, action, and reward at the current time step t (or current stage of the process).

It's worth noting that the agent may take a considerable amount of time before deciding to take action A_t when it observes the environment state S_t. As long as the agent does not violate the rules of the environment, it has the flexibility to take action at its own pace. Therefore, a regular time interval is not applicable in this case.

In this book, we adapt the mathematical notation employed by Professor Emma Brunskill in her remarkable course on reinforcement learning [1]. We generally assume that the reward only depends on the state S_t and the action A_t taken by the agent. To keep things simple, we use the same index t for the reward, which is expressed as $R_t = R(S_t, A_t)$, while some may prefer to use $R_{t+1} = R(S_t, A_t, S_{t+1})$ instead of R_t to emphasize that a reward also depends on the successor state S_{t+1}, as Sutton and Barto discussed in their book.[1] However, this alternative expression can sometimes lead to confusion, particularly in simple cases like the ones we present in this book, where the reward does not depend on the successor state.

It is important to remember that in practical implementation, the reward is typically received one time step later, along with the successor state, as illustrated in Fig. 2.1.

We use upper case in S_t, A_t, R_t because these are random variables, and the actual outcome of these random variables could vary. When we are talking about the specific outcome of these random variables, we often use the lower case s, a, r.

Taken together, the state space, action space, transition model, and reward function provide a complete description of the environment and the agent's interactions with it. In the following sections, we'll explore how these elements interact to form the foundation for solving reinforcement learning problems.

To better understand the interactions between the agent and the environment, we can unroll the interaction loop as follows, as shown in Fig. 2.2:

- The agent observes the state S_0 from the environment.
- The agent takes action A_0 in the environment.

---

[1] In their book, Sutton and Barto briefly discussed why they chose to use R_{t+1} instead of R_t as the immediate reward (on page 48). However, they also emphasized that both conventions are widely used in the field.

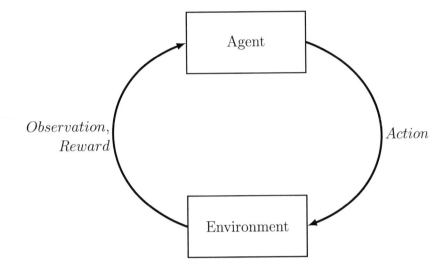

Fig. 2.1 Agent-environment interaction

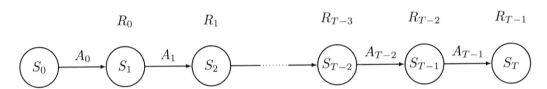

Fig. 2.2 Example of the agent-environment iteration loop unrolled by time for episodic problem

- The environment transition into a new state S_1 and also generate a reward signal R_0, where R_0 is conditioned on S_0 and A_0.
- The agent receives reward R_0 along with the successor state S_1 from the environment.
- The agent decides to take action A_1.
- The interaction continues to the next stage until the process reaches the terminal state S_T. Once it reaches terminal state, no further action is taken and a new episode can be started from the initial state S_0.

The reason why there is no reward when the agent observes the first environment state S_0 is due to the fact that the agent has not yet interacted with the environment. As mentioned earlier, in this book, we assume that the reward function $R_t = R(S_t, A_t)$ is conditioned on the current state of the environment S_t and the action A_t taken by the agent. Since no action is taken in state S_0, there is no reward signal associated with the initial state. However, in practice, it can sometimes be convenient to include a "fake" initial reward, such as using 0, to simplify the code.

In summary, the MDP provides a framework for modeling reinforcement learning problems, where the agent interacts with the environment by taking actions based on the current state and receives rewards that are conditioned on the current state and action. By understanding the interactions between the agent and environment, we can develop algorithms that learn to make good decisions and maximize the cumulative reward over time.

Markov Property

Not every problem can be modeled using the MDP framework. The Markov property is a crucial assumption that must hold for the MDP framework to be applicable. This property states that the future state of the system is independent of the past given the present. More specifically, the successor state S_{t+1} depends only on the current state S_t and action A_t, and not on any previous states or actions.

In other words, the Markov property is a restriction on the environment state, which must contain sufficient information about the history to predict the future state. Specifically, the current state must contain enough information to make predictions about the next state, without needing to know the complete history of past states and actions.

$$P\left(S_{t+1} \mid S_t, A_t\right) = P\left(S_{t+1} \mid S_t, A_t, S_{t-1}, A_{t-1}, \cdots, S_1, A_1, S_0, A_0\right) \tag{2.1}$$

For example, a robot trying to navigate a room can be modeled using the MDP framework only if it satisfies the Markov property. If the robot's movement depends on its entire history, including its past positions and actions, the Markov property will be violated, and the MDP framework will no longer be applicable. This is because the robot's current position would not contain sufficient information to predict its next position, making it difficult to model the environment and make decisions based on that model.

Service Dog Example

In this example, we imagine training a service dog to retrieve an object for its owner. The training is conducted in a house with three rooms, one of which contains a personal object that the dog must retrieve and bring to its owner or trainer. The task falls into the episodic reinforcement learning problem category, as the task is considered finished once the dog retrieves the object. To simplify the task, we keep placing the object in the same (or almost the same) location and initialize the starting state randomly. Additionally, one of the rooms leads to the front yard, where the dog can play freely. This scenario is illustrated in Fig. 2.3.

We will use this service dog example in this book to demonstrate how to model a reinforcement learning problem as an MDP (Markov decision process), explain the dynamics function of the environment, and construct a policy. We will then introduce specific algorithms, such as dynamic programming, Monte Carlo methods, and temporal difference methods, to solve the service dog reinforcement learning problem.

2.3   Markov Process or Markov Chain

To start our discussion on Markov decision processes (MDPs), let's first define what a Markov process (Markov chain) is. A Markov process is a memoryless random process where the probability of transitioning to a new state only depends on the current state and not on any past states or actions. It is the simplest case to study in MDPs, as it involves only a sequence of states without any rewards or actions. Although it may seem basic, studying Markov chains is important as they provide fundamental insights into how states in a sequence can influence one another. This understanding can be beneficial when dealing with more complex MDPs.

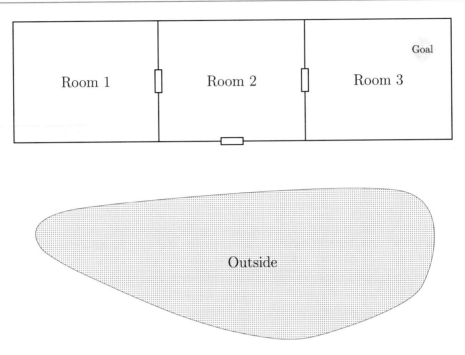

Fig. 2.3 Simple drawing to illustrate the service dog example

A Markov chain can be defined as a tuple of $(\mathcal{S}, \mathcal{P})$, where $\mathcal{S}$ is a finite set of states called the state space, and $\mathcal{P}$ is the dynamics function (or transition model) of the environment, which specifies the probability of transitioning from a current state s to a successor state s'. Since there are no actions in a Markov chain, we omit actions in the dynamics function $\mathcal{P}$. The probability of transitioning from state s to state s' is denoted by $P(s'|s)$.

For example, we could model a Markov chain with a graph, where each node represents a state, and each edge represents a possible transition from one state to another, with a transition probability associated with each edge. This can help us understand how states in a sequence can affect one another. We've modeled our service dog example as a Markov chain, as shown in Fig. 2.4. The open circle represents a non-terminal state, while the square box represents the terminal state. The straight and curved lines represent the transitions from the current state s to its successor state s', with a transition probability $P(s'|s)$ associated with each possible transition. For example, if the agent is currently in state *Room 1*, there is a 0.8 probability that the environment will transition to its successor state *Room 2* and a 0.2 probability of staying in the same state *Room 1*. Note that these probabilities are chosen randomly to illustrate the idea of the dynamics function of the environment.

Transition Matrix for Markov Chain

The transition matrix $\mathcal{P}$ is a convenient way to represent the dynamics function (or transition model) of a Markov chain. It lists all the possible state transitions in a single matrix, where each row represents a current state s, and each column represents a successor state s'. The transition probability for transitioning from state s to state s' is denoted by $P(s'|s)$. Since we are talking about probability,

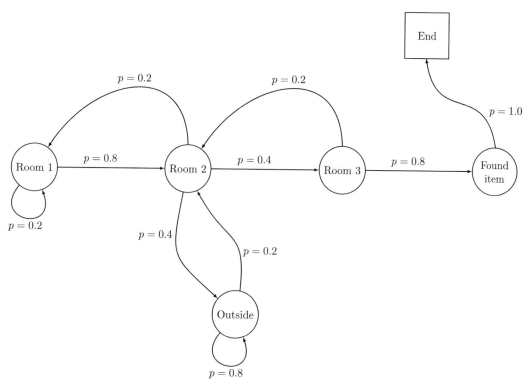

Fig. 2.4 Service dog Markov chain

the sum of each row is always equal to 1.0. Here, we list the transition matrix for our service dog Markov chain:

$$
\mathcal{P} = \begin{matrix}
 & \text{Room 1} & \text{Room 2} & \text{Room 3} & \text{Outside} & \text{Found item} & \text{End} \\
\text{Room 1} & 0.2 & 0.8 & 0 & 0 & 0 & 0 \\
\text{Room 2} & 0.2 & 0 & 0.4 & 0.4 & 0 & 0 \\
\text{Room 3} & 0 & 0.2 & 0 & 0 & 0.8 & 0 \\
\text{Outside} & 0 & 0.2 & 0 & 0.8 & 0 & 0 \\
\text{Found item} & 0 & 0 & 0 & 0 & 0 & 1.0 \\
\text{End} & 0 & 0 & 0 & 0 & 0 & 1.0
\end{matrix}
$$

With access to the dynamics function of the environment, we can sample some state transition sequences $S_0, S_1, S_2, \cdots$ from the environment. For example:

- Episode 1: (Room 1, Room 2, Room 3, Found item, End)
- Episode 2: (Room 3, Found item, End)
- Episode 3: (Room 2, Outside, Room 2, Room 3, Found item, End)
- Episode 4: (Outside, Outside, Outside, ...)

We now have a basic understanding about the state transition in the environment; let's move on to add rewards into the process.

2.4   Markov Reward Process

As we've said before, the goal of a reinforcement learning agent is to maximize rewards, so the next natural step is to add rewards to the Markov chain process. The Markov reward process (MRP) is an extension of the Markov chain, where rewards are added to the process. In a Markov reward process, the agent not only observes state transitions but also receives a reward signal along the way. Note, there are still no actions involved in the MRPs. We can define the Markov reward process as a tuple $(\mathcal{S}, \mathcal{P}, \mathcal{R})$, where

- $\mathcal{S}$ is a finite set of states called the state space.
- $\mathcal{P}$ is the dynamics function (or transition model) of the environment, where $P(s'|s) = P\left[S_{t+1} = s' \mid S_t = s\right]$ specify the probability of environment transition into successor state s' when in current state s.
- $\mathcal{R}$ is a reward function of the environment. $R(s) = \mathbb{E}\left[R_t \mid S_t = s\right]$ is the reward signal provided by the environment when the agent is in state s.

As shown in Fig. 2.5, we added a reward signal to each state in our service dog example. As we've talked briefly in Chap. 1, we want the reward signals to align with our desired goal, which is to find the object, so we decided to use the highest reward signal for the state *Found item*. For the state

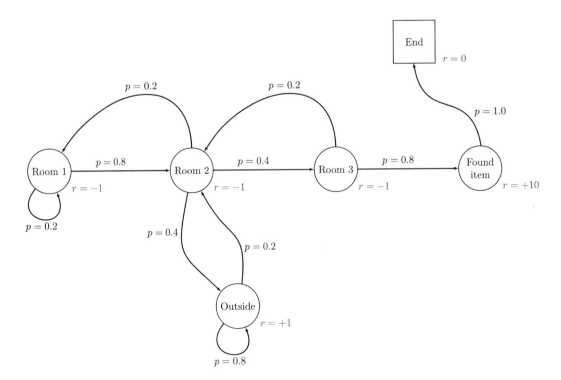

Fig. 2.5 Service dog MRP

Outside, the reward signal is +1, because being outside playing might be more enjoyable for the agent compared to wandering between different rooms.

With reward signals, we can compute the total rewards the agent could get for these different sample sequences, where the total rewards are calculated for the entire sequence:

- Episode 1: (Room 1, Room 2, Room 3, Found item, End)
  Total rewards $= -1 - 1 - 1 + 10 + 0 = 7.0$
- Episode 2: (Room 3, Found item, End)
  Total rewards $= -1 + 10 = 9.0$
- Episode 3: (Room 2, Outside, Room 2, Room 3, Found item, End)
  Total rewards $= -1 + 1 - 1 - 1 + 10 + 0 = 8.0$
- Episode 4: (Outside, Outside, Outside . . .)
  Total rewards $= 1 + 1 + \cdots = \infty$

Return

To quantify the total rewards the agent can get in a sequence of states, we use a different term called *return*. The return G_t is simply the sum of rewards from time step t to the end of the sequence and is defined by the mathematical equation in Eq. (2.2), where T is the terminal time step for episodic reinforcement learning problems.[2]

$$G_t = R_t + R_{t+1} + R_{t+2} + \cdots + R_{T-1} \tag{2.2}$$

One issue with Eq. (2.2) is that the return G_t could become infinite in cases where there are recursion or loops in the process, such as in our sample sequence episode 4. For continuing reinforcement learning problems, where there is no natural end to the task, the return G_t could also easily become infinite. To resolve this issue, we introduce the discount factor.

The discount factor γ is a parameter, where $0 \leq \gamma \leq 1$, that helps us to solve the infinite return problem. By discounting rewards received at future time steps, we can avoid the return becoming infinite. When we add the discount factor to the regular return in Eq. (2.2), the return becomes the *discounted sum of rewards* from time step t to a horizon H, which can be the length of the episode or even infinity for continuing reinforcement learning problems. We'll be using Eq. (2.3) as the definition of return for the rest of the book. Notice that we omit the discount γ for the immediate reward R_t, since $\gamma^0 = 1$.

$$G_t = R_t + \gamma R_{t+1} + \gamma^2 R_{t+2} + \cdots + \gamma^{H-1-t} R_{H-1} \tag{2.3}$$

We want to emphasize that the discount factor γ not only helps us to solve the infinite return problem, but it can also influence the behavior of the agent (which will make more sense when we

---

[2] The notation used in Eq. (2.2), as well as Eqs. (2.3) and (2.4), may seem unfamiliar to readers familiar with the work of Sutton and Barto. In their book, they utilize a different expression denoted as $G_t = R_{t+1} + R_{t+2} + R_{t+3} + \cdots + R_T$ for the nondiscounted case. In their formulation, the immediate reward is denoted as R_{t+1}. However, in our book, we adopt a simpler reward function and notation, as explained earlier. We represent the immediate reward as R_t, assuming it solely depends on the current state S_t and the action A_t taken in that state. Therefore, in Eq. (2.2), as well as Eqs. (2.3) and (2.4), we start with R_t instead of R_{t+1}, and we use R_{T-1} instead of R_T for the final reward. It is important to note that despite this slight time step shift, these equations essentially compute the same result: the sum of (or discounted) rewards over an episode.

later talk about value functions and policy). For example, when $\gamma = 0$, the agent only cares about the immediate reward, and as γ gets closer to 1, future rewards become as important as immediate rewards. Although there are methods that do not use discount, the mathematical complexity of such methods is beyond the scope of this book.

There is a useful property about the return G_t in reinforcement learning: the return G_t is the sum of the immediate reward R_t and the discounted return of the next time step γG_{t+1}. We can rewrite it recursively as shown in Eq. (2.4). This recursive property is important in MRPs, MDPs and reinforcement learning because it forms the foundation for a series of essential mathematical equations and algorithms, which we will introduce later in this chapter and in the next few chapters.

$$
\begin{aligned}
G_t &= R_t + \gamma R_{t+1} + \gamma^2 R_{t+2} + \gamma^3 R_{t+3} + \cdots \\
&= R_t + \gamma \left(R_{t+1} + \gamma R_{t+2} + \gamma^2 R_{t+3} + \cdots \right) \\
&= R_t + \gamma G_{t+1}
\end{aligned}
\tag{2.4}
$$

We can compute the return G_t for the sample sequences of a particular MRP or even MDP. The following shows the returns for some sample episodes in our service dog example, where we use discount factor $\gamma = 0.9$:

- Episode 1: (Room 1, Room 2, Room 3, Found item, End)
  $G_0 = -1 - 1 * 0.9 - 1 * 0.9^2 + 10 * 0.9^3 = 4.6$
- Episode 2: (Room 3, Found item, End)
  $G_0 = -1 + 10 * 0.9 = 8.0$
- Episode 3: (Room 2, Outside, Room 2, Room 3, Found item, End)
  $G_0 = -1 + 1 * 0.9 - 1 * 0.9^2 - 1 * 0.9^3 + 10 * 0.9^4 = 4.9$
- Episode 4: (Outside, Outside, Outside, ...)
  $G_0 = 1 + 1 * 0.9 + 1 * 0.9^2 + \cdots = 10.0$

As long as the discount factor is not 1, we won't have the infinite return problem even if there's a loop in the MRP process. Comparing the return for these different sample sequences, we can see that episode 4 has the highest return value. However, if the agent gets stuck in a loop staying in the same state *Outside*, it has no way to achieve the goal, which is to find the object in *Room 3*. This does not necessarily mean that the reward function is flawed. To prove this, we need to use the value function, which we will explain in detail in the upcoming sections.

Value Function for MRPs

In Markov reward processes (MRPs), the return G_t measures the total future reward from time step t to the end of the episode. However, comparing returns for different sample sequences alone has its limits, as it only measures returns starting from a particular time step. This is not very helpful for an agent in a specific environment state who needs to make a decision. The value function can help us overcome this problem.

Formally, the state value function $V(s)$ for MRP measures the *expected return* starting from state s and to a horizon H. We call it the expected return because the trajectory starting from state s to a horizon H is often a random variable. In simple words, $V(s)$ measures the average return starting from state s and up to a horizon H. For episodic problems, the horizon is just the terminal time step, that is, $H = T$.

$$V(s) = \mathbb{E}\left[G_t \mid S_t = s\right] \tag{2.5}$$

The state value function $V(s)$ for MRPs also shares the recursive property as shown in Eq. (2.6). Equation (2.6) is also called the Bellman expectation equation for $V(s)$ (for MRPs). We call it the Bellman expectation equation because it's written in a recursive manner but still has the expectation sign $\mathbb{E}$ attached to it.

$$
\begin{aligned}
V(s) &= \mathbb{E}\left[G_t \mid S_t = s\right] \\
&= \mathbb{E}\left[R_t + \gamma R_{t+1} + \gamma^2 R_{t+2} + \cdots \mid S_t = s\right] \\
&= \mathbb{E}\left[R_t + \gamma\left(R_{t+1} + \gamma R_{t+2} + \cdots\right) \mid S_t = s\right] \\
&= \mathbb{E}\left[R_t + \gamma G_{t+1} \mid S_t = s\right] \\
&= \mathbb{E}\left[R_t + \gamma V(S_{t+1}) \mid S_t = s\right] \tag{2.6} \\
&= R(s) + \gamma \sum_{s' \in \mathcal{S}} P(s'|s)V(s'), \quad \text{for all } s \in \mathcal{S} \tag{2.7}
\end{aligned}
$$

Equation (2.7) is called the Bellman equation for $V(s)$ (for MRPs). We can see how the step from Eqs. (2.6) to (2.7) removed the expectation sign $\mathbb{E}$ from the equation, by considering the values of all the possible successor states s' and weighting each by the state transition probability $P(s'|s)$ from the environment for the MRPs.

Worked Example

As an example, we can use Eq. (2.7) to compute the expected return for a particular state. Let's say our initial values for state *Room 2* are 5.0 and for state *Found item* are 10.0, as shown in Fig. 2.6 (these initial values were chosen randomly), and we use no discount $\gamma = 1.0$. Now let's compute the expected return for state *Room 3*. Since we already know the model (dynamics function and reward function) of the Markov reward process (MRP), we know the immediate reward is -1 no matter what successor state s' will be. There's a 0.8 probability the environment will transition to successor state *Found item*, and the value for this successor state is $V(\textit{Found item}) = 10.0$. And there's also a 0.2 probability the environment will transition to successor state *Room 2*, and the value for this successor state is $V(\textit{Room 2}) = 5.0$. So we can use the Bellman equation to compute our estimated value for state $V(\textit{Room 3})$ as follows:

$$V(\textit{Room 3}) = -1 + 0.8 * 10.0 + 0.2 * 5.0 = 8.0$$

Of course, this estimated value for state *Room 3* is not accurate, since we started with randomly guessed values, and we didn't include other states. But if we include all the states in the state space and repeat the process over a large number of times, in the end, the estimated values would be very close to the true values.

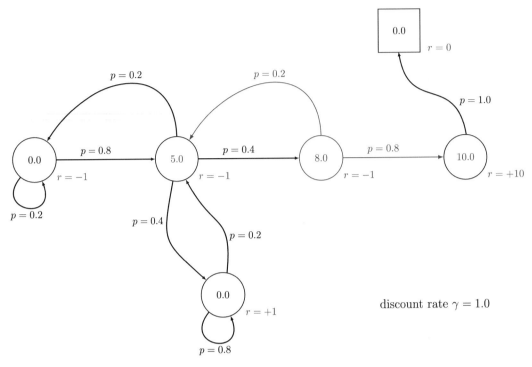

Fig. 2.6 Example of how to compute the value of a state for the service dog MRP, $\gamma = 1.0$; the numbers are chosen randomly

Figure 2.7 shows the true values of the states for our service dog MRP. The values are computed using Eq. (2.7) and dynamic programming, which is an iterative method and can be used to solve MDP. We will introduce dynamic programming methods in the next chapter. For this experiment, we use a discount factor of $\gamma = 0.9$. We can see that the state *Found item* has the highest value among all states.

Why do we want to estimate the state values? Because it can help the agent make better decisions. If the agent knows which state is better (in terms of expected returns), it can choose actions that may lead it to those better states. For example, in Fig. 2.7, if the current state is *Room 2*, then the best successor state is *Room 3* since it has the highest state value of 7.0 among all the possible successor states for *Room 2*. By selecting actions that lead to high-value states, the agent can maximize its long-term expected return.

2.5   Markov Decision Process

Now we're ready to discuss the details of the MDP. Similar to how the MRP extends the Markov chain, the Markov decision process (MDP) extends the MRP by including actions and policy into the process. The MDP contains all the necessary components, including states, rewards, actions, and policy. We can define the MDP as a tuple $(\mathcal{S}, \mathcal{A}, \mathcal{P}, \mathcal{R})$:

- $\mathcal{S}$ is a finite set of states called the state space.
- $\mathcal{A}$ is a finite set of actions called the action space.

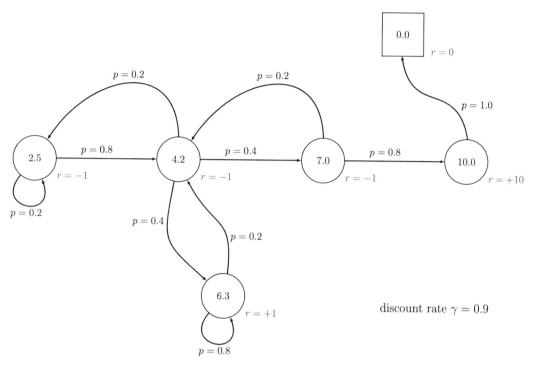

Fig. 2.7 State values for the service dog MRP, $\gamma = 0.9$

- $\mathcal{P}$ is the dynamics function (or transition model) of the environment, where $P(s'|s, a) = P\big[$ $S_{t+1} = s' \mid S_t = s, A_t = a\big]$ specify the probability of environment transition into successor state s' when in current state s and take action a.
- $\mathcal{R}$ is a reward function of the environment; $R(s, a) = \mathbb{E}\big[R_t \mid S_t = s, A_t = a\big]$ is the reward signal provided by the environment when the agent is in state s and taking action a.

Note that the dynamics function $\mathcal{P}$ and reward function $\mathcal{R}$ are now conditioned on the action A_t chosen by the agent at time step t.

In Fig. 2.8, we have modeled our service dog example using an MDP. Before we move on, we want to explain the small changes we've made to Fig. 2.8. First, we have merged states *Found item* and *End* into a single terminal state *Found item*. This makes sense because the reward is now conditioned on (s, a), and there are no additional meaningful states after the agent has reached the state *Found item*. Second, the straight and curly lines in Fig. 2.8 represent valid actions that the agent can choose in a state, rather than state transition probabilities. Finally, the reward now depends on the action chosen by the agent, and the reward values are slightly different.

In fact, our service dog example is now modeled as a deterministic (stationary) reinforcement learning environment. This means that if the agent takes the same action a in the same state s, the successor state s' and reward r will always be the same, regardless of whether we repeat it 100 times or 1 million times. The transition to the successor state s' is guaranteed to happen with 1.0 probability. For example, if the agent is in state *Room 2* and chooses to *Go outside*, the successor state will always be *Outside*; there is zero chance that the successor state will be *Room 3* or *Room 1*.

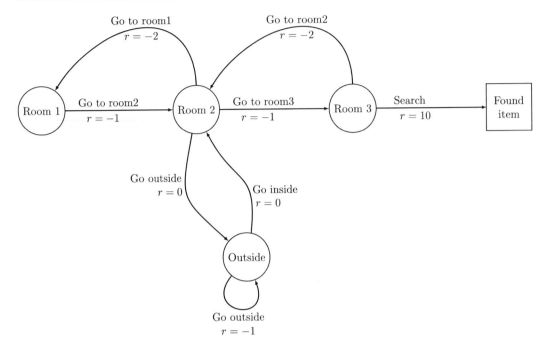

Fig. 2.8 Service dog MDP

We can list all the elements in the set of $\mathcal{S}, \mathcal{A}, \mathcal{R}$ for our service dog MDP; notice that not all actions are available (or legal) in each state:

- $\mathcal{S} = \{\text{Room 1, Room 2, Room 3, Outside, Found item}\}$
- $\mathcal{A} = \{\text{Go to room1, Go to room2, Go to room3, Go outside, Go inside, Search}\}$
- $\mathcal{R} = \{-1, -2, +1, 0, +10\}$

As we have explained before, our service dog MDP is a deterministic (stationary) reinforcement learning environment. The dynamics function is slightly different compared to our previous example. For MDPs, the transition from the current state s to its successor state s' depends on the current state s and the action a chosen by the agent. For a deterministic environment, the transition probability is always 1.0 for legal actions $a \in \mathcal{A}(s)$, and 0 for illegal actions $a \notin \mathcal{A}(s)$, which are actions not allowed in the environment. Illegal actions should never be chosen by the agent, since most environments have enforced checks at some level. For example, in the game of Go, if a player makes an illegal move, they automatically lose the game.

We can still construct a single matrix for the dynamics function for our service dog MDP, but this time it needs to be a 3D matrix. Since it's not easy for us to draw a 3D matrix, we chose to use a 2D plane to explain the concept for a single state. Assume the current state of the environment is *Room 2*, each row of the plane represents an action, and each column of the plane represents a successor state s'. For consistency purposes, we set the transition probability to 1.0 for the successor state *Room 2*, for all illegal actions (*Go to room2, Go inside, Search*). This just means that these illegal actions won't affect the state of the environment.

$$
\mathcal{P} = \begin{array}{r} \text{Go to room1} \\ \text{Go to room2} \\ \text{Go to room3} \\ \text{Go outside} \\ \text{Go inside} \\ \text{Search} \end{array}
\begin{pmatrix}
1.0 & 0 & 0 & 0 & 0 \\
0 & 1.0 & 0 & 0 & 0 \\
0 & 0 & 1.0 & 0 & 0 \\
0 & 0 & 0 & 1.0 & 0 \\
0 & 1.0 & 0 & 0 & 0 \\
0 & 1.0 & 0 & 0 & 0
\end{pmatrix}
$$

	Room 1	Room 2	Room 3	Outside	Found item
Go to room1	1.0	0	0	0	0
Go to room2	0	1.0	0	0	0
Go to room3	0	0	1.0	0	0
Go outside	0	0	0	1.0	0
Go inside	0	1.0	0	0	0
Search	0	1.0	0	0	0

For illustration purposes, let's imagine our service dog MDP is not deterministic but a stochastic reinforcement learning environment. For example, every legal action could also lead to the successor state *Outside* randomly. In this case, the transition matrix for the stochastic environment would be something like this:

	Room 1	Room 2	Room 3	Outside	Found item
Go to room1	0.6	0	0	0.4	0
Go to room2	0	1.0	0	0	0
Go to room3	0	0	0.2	0.8	0
Go outside	0	0	0	1.0	0
Go inside	0	1.0	0	0	0
Search	0	1.0	0	0.0	0

Policy

In reinforcement learning and MDPs, the agent has the ability to decide how to act in the environment, but it is not always clear how to choose an action in a given state or what rules to use for making decisions. To address this problem, we introduce the concept of a policy.

Formally, a policy π is a function that maps each state s to a probability distribution over the set of all possible actions in that state s. A policy can be deterministic, meaning that it always chooses the same action for a given state, or it can be stochastic, meaning that it chooses actions according to a probability distribution. For example, if a particular state has four actions a_1, a_2, a_3, and a_4, with a_1 being the best action in terms of rewards, a_4 being the worst action, and a_2 being better than a_3, a deterministic policy would always choose a_1, while a stochastic policy might choose a_1 60% of the time, a_2 20% of the time, a_3 15% of the time, and a_4 5% of the time.

Mathematically, we can represent a policy as a collection of functions, each of which maps a specific environment state s to a probability distribution over all actions $a \in \mathcal{A}(s)$. This can be written as

$$
\pi(a|s) = P\left[A_t = a \mid S_t = s\right], \quad \text{for all } s \in \mathcal{S}, a \in \mathcal{A} \tag{2.8}
$$

For any state s, the sum of probabilities over all possible actions must equal 1:

$$
\sum_{a \in A} \pi(a|s) = 1.0
$$

The number of possible deterministic policies for a finite MDP is $N(A)^{N(S)}$, where $N(S)$ is the size of the state space, and $N(A)$ is the size of the action space. For our service dog MDP, this number would be $6^5 = 7776$. However, this estimate includes illegal actions for each state, so a more accurate estimate that only includes legal actions is $3^5 = 243$.

Value Functions for MDPs

The goal of modeling the reinforcement learning problem using MDPs is that we want to find an optimal policy so that the agent can get maximum rewards when interacting with the environment. The way we can do that within the context of MDP is by using the value functions.

A value function measures the expected return when following some policy. There are two different value functions for an MDP: the state value function V_π and state-action value function Q_π.

The state value function V_π for MDP measures the expected return starting in state s, then following policy π afterward to a horizon H, where H could be the length of an episode, or it could be infinite for the continuing problems, while the state-action value function Q_π for a policy π measures the expected return starting from state s, taking action a, *then* following policy π.

Equation (2.9) is the definition for the state value function V_π, which looks almost identical to Eq. (2.5), except it now has a small subscript π attached to it, which means the values are measured for a specific policy. This should make sense since the rewards the agent receives will depend on how it acts in the environment, and the agent relies on the policy π to act in the environment. Even for the same state s, different policies often will have different values.

$$V_\pi(s) = \mathbb{E}_\pi\Big[G_t \mid S_t = s\Big], \quad \text{for all } s \in S \tag{2.9}$$

Equation (2.10) is the definition of state-action value function Q_π for a policy π.

$$Q_\pi(s, a) = \mathbb{E}_\pi\Big[G_t \mid S_t = s, A_t = a\Big], \quad \text{for all } s \in S, a \in A \tag{2.10}$$

There is a nice relationship between the $V_\pi(s)$ and $Q_\pi(s, a)$. The value of a state s under a policy π is just the weighted sum over all state-action values $Q_\pi(s, a)$ for all possible actions $a \in A$, as shown in Eq. (2.11).

$$V_\pi(s) = \sum_{a \in A} \pi(a|s) Q_\pi(s, a) \tag{2.11}$$

Here, $\pi(a|s)$ is the probability of taking action a in state s when following policy π.

With this concept in mind, we can define the Bellman equations for the V_π and Q_π. Equation (2.12) is called the Bellman expectation equation for V_π; in this equation, we use the immediate reward R_t and the discounted value of successor state(s) $\gamma V_\pi(S_{t+1})$ to compute the value of a state $V_\pi(s)$ under some policy π.

$$V_\pi(s) = \mathbb{E}_\pi\Big[G_t \mid S_t = s\Big]$$

$$= \mathbb{E}_\pi\Big[R_t + \gamma R_{t+1} + \gamma^2 R_{t+2} + \cdots \mid S_t = s\Big]$$

$$= \mathbb{E}_\pi\Big[R_t + \gamma\big(R_{t+1} + \gamma R_{t+2} + \cdots\big) \mid S_t = s\Big]$$

$$= \mathbb{E}_\pi \Big[R_t + \gamma G_{t+1} \, \Big| \, S_t = s \Big]$$

$$= \mathbb{E}_\pi \Big[R_t + \gamma V_\pi (S_{t+1}) \, \Big| \, S_t = s \Big] \tag{2.12}$$

$$= \sum_{a \in A} \pi(a|s) \Big[R(s,a) + \gamma \sum_{s' \in S} P(s'|s,a) V_\pi(s') \Big], \quad \text{for all } s \in S \tag{2.13}$$

Equation $(2.13)^3$ further expands the Bellman expectation equation for V_π by weighting probability of taking action a when in state s under the policy π; it also weights the value of each successor state $V_\pi(s')$ by using the probability of transitioning into that particular successor state when the agent is in state s and taking action a. This equation is also called the Bellman equation for V_π.

Similarly, Eq. (2.14) is the Bellman expectation equation for Q_π, and Eq. (2.16) is the Bellman equation for Q_π. Note that Eq. (2.16) involves conditioning on the value of the next state-action pair, whereas the other equations do not. This may seem strange at first, but it is necessary because the next state and action are random variables, and the Bellman equation needs to express the expected value of the state-action pair, taking into account all possible outcomes. Specifically, Eq. (2.16) involves a new expectation over the value of the successor state-action pair (s', a'), because this pair depends on the agent's policy.

$$Q_\pi(s,a) = \mathbb{E}_\pi \Big[G_t \, \Big| \, S_t = s, A_t = a \Big]$$

$$= \mathbb{E}_\pi \Big[R_t + \gamma R_{t+1} + \gamma^2 R_{t+2} + \cdots \, \Big| \, S_t = s, A_t = a \Big]$$

$$= \mathbb{E}_\pi \Big[R_t + \gamma G_{t+1} \, \Big| \, S_t = s, A_t = a \Big]$$

$$= \mathbb{E}_\pi \Big[R_t + \gamma Q_\pi(S_{t+1}, A_{t+1}) \, \Big| \, S_t = s, A_t = a \Big] \tag{2.14}$$

$$= R(s,a) + \gamma \mathbb{E}_\pi \Big[Q_\pi(s',a') \, \Big| \, S_{t+1} = s', A_{t+1} = a' \Big] \tag{2.15}$$

$$= R(s,a) + \gamma \sum_{s' \in S} P(s'|s,a) \sum_{a' \in A} \pi(a'|s') Q_\pi(s',a'), \quad \text{for all } s \in S, a \in A \tag{2.16}$$

Because of the relationship between V_π and Q_π as shown in Eq. (2.11), we can also rewrite the Bellman equation for Q_π as

$$Q_\pi(s,a) = R(s,a) + \gamma \sum_{s' \in S} P(s'|s,a) V_\pi(s'), \quad \text{for all } s \in S, a \in A \tag{2.17}$$

---

[3] The notation used in Eq. (2.13) and later in Eq. (2.16) may appear unfamiliar to those who are familiar with the work of Sutton and Barto. In their book, they employ a different expression denoted as $\sum_{a \in A} \pi(a|s) \sum_{s',r} P(s',r|s,a) \big[r +$
$\gamma V_\pi(s') \big]$. The reason for this difference is that they combine the reward function and dynamics function together to form a single function which takes in four arguments $p(s',r|s,a)$. By doing so, they are able to compute the expected reward or transition probabilities based on this single function. However, in our book, we adopt a slightly different and simpler notation. We use separate functions for reward and dynamics, namely, $r = R(s,a)$ and $p(s'|s,a)$, respectively. Moreover, we assume that the reward only depends on the state-action pair and not the successor state. Consequently, this allows us to separate the immediate reward $R(s,a)$ from the inner summation in Eq. (2.13). We further elaborate on this distinction later in this chapter when we introduce the alternative Bellman equations.

Fig. 2.9 Backup diagram
for V_π

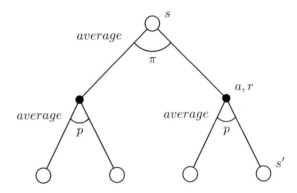

The additional step to derive Eq. (2.17) is necessary because in some cases, we don't have an accurate estimate of $Q_\pi(s, a)$ as we'll see later in the book when we introduce dynamic programming algorithms.

The Bellman equations are at the core of solving V_π and Q_π. That means they are also at the core of solving reinforcement learning problems if we are using value function methods, which estimate the value of states or state-action pairs. For example, the Bellman equation for V_π says that the value of a state can be decomposed into two parts: the immediate reward $R(s, a)$ and the discounted value of the successor states $\gamma \sum_{s' \in S} P(s'|s, a) V_\pi(s')$, which are the states that can be reached from s under policy π. In some environment, there could be multiple successor states s', and we compute the expected value over all of them. The similar concept also applies to Q_π, which estimates the value of taking action a in state s and following policy π thereafter.

The Bellman equations form the basis for a series of algorithms we will introduce later in the book to compute or approximate V_π and Q_π, which are at the core of so-called value-based methods for reinforcement learning. The Bellman equations help us define the rules for transferring value back to a state from its successor states, which we often call the update or backup operation. Sometimes, we also call them the update (backup) rule for the value functions V_π and Q_π. Value function methods are one class of reinforcement learning algorithms, and there are many others we will discuss later in the book.

Figure 2.9 is a backup diagram that can help us better understand Eq. (2.13). In this diagram, open circles represent states, and black dots represent actions. At the root is a state node s, and in state s, there may be multiple actions that the agent can choose from (we only show two actions). The exact action a that is chosen will depend on the policy π that the agent is following. From each action a, the environment could transition into one of several successor states s' (we show two such successor states), with a transition probability p that depends on (s, a). The Bellman equation takes the expectation (as indicated by the little arc under the root state node s and action node a) over all possible outcomes, weighted by the probability of their occurrence. The term $\sum_{a \in A} \pi(a|s)$ represents the expectation over all actions for state s, weighted by the probability of each action being selected according to the policy π. The term $\sum_{s' \in S} P(s'|s, a)$ represents the expectation over all possible successor states s', weighted by the state transition probability from the environment.

We can also draw a backup diagram for Eq. (2.16). The idea of taking an expectation over all possible outcomes, which we explained when introducing the Bellman equation for V_π, also applies here. As we can see from Fig. 2.10, we start from the first action node, which represents the action a taken by the agent. When the agent takes action a in state s (not necessarily following policy π when taking the action), the environment could transition into one of several successor states s' (we show two such successor states), with a transition probability p that depends on (s, a). In each of

Fig. 2.10 Backup
diagram for Q_π

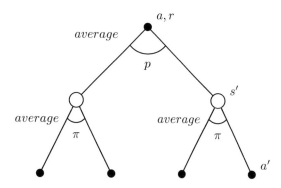

the successor states s', there may be multiple actions that the agent can choose from (we only show two actions). The exact action a' chosen will depend on the policy π that the agent is following. The Bellman equation takes the expectation (as indicated by the little arc under the successor state s') over all possible outcomes, weighted by the probability of their occurrence.

Worked Example for V_π

To better understand how to compute the value for a state under policy π, let's look at an example using our service dog MDP. We chose the random policy π, which means the probability of selecting each action was evenly distributed across all (legal) actions. In this case, we're going to focus on computing the value for state *Room 2*. There are three legal actions *Go to room1, Go to room3, Go outside* the agent can choose in this state. And our initial guessed values for each of the successor states are $V_\pi(Room\ 1) = 1.0$, $V_\pi(Room\ 3) = 10.0$, $V_\pi(Outside) = 5.0$. Assuming we know the model of the environment and we use discount $\gamma = 0.9$, how might we compute the value for state *Room 2* for this random policy π?

We know from the model of the environment the immediate reward for each of the legal actions is $-2, -1, 0$, and the state transition probability from the current state to its successor states is always 1.0 for all legal actions. And we also know the probability of the agent selecting each of the legal actions is $0.33, 0.33, 0.33$. Finally, we know (guessed) the value for the different successor states s'. So we just need to plug those numbers into Eq. (2.13), and we can compute the values for state *Room 2* as

$$V_\pi(Room\ 2) = 0.33 * (-2 + 0.9 * 1.0) + 0.33 * (-1 + 0.9 * 10.0)$$
$$+ 0.33 * (0 + 0.9 * 5.0)$$
$$= 0.33 * -1.1 + 0.33 * 8 + 0.33 * 4.5$$
$$= 3.76$$

The above results are shown in Fig. 2.11. Of cause, the estimated value for state *Room 2* is not accurate, just like the worked example for MRP. But if we include all the states in the state space and repeat the process over a large number of times, in the end the estimated values would be very close to the true values. This idea is essentially what the dynamic programming algorithms do, which we'll introduce in Chap. 3.

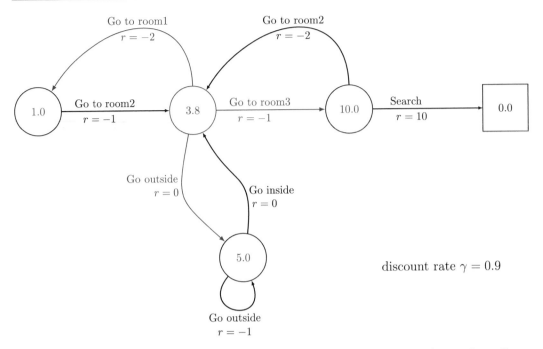

Fig. 2.11 Example of how to compute the value of state *Room 2* for the service dog MDP, for a random policy π, $\gamma = 0.9$; the initial state values are chosen randomly

Worked Example for Q_π

Let's look at an example of how to compute the state-action values $Q_\pi(s, a)$ using our service dog MDP. The conditions were almost identical to the worked example where we compute the value for $V_\pi(Room\ 2)$. But in this case, we want to compute the state-action value for all legal actions in state *Room 2*.

We still use the random policy π, which means the probability of selecting each action was evenly distributed across all (legal) actions. Our initial guessed values for each of the states are $V_\pi(Room\ 1) = 1.0$, $V_\pi(Room\ 3) = 10.0$, $V_\pi(Outside) = 5.0$. Assuming we know the model of the environment and we use discount $\gamma = 0.9$, how might we compute the state-action value for each of the (legal) actions in state *Room 2* for this random policy?

We know from the model of the environment the immediate reward for each of the legal actions is $-2, -1, 0$, and the state transition probability from the current state to its successor state is always 1.0 for each state-action pair (s, a). And we know (guessed) the value for the different successor states s'. So we just need to plug those numbers into Eq. (2.17), and we can compute the values for each action as

$$Q_\pi(Room\ 2,\ Go\ to\ room1) = -2 + 0.9 * 1.0$$
$$= -1.1$$

$$Q_\pi(Room\ 2,\ Go\ to\ room3) = -1 + 0.9 * 10.0$$
$$= 8.0$$

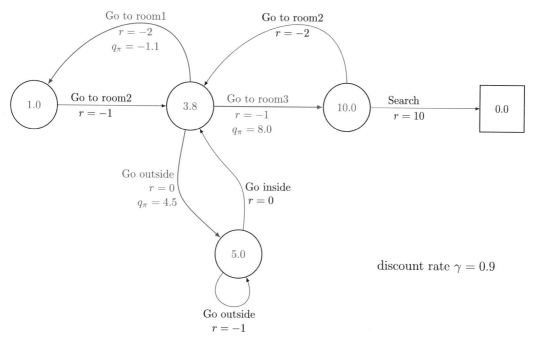

Fig. 2.12 Example of how to compute state-action values for all legal actions in state *Room2*, for a random policy to the service dog MDP. $\gamma = 0.9$; the initial state values are chosen randomly

$$Q_\pi(Room\ 2,\ Go\ outside) = 0 + 0.9 * 5.0$$
$$= 4.5$$

We can use Eq. (2.11) to compute the value for state *Room 2* under this random policy π; the outcome is exactly the same as we've done before using Eq. (2.13):

$$V_\pi(Room\ 2) = 0.33 * -1.1 + 0.33 * 8 + 0.33 * 4.5$$
$$= 3.76$$

The above results are shown in Fig. 2.12. Once we have an accurate estimate of Q_π for the policy π, then it's fairly easy to use it to help the agent to make better decisions. All the agent needs to do is select the action that has the highest state-action value $\arg\max_a Q_\pi(s, a)$ for all states in the state space. And it doesn't need to do a one-step look-ahead anymore.

2.6   Alternative Bellman Equations for Value Functions

One limitation Eq. (2.13) has is that we assume the reward signal from the environment is only conditioned on the current environment state s and the action a taken by the agent. However, there are cases the reward function is also conditioned on the successor state s', like $R(s, a, s')$, then the

Bellman equation for V_π needs to be rewritten to support this scenario as shown in Eq. (2.18):

$$V_\pi(s) = \mathbb{E}_\pi\left[R_t + \gamma V_\pi(S_{t+1}) \mid S_t = s\right]$$

$$= \sum_{a \in A} \pi(a|s)\left[\sum_{s' \in S} P(s'|s,a)R(s,a,s') + \gamma \sum_{s' \in S} P(s'|s,a)V_\pi(s')\right]$$

$$= \sum_{a \in A} \pi(a|s)\sum_{s' \in S} P(s'|s,a)\left[R(s,a,s') + \gamma V_\pi(s')\right], \quad \text{for all } s \in S \qquad (2.18)$$

Similarly, we can rewrite the Bellman equation for Q_π for the case where the reward function is conditioned on the successor state s' like shown in Eq. (2.19):

$$Q_\pi(s,a) = \mathbb{E}_\pi\left[R_t + \gamma Q_\pi(S_{t+1}, A_{t+1}) \mid S_t = s, A_t = a\right]$$

$$= \sum_{s' \in S} P(s'|s,a)R(s,a,s') + \gamma \mathbb{E}_\pi\left[Q_\pi(s',a') \mid S_{t+1} = s', A_{t+1} = a'\right]$$

$$= \sum_{s' \in S} P(s'|s,a)R(s,a,s') + \gamma \sum_{s' \in S} P(s'|s,a)\sum_{a' \in A} \pi(a'|s')Q_\pi(s',a')$$

$$= \sum_{s' \in S} P(s'|s,a)\left[R(s,a,s') + \gamma \sum_{a' \in A} \pi(a'|s')Q_\pi(s',a')\right], \quad \text{for all } s \in S, a \in A \quad (2.19)$$

However, both of these equations assume the reward of the environment is deterministic. That means the same (s,a) or (s,a,s') will always yield the same reward signal. But there are cases the reward could also be stochastic. Then in that case, they will fail to capture this randomness.

In their famous book *Reinforcement Learning: An Introduction* [2], Sutton and Barto used Eq. (2.20) to cover this scenario. They consolidated the dynamics function P and the reward function R into a single four-parameter function $P(s', r|s, a)$, which covers the problem where the reward could be conditioned on the successor state s', or the reward could be stochastic all at the same time.

$$V_\pi(s) = \mathbb{E}_\pi\left[R_t + \gamma V_\pi(S_{t+1}) \mid S_t = s\right]$$

$$= \sum_{a \in A} \pi(a|s)\sum_{s'}\sum_{r} P(s', r|s, a)\left[r + \gamma V_\pi(s')\right]$$

$$= \sum_{a \in A} \pi(a|s)\sum_{s',r} P(s', r|s, a)\left[r + \gamma V_\pi(s')\right], \quad \text{for all } s \in S \qquad (2.20)$$

And similarly in this case, the Bellman equation for Q_π becomes

$$Q_\pi(s,a) = \mathbb{E}_\pi\left[R_t + \gamma Q_\pi(S_{t+1}, A_{t+1}) \mid S_t = s, A_t = a\right]$$

$$= \sum_{s'}\sum_{r} P(s', r|s, a)\left[r + \gamma \mathbb{E}_\pi\left[Q_\pi(s',a') \mid S_{t+1} = s', A_{t+1} = a'\right]\right]$$

$$= \sum_{s',r} P(s', r|s, a)\left[r + \gamma \sum_{a' \in A} \pi(a'|s')Q_\pi(s',a')\right], \quad \text{for all } s \in S, a \in A \qquad (2.21)$$

While Eqs. (2.20) and (2.21) are more general, they might be much harder to fully understand. To keep things simple, we will stick to the Bellman equations (2.13), (2.16), and (2.17) in this book.

2.7 Optimal Policy and Optimal Value Functions

The goal of using reinforcement learning to solve a problem is to find an optimal policy. If we can find a policy that yields the highest return for every single state in the entire state space, then we can consider the MDP is solved. This makes sense, because the only way the agent can get these maximum returns is by choosing the best actions in every single state. Formally, the optimal policy π_* is defined as the policy (or policies) which has the maximum state value V_π over all other policies, for all states in the state space, as shown in Eq. (2.22).

$$\pi_*(s) = \arg\max_\pi V_\pi(s), \quad \text{for all } s \in \mathcal{S} \tag{2.22}$$

In theory, for every MDP there is an optimal policy π_* that is better than or equal to other policies. It's possible for the MDP to have multiple optimal policies. For example, if in a state, there are two or more actions that lead to the same successor state, and the rewards for these different actions are the same. Figure 2.13 is a small example MDP to illustrate this case. In this small example, the agent needs to drive a toy car from state A to state C. But the only way to get to state C is through state B. There are two possible actions (routes) that the agent can choose in state A; both actions will lead to state B. The rewards for these two actions are the same, so the agent can choose any one of these actions.

We should not confuse optimal policy with a deterministic policy. A deterministic policy always chooses the same action in the same state, but that doesn't mean the chosen action is the best action. But the optimal policy will always choose the best action.

The optimal policy (or policies) π_* of an MDP shares the same optimal state value function V_*:

$$V_*(s) = \max_\pi V_\pi(s), \quad \text{for all } s \in \mathcal{S} \tag{2.23}$$

Similarly, the optimal policy (or policies) π_* of an MDP shares the same optimal state-action value function Q_*:

$$Q_*(s, a) = \max_\pi Q_\pi(s, a), \quad \text{for all } s \in \mathcal{S}, a \in \mathcal{A} \tag{2.24}$$

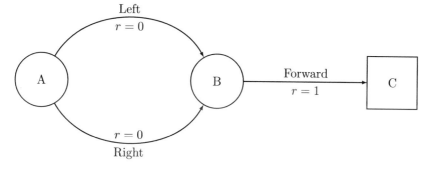

Fig. 2.13 Example of a toy car MDP with multiple optimal policies

There's a nice relationship between the optimal state value function V_* and the optimal state-action value function Q_*. It says the optimal value of a state $V_*(s)$ is equal to $Q_*(s, a)$ where a is the action that has the maximum state-action value over all other actions in this state s. This should make sense, because if the agent is following the optimal policy, then there's a 1.0 probability it will select the best action and a 0 probability of selecting nonbest actions (at least for the case of deterministic policy).

$$V_*(s) = \max_{a \in A} Q_*(s, a) \tag{2.25}$$

Like the regular value functions, we can also rewrite Eq. (2.23) in a recursive manner. Here, we only care about the action that has the maximum immediate reward and the discount value of successor states. Equation (2.26) is the Bellman optimality equation for V_*:

$$
\begin{aligned}
V_*(s) &= \max_{a \in A} Q_*(s, a) \\
&= \max_{a} \mathbb{E}\Big[R_t + \gamma G_{t+1} \;\Big|\; S_t = s, A_t = a \Big] \\
&= \max_{a} \mathbb{E}\Big[R_t + \gamma V_*(S_{t+1}) \;\Big|\; S_t = s, A_t = a \Big] \\
&= \max_{a} \Big[R(s, a) + \gamma \sum_{s' \in \mathcal{S}} P(s'|s, a) V_*(s') \Big], \quad \text{for all } s \in \mathcal{S}
\end{aligned}
\tag{2.26}
$$

And similarly, we can also derive the Bellman optimality equation for Q_*, as shown in Eq. (2.27):

$$
\begin{aligned}
Q_*(s, a) &= \mathbb{E}\Big[R_t + \gamma \max_{a'} Q_*(S_{t+1}, a') \;\Big|\; S_t = s, A_t = a \Big] \\
&= R(s, a) + \gamma \sum_{s' \in \mathcal{S}} P(s'|s, a) \max_{a' \in A} Q_*(s', a'), \quad \text{for all } s \in \mathcal{S}, a \in A(s)
\end{aligned}
\tag{2.27}
$$

Since $V_*(s) = \max_a Q_*(s, a)$, we can also rewrite the Bellman optimality equation for Q_* using V_*:

$$
\begin{aligned}
Q_*(s, a) &= \mathbb{E}\Big[R_t + \gamma V_*(S_{t+1}) \;\Big|\; S_t = s, A_t = a \Big] \\
&= R(s, a) + \gamma \sum_{s' \in \mathcal{S}} P(s'|s, a) V_*(s')
\end{aligned}
\tag{2.28}
$$

We can draw the backup diagram for Bellman optimality equation. Instead of averaging over all the actions, the Bellman optimality equation for V_* only considers the action with the highest value, as shown in Fig. 2.14. Notice here it still averages over all the possible successor states s'; however, these averaged values from different action nodes (two are shown here) are not summed together.

And Fig. 2.15 shows the backup diagram for Eq. (2.27). Similarly, here it only considers the action that yields the highest value for the successor state s'.

In practice, we are more interested in the optimal state-action value function Q_*. The reason is very simple; if we have Q_*, we can easily construct an optimal policy based on Q_* like shown in

Fig. 2.14 Backup
diagram for V_*

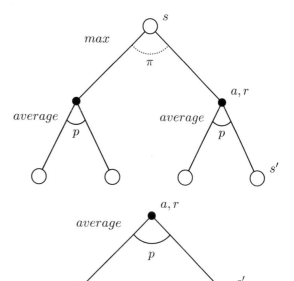

Fig. 2.15 Backup
diagram for Q_*

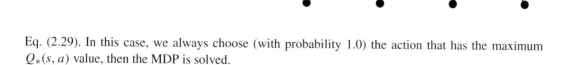

Eq. (2.29). In this case, we always choose (with probability 1.0) the action that has the maximum $Q_*(s, a)$ value, then the MDP is solved.

$$\pi_*(a|s) = \begin{cases} 1, & \text{if} \quad a = \arg\max_{a \in A} Q_*(s, a) \\ 0, & \text{otherwise} \end{cases} \tag{2.29}$$

The situation becomes a little bit more complex if we only know the optimal state value function V_*. Because the values are averaged over all actions, we need to do a one-step look-ahead to decide which action is the best one. And we'll need to use the model of the environment to do that.

$$\pi_*(a|s) = \begin{cases} 1, & \text{if} \quad a = \arg\max_{a \in A} \sum_{s' \in \mathcal{S}} \left[R(s, a) + \gamma V_*(s') \right] \\ 0, & \text{otherwise} \end{cases} \tag{2.30}$$

2.8 Summary

This chapter provided a comprehensive introduction to Markov decision processes (MDPs) which form the foundation of reinforcement learning. The chapter began with an overview of MDPs, followed by an explanation of how they can be used to model the RL problem. This chapter holds significant importance as it laid the groundwork for the subsequent content in the book. The concepts and mathematical equations presented here are so crucial that they are extensively utilized throughout the rest of the book, making it a key chapter deserving of our highest recognition.

The chapter delved into the details of Markov processes, Markov reward processes, and Markov decision processes, providing a clear understanding of the components that make up an MDP. The state value function and state-action value function for MDPs were examined, along with their computation utilizing the powerful Bellman equations. Furthermore, alternative Bellman equations for value functions were explored, fostering a more profound insight into the calculation of these essential functions for more complex problems.

Finally, the chapter covered optimal policies and optimal value functions, discussing how they can be derived using various methods. Overall, this chapter provided a solid foundation for understanding MDPs and their role in RL.

The next chapter of the book will focus on using dynamic programming (DP) methods to solve tabular RL problems. By leveraging the knowledge gained in this chapter, readers will be able to better understand the DP methods and their applications in RL.

References

[1] Professor Emma Brunskill and Stanford University. Cs234: Reinforcement learning. http://web.stanford.edu/class/cs234/, 2021.
[2] Richard S. Sutton and Andrew G. Barto. *Reinforcement Learning: An Introduction*. The MIT Press, second edition, 2018.

Dynamic Programming

3

The ultimate goal of modeling a reinforcement learning problem using Markov decision processes (MDPs) is that we can use the Bellman equations to find an optimal policy π_* that maximizes the expected cumulative reward. However, finding such a policy is not always straightforward. In this chapter, we'll introduce dynamic programming (DP) algorithms as a way to find the optimal policy π_* when we have access to a perfect model of the environment.

Dynamic programming (DP) is a mathematical optimization method developed by Richard Bellman in the 1950s [1]. The core idea of dynamic programming is to break down a complex problem into simpler subproblems and solve them in a recursive manner. This approach is particularly useful for problems that exhibit overlapping subproblems and optimal substructure, meaning that the solution to a subproblem can be used to solve the overall problem.

In the context of reinforcement learning, dynamic programming algorithms can help an agent find the optimal value functions V_* and Q_*.

The Bellman equations, which we introduced in Chap. 2, provide a recursive way to compute the value functions V_π and Q_π. Specifically, the Bellman equations express the value of a state or state-action pair in terms of the expected immediate reward and the expected value of the next state or state-action pair under the current policy π. This recursive formulation allows us to compute the value functions iteratively, starting with an initial guess and updating the values until convergence to V_* and Q_*.

However, to find the true optimal value functions using DP algorithms, the agent needs to have access to the perfect model of the environment, which means that DP algorithms are model-based reinforcement learning methods.

Additionally, DP algorithms can be computationally expensive, especially for problems with large state and action spaces. As a result, other RL methods, such as Monte Carlo methods and temporal difference learning which we'll introduce later in this book, have been developed to overcome these limitations.

Overall, DP algorithms provide a powerful framework for solving reinforcement learning problems when a perfect model of the environment is available. By breaking down the problem into simpler subproblems and using the Bellman equations to compute the value functions, we can find the optimal policy π_* that maximizes the expected cumulative reward.

© The Author(s), under exclusive license to APress Media, LLC, part of Springer Nature 2023   47
M. Hu, *The Art of Reinforcement Learning*,
https://doi.org/10.1007/978-1-4842-9606-6_3

3.1   Use DP to Solve MRP Problem

To get us started with DP algorithms, let's first focus on the Markov reward process (MRP) case, which is much simpler than MDP as it does not involve any actions in the process. We still focus on our service dog MRP problem, which we've introduced in Chap. 2.

The Bellman equation for the state value function in the case of Markov reward processes (MRPs) is expressed as

$$V(s) = R(s) + \gamma \sum_{s' \in \mathcal{S}} P(s'|s)V(s'), \quad \text{for all } s \in \mathcal{S}$$

Here, γ is the discount rate; $R(s)$ is the reward obtained from the environment; $P(s'|s)$ is the dynamics function of the environment, which specifies the probability of transitioning from state s to s'; and $V(s')$ is the value of the successor state.

If we have access to the true model of the MRP, we can use the DP method to compute the state values for the different states of MRP. The DP algorithm works by sweeping over the entire state space of the MRP in each iteration, computing and updating the state value for each state using the reward and discounted values of the successor state. This process is repeated for a large number of iterations until the true state values are obtained. The pseudocode for the algorithm is shown in Algorithm 1, which is the algorithm we used to compute the state values for our service dog MRP back in Chap. 2 as shown in Fig. 2.7.

Algorithm 1: DP compute state value function for MRP

Input: Discount rate γ
Initialize: $V^0(s) = 0$, for all $s \in \mathcal{S}$
Output: The state value function V for an MRP
1 **for** $k = 1$, until convergence **do**
2   **for** $s \in \mathcal{S}$ **do**
3           $V^k(s) = R(s) + \gamma \sum_{s' \in \mathcal{S}} P(s' \mid s)V^{k-1}(s')$

But reinforcement learning is primarily concerned with solving Markov decision processes (MDPs), which involve agents making decisions during the learning process. In the remainder of this chapter, we will focus on the use of DP algorithms to solve MDPs.

3.2   Policy Evaluation

The goal of solving a reinforcement learning or MDP problem is to find the optimal policy π_*. However, before we start searching for the optimal policy, let's first consider a simpler problem: given two different policies π and π', how can we tell which one is better?

To answer this question, we can use the state value function V_π. The value function measures how good or bad it is for the agent to be in a state by measuring the expected return starting from that state and following a policy π. If we have an accurate estimate of the state value function for each of these policies π and π', we can then measure the difference between these two policies. We say a policy π'

is as good as or better than policy π when the expected return $V_{\pi'}(s)$ is greater than or equal to $V_\pi(s)$, for all states s in the state space:

$$\pi' \geq \pi \quad \text{if } V_{\pi'}(s) \geq V_\pi(s), \quad \text{for all } s \in \mathcal{S}$$

We can estimate the value of a single state s for a policy π using the Bellman equation for V_π and the model of the environment. To compute the value of each state in the state space, we can generalize this process by iteratively computing the value of each state in the state space using the Bellman equation for V_π and the model of the environment. This process is known as *policy evaluation*. The algorithm estimates the state value function V_π for an arbitrary policy π, Policy evaluation is also known as prediction, as it predicts the state value function for a policy π.

The core of the policy evaluation algorithm is the Bellman equation for V_π:

$$V_\pi(s) = \mathbb{E}_\pi\left[R_t + \gamma V_\pi(S_{t+1}) \mid S_t = s\right]$$

$$= \sum_{a \in A} \pi(a|s)\left[R(s,a) + \gamma \sum_{s' \in \mathcal{S}} P(s'|s,a)V_\pi(s')\right], \quad \text{for all } s \in \mathcal{S} \qquad (3.1)$$

The algorithm uses the Bellman equation as an update rule and also uses the model (dynamics function and reward function) of the environment during the value update. In theory, the computed values will converge to the true values for the policy π if we run the algorithm forever. It is important to note that the convergence is guaranteed as long as the algorithm satisfies certain conditions, such as being run for an infinite number of iterations and the discount factor being less than one. The convergence rate can depend on the policy and the problem instance.

For example, consider a simple grid world with a goal state and a pitfall state, as shown in Fig. 3.1. The agent receives a reward of +1 for reaching the goal state and a reward of −1 for falling into the pitfall state. All other state transitions have a reward of 0. The agent has four actions available in each state: up, down, left, and right. Suppose we have an arbitrary policy that always moves up, except in the goal state where it stops. We can use policy evaluation to estimate the value function for this policy and then compare it with other policies. By repeating the Bellman update equation for each state in the grid world, we can estimate the value of each state for the given policy. This process can be repeated iteratively until convergence is achieved. The resulting state value function can help us to measure the quality of the policy and identify areas where the policy can be improved. For example, in the grid world problem, if we find that the value of the state immediately above the pitfall is higher than the value of the state immediately below it, we can infer that it would be better to move down instead of up in this particular state, to avoid the pitfall.

Overall, policy evaluation is an important algorithm in reinforcement learning as it allows us to estimate the value function of a given policy, which is a crucial step in solving reinforcement learning problems. It is worth noting that there are other methods, such as Monte Carlo and temporal difference learning, that can be used to estimate the value function, but policy evaluation is a fundamental approach that forms the basis of many other reinforcement learning algorithms.

The algorithm takes an arbitrary policy π, a discount factor γ, and a small threshold ϵ which determines the stopping criterion. The algorithm starts with an initial estimated value of zero for all states and then iterates until the estimated values converge to the true state value for the input policy π. Specifically, the algorithm updates the estimated value of each state by taking a weighted average

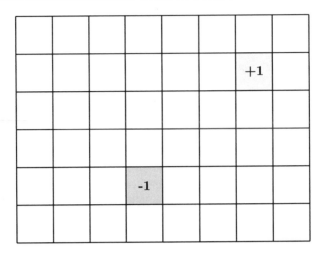

Fig. 3.1 Example of a grid world

of the immediate reward obtained in the next time step and the estimated value of the next state, using the current policy π to select the next action, which is captured in Eq. (3.2).

$$V_\pi^{k+1}(s) \leftarrow \sum_{a\in A} \pi(a|s)\left[R(s,a) + \gamma \sum_{s'\in\mathcal{S}} P(s'|s,a)V_\pi^k(s')\right] \tag{3.2}$$

where $V_\pi^k(s)$ is the estimated value of state s at iteration k, $\pi(a|s)$ is the probability of selecting action a in state s under policy π, $P(s'|s,a)$ is the probability of transitioning to successor state s', and $R(s,a)$ is the reward provided by the environment when the agent takes action a in state s. The algorithm repeats this update for all states until the difference between the updated and previous estimated values for all states is less than ϵ. This threshold determines the stopping criterion and ensures that the algorithm terminates when the estimates have converged to the true values.

Algorithm 2: Policy evaluation, iterative algorithm

Input: Policy π to be evaluated, discount rate γ, a small threshold ϵ determining the accuracy of the estimation
Initialize: $V_\pi(s) = 0$, for all $s \in S$
Output: The estimated state value function V_π for the input policy π

1 **while** not converge **do**
2 $\quad$ $\delta = 0$
3 $\quad$ **for** $s \in \mathcal{S}$ **do**
4 $\quad\quad$ $v = V_\pi(s)$
5 $\quad\quad$ $V_\pi(s) \leftarrow \sum_a \pi(a|s)\left[R(s,a) + \gamma \sum_{s'\in\mathcal{S}} P(s'|s,a)V_\pi(s')\right]$
6 $\quad\quad$ $\delta = \max\left(\delta, |v - V_\pi(s)|\right)$
7 $\quad$ **if** $\delta < \epsilon$ **then**
8 $\quad\quad$ Break

For each iteration of the policy evaluation algorithm, the agent sweeps over all states in the state space. For each state, the agent may choose from multiple actions, each with a different probability of being selected according to the policy π. From each action a, the environment may transition to one of several successor states s', with a transition probability p that depends on the current state s and the action a taken by the agent. This is possible since the agent has access to the true model of the environment, that's the reward signal $R(s, a)$ and transition probability $P(s'|s, a)$ for each state-action pair. The algorithm uses Eq. (3.1) as a backup rule, which takes the average over all possible outcomes, weighted by their respective probabilities.

Once the agent has computed new estimates for the value of each state, it begins a new iteration with these updated values. While the estimated values will converge to the true values for the policy π in the limit, in practice we must decide when to stop the algorithm before this point. To determine when the algorithm has converged to an accurate estimate of the value function, we compare the absolute difference between the updated value $V_\pi(s)$ and the previous estimate for each state s and keep track of the maximum absolute difference across all states in each iteration. After completing one full iteration over all states, we compare the maximum absolute difference to a small threshold ϵ. If this maximum absolute difference is smaller than ϵ, we consider the algorithm to have converged and terminate the loop.

Storing the estimated values for all states in a single array can speed up the evaluation process because it allows us to update the values in place. This means that we can update the estimate of the value function for a particular state and then immediately use the updated estimate in subsequent calculations without having to copy it to a new data structure.

In contrast, if we stored the estimated values for each state in a separate data structure, we would need to copy the updated value from one data structure to another for each update. This copying operation could be computationally expensive, especially if we have a large number of states.

In addition, storing the values for all states in a single array can also make the code simpler and easier to read. It is easier to iterate over a single array than to keep track of multiple data structures for each state.

Overall, storing the estimated values for all states in a single array can be a practical optimization that improves both the speed and clarity of the code.

Worked Example

Let's use the policy evaluation algorithm to estimate the state value function V_π for a (fixed) random policy π for the service dog MDP. The values shown in Fig. 3.2 were computed using a discount factor of 0.9 and a threshold of $1e - 5$.

It's worth noting that the value for the terminal state is always zero, regardless of whether the input policy is optimal or random. This is because the state value function measures the expected return starting from state s under policy π. As there are no immediate rewards or successor states after the environment reaches the terminal state, its value is always zero.

3.3   Policy Improvement

The ultimate goal of reinforcement learning is to find the optimal policy π_* that maximizes the expected return for all states in the entire state space. However, searching the entire policy space one by one to find π_* is infeasible for MDPs with large state and action spaces. Fortunately, there is a more efficient way to improve a given policy, which is called policy improvement.

The idea behind policy improvement is simple. Instead of searching for the optimal policy in one go, we iteratively make small and incremental improvements to the current policy by modifying the policy to be greedier with respect to the estimated state-action value function Q_π for the current

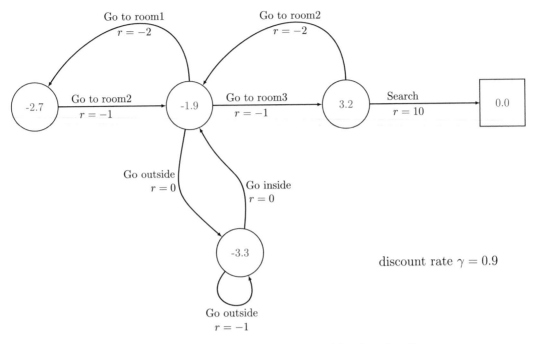

Fig. 3.2 State value for random policy π for the service dog MDP, $\gamma = 0.9$ and $\epsilon = 1e - 5$

policy. If a modified policy has higher expected returns for at least one state, it is guaranteed to be better than the original policy. This is the key concept behind policy improvement, and it is based on the following theorem:

For any two deterministic policies π and π', π' is at least as good as π if and only if $V_{\pi'}(s) \geq V_\pi(s)$ for all states $s \in \mathcal{S}$.

Here, $V_\pi(s)$ is the estimated value of state s under policy π. In other words, a modified policy is better than the original policy if it increases the expected return for at least one state.

This is where we need to use the state-action value function $Q_\pi(s, a)$, because the state value function $V_\pi(s)$ doesn't tell us which action is better when the agent is in state s and following policy π. We've already introduced the Bellman equation for Q_π in Chap. 2. Here's a quick review.

$$Q_\pi(s, a) = R(s, a) + \gamma \mathbb{E}_\pi \left[Q_\pi(s', a') \mid S_{t+1} = s', A_{t+1} = a' \right]$$

$$= R(s, a) + \gamma \sum_{s' \in \mathcal{S}} P(s'|s, a) \sum_{a' \in A} \pi(a'|s') Q_\pi(s', a'), \quad \text{for all } s \in \mathcal{S}, a \in A \quad (3.3)$$

Equation (3.3) tells us how to update values for a state-action pair (s, a) using Q_π, but we don't have an accurate estimate of Q_π yet. Luckily, there's a relationship between Q_π and V_π, where $V_\pi(s) = \sum_{a \in A} \pi(a|s) Q_\pi(s, a)$. We've already know how to use the policy evaluation algorithm to compute the state value function V_π for a policy π. And we can simply plug V_π into Eq. (3.4) which will help us finish the value update rule for estimated state-action value function Q_π.

$$Q_\pi(s, a) = R(s, a) + \gamma \sum_{s' \in S} P(s'|s, a) V_\pi(s'), \quad \text{for all } s \in S, a \in A \qquad (3.4)$$

This equation tells us that the value of taking action a in state s and then following policy π is the expected reward R_t plus the discounted value of the next state S_{t+1}, which is given by the state value function $V_\pi(St + 1)$. We can compute Q_π for all state-action pairs using the policy evaluation algorithm.

Once we have an estimate of the state-action value function Q_π, we can perform policy iteration, which involves improving the policy by making it greedier with respect to Q_π. Specifically, we can construct a new policy π' by selecting the action that maximizes the state-action value function at each state.

We now introduce the *policy improvement* algorithm as shown in Algorithm 3. It takes in the estimated state value function V_π for an arbitrary policy π (which is coming from policy evaluation algorithm). It then sweeps over all the states in the state space; for each state, it goes through all (legal) actions in this state to compute state-action value function Q_π based on the model of the environment and V_π. After the sweep is done, it then uses Q_π to compute a better, improved deterministic policy π'. It does this by going through all the states in the state space; for every state, it updates the policy probability distribution for each action, so that the probability of choosing the action with the highest estimated value $Q_\pi(s, a)$ is 1, and the probability of choosing other actions is 0.

Algorithm 3: Policy improvement, iterative algorithm

Input: Estimated state value function V_π for π, discount rate γ
Initialize: $Q_\pi(s, a) = 0$, for all $s \in S, a \in A$
Output: Improved deterministic policy π'

```
/* Compute state-action value function using estimated state value
   function                                                          */
```
1 **for** $s \in S$ **do**
2   **for** $a \in A(s)$ **do**
3          $Q_\pi(s, a) \leftarrow R(s, a) + \gamma \sum_{s' \in S} P(s'|s, a) V_\pi(s')$

```
/* Compute an improved deterministic policy                          */
```
4 **for** $s \in S$ **do**
5      $A_* = \arg\max_a Q_\pi(s, a)$
6   **for** $a \in A(s)$ **do**
7   **if** $a = A_*$ **then**
8              $\pi'(a|s) = 1$
9   **else**
10             $\pi'(a|s) = 0$

The policy improvement algorithm consists of two steps: the first step is to use Eq. (3.3) and the model of the environment to compute the estimated state-action value function Q_π for the current policy π; the second step is to improve the policy π by constructing a new, better deterministic policy π' based on Q_π. Unlike the policy evaluation algorithm, the policy improvement algorithm does not run forever. It only goes through these two steps once. This makes sense as we now have a new, better policy π'; the old state value function V_π does not represent the values of this new policy anymore because the policy has changed.

To implement Algorithm 3, we could use a 2D matrix with the shape of (size of state space, size of action space) to represent the state-action value function Q_π. One thing to pay attention to is during the construction of the improved policy, we should only consider the legal actions for the specific state; legal actions are those that are available in the current state. Excluding illegal actions is important because it ensures that the action selection process is performed correctly and does not lead to suboptimal or even dangerous behavior.

In reinforcement learning, excluding illegal actions is particularly important when constructing a policy based on the estimated state-action value function. Without excluding illegal actions, the policy may select suboptimal or even illegal actions leading to poor performance or failure to achieve the desired goal. By considering only legal actions in the action selection process, the policy can ensure that it selects the best action possible for the given state while still satisfying any constraints or safety requirements.

For example, consider a robot in a factory environment with certain safety constraints, such as not colliding with other robots or objects. If the robot's action selection process includes illegal actions, such as moving in a direction that would result in a collision, it could cause damage to itself, other robots, or the factory environment.

Another concrete example is this: let's say the values for the legal actions are all negative $Q_\pi(s, a) < 0$ for some state s; if we don't exclude the illegal actions, then a standard $\arg\max$ will always choose one of the illegal actions, since zero is greater than negative values.

3.4   Policy Iteration

Policy iteration [2] is an algorithm for finding the optimal policy in a Markov decision process (MDP). The algorithm consists of two main steps, policy evaluation and policy improvement, which are iteratively applied to refine the policy until it converges to the optimal policy.

The key idea behind policy iteration is to use the value function to greedily select better actions and then update the value function using the new policy. The algorithm starts with an initial policy, which is typically chosen randomly, and then iteratively improves the policy by estimating the value function and selecting better actions.

The policy iteration algorithm runs these two steps iteratively in a loop. At each iteration, we first estimate the state value function V_π^i for the current policy π^i, using the policy evaluation algorithm. The value function represents the expected cumulative reward that can be obtained by following the policy from a given state.

$$\pi^0 \xrightarrow{\text{evaluation}} V_\pi^0 \xrightarrow{\text{improvement}} \pi^1 \xrightarrow{\text{evaluation}} V_\pi^1 \xrightarrow{\text{improvement}} \pi^2 \cdots \xrightarrow{\text{improvement}} \pi_*$$

Once we have estimated the value function, we use it to select better actions by running the policy improvement algorithm. The policy improvement algorithm greedily selects the action that maximizes the expected cumulative reward, based on the estimated value function. This new policy is denoted by π^{i+1} and is an improvement over the previous policy π^i.

We repeat these two steps, policy evaluation and policy improvement, until convergence. Convergence is achieved when the old policy π^i is the same as π^{i+1} for every state in the state space, meaning there's no improvement that can be made anymore.

The policy iteration algorithm is summarized in Algorithm 4. Note that this algorithm can be computationally expensive, especially for large MDPs, but it is guaranteed to converge to the optimal policy.

Algorithm 4: Policy iteration, iterative algorithm

Input: Discount rate γ, a small threshold ϵ determining the accuracy of the estimation
Initialize: $i = 0$, random policy π^0
Output: Optimal policy π_* and the optimal value functions V_*, Q_*

1 **while** not converge **do**
2   V_π^i = policy **evaluation**$(\pi^i, \gamma, \epsilon)$
3   π^{i+1} = policy **improvement**(V_π^i, γ)
4       $i = i + 1$
        /* L1-norm to measure if the policy changed for any state   */
5   **if** $|\pi^i - \pi^{i-1}|_1 = 0$ **then**
6 | Break

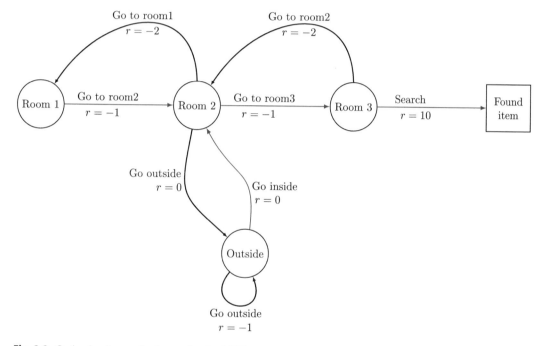

Fig. 3.3 Optimal policy π_* for the service dog MDP

Overall, policy iteration is a powerful algorithm for finding the optimal policy in an MDP. By iteratively refining the policy and updating the value function, we can converge to the optimal policy. However, it is important to note that the policy iteration algorithm may not be suitable for large MDPs, as it can be computationally expensive. In such cases, other algorithms such as value iteration (which we will introduce later in this chapter) may be more appropriate.

Worked Example

We can use the policy iteration algorithm to solve our service dog MDP. The computed optimal policy π_* for the MDP is shown in Fig. 3.3. Along with the optimal policy, we obtain the optimal value functions V_* and Q_*, which are shown in Figs. 3.4 and 3.5, respectively. These values were computed using a discount factor of $\gamma = 0.9$.

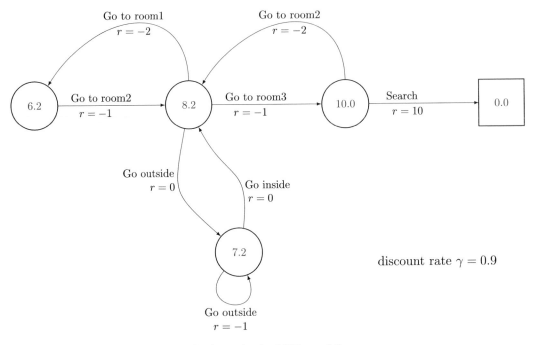

Fig. 3.4 Optimal state value function V_* for the service dog MDP, $\gamma = 0.9$

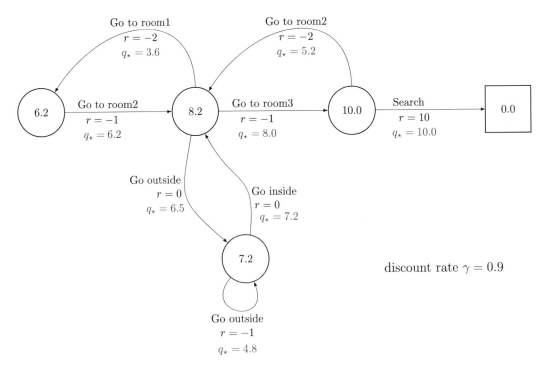

Fig. 3.5 Optimal state-action value function Q_* for the service dog MDP, $\gamma = 0.9$

Table 3.1 Optimal state value for the service dog MDP, with different discount factors γ

Discount	Room 1	Room 2	Room 3	Outside	Found item
$\gamma = 0.0$	−1.0	−1.0	10.0	0.0	0
$\gamma = 0.3$	−0.4	2.0	10.0	0.6	0
$\gamma = 0.5$	1.0	4.0	10.0	2.0	0
$\gamma = 0.7$	3.2	6.0	10.0	4.2	0
$\gamma = 0.9$	6.2	8.0	10.0	7.2	0
$\gamma = 1.0$	8.0	9.0	10.0	9.0	0

By running the policy iteration algorithm with different discount factors γ, we can observe how the discount affects the state values, as shown in Table 3.1. Notably, the value of the terminal state is always zero.

3.5   General Policy Iteration

The policy iteration algorithm is a widely used and effective method in reinforcement learning. It is characterized by running policy evaluation and policy improvement processes concurrently. All policy iteration algorithms share this basic structure and concept, but we have the flexibility to modify the specific implementations of the policy evaluation and policy improvement steps.

In policy iteration, we start with an initial policy, which is a mapping from states to actions. We then perform policy evaluation to estimate the value function for the policy. This involves iterating over the state space, computing the expected value of each state under the policy, and updating the value function accordingly. After policy evaluation, we perform policy improvement by constructing a new policy that is greedier than the old policy with respect to the value function. This new policy is obtained by choosing the action that maximizes the expected value of the next state, starting from the current state, and repeating this process until the end of the episode. The new policy is then used for the next round of policy evaluation, and the process is repeated until convergence.

Algorithm 5: General policy iteration

    Initialize: $i = 0$, random policy π^0
    Output: The (optimal) policy π after the search
1 **while** not converge **do**
2 $V_\pi^i = $ policy **evaluation**(π^i)
3 $\pi^{i+1} = $ policy **improvement**(V_π^i)
4 $i = i + 1$

Although dynamic programming is generally considered a model-based reinforcement learning method, policy iteration can be used for both model-based and model-free methods in practice. For model-based methods, the policy evaluation step involves computing the state transition probabilities and rewards from the model, while for model-free methods, these values are estimated from experience using Monte Carlo or TD methods, which we'll discuss later in the book.

3.6   Value Iteration

Value iteration is a powerful algorithm for finding the optimal policy in reinforcement learning. While policy iteration is effective, one downside is that it requires multiple iterations of policy evaluation and policy improvement to find the optimal policy π_*, which can be time-consuming and computationally expensive. Value iteration addresses this issue by combining policy evaluation and policy improvement into one step.

To find the optimal policy, the value iteration algorithm computes the optimal state value function V_* directly instead of the estimated state value function V_π for an arbitrary policy π. The optimal state value function V_* is the one that has the highest state values across the entire policy space.

The value iteration algorithm uses the Bellman optimality equation for V_* as an update rule to update the values for different states. The equation computes the expected value of the reward plus the discounted value of the next state, given the current state and action. The maximum value of this expression over all possible actions is used to update the value of the current state. This process is repeated until the values of all states converge to their optimal values.

$$
\begin{aligned}
V_*(s) &= \max_a \mathbb{E}\left[R_t + \gamma V_*(S_{t+1}) \mid S_t = s, A_t = a \right] \\
&= \max_a \left[R(s, a) + \gamma \sum_{s' \in \mathcal{S}} P(s'|s, a) V_*(s') \right], \quad \text{for all } s \in \mathcal{S}
\end{aligned} \tag{3.5}
$$

Here, s represents the current state, a is an action, R_t is the reward at time t, γ is the discount factor, S_{t+1} is the next state, $P(s'|s, a)$ is the probability of transitioning to state s' from state s under action a, and $V_*(s')$ is the optimal value of the next state.

The value iteration algorithm sweeps over all states in the state space, and for each state, it uses Eq. (3.5) as an update rule to compute the estimated optimal value for this state. This is done using the model (reward function and dynamics function) of the environment. Once the current sweep is complete, the algorithm repeats this process for all states until convergence to the true optimal state value function V_*.

Once we have V_*, we can easily compute an optimal policy π_*. For each state in the state space, we select the action that yields the highest value of its successor state. We then set the probability of selecting this action to one and zero for all other actions. This process results in a deterministic policy that will always select the action that leads to the highest expected reward.

In summary, value iteration is an efficient algorithm for finding the optimal policy in reinforcement learning. It combines policy evaluation and policy improvement into a single step by computing the optimal state value function directly using the Bellman optimality equation. Once the optimal state value function is computed, it's easy to determine the optimal policy by selecting the action that leads to the highest expected reward for each state in the state space.

Algorithm 6 shows the pseudocode for the value iteration algorithm, which uses Eq. (3.5) as the value update rule. To illustrate the application of the value iteration algorithm, we can use the example of the service dog MDP. Applying the value iteration algorithm to the service dog MDP should yield the same result as shown in Figs. 3.3 and 3.4. However, it should be noted that the speedup of the value iteration algorithm over the policy iteration algorithm may not be noticeable for this simple example. This could be due to the relatively small number of states and actions in the problem or because the computational complexity of the two algorithms is similar for this particular problem.

Algorithm 6: Value iteration, iterative algorithm

Input: Discount rate γ, a small threshold ϵ determining the accuracy of the estimation
Initialize: $V(s) = 0$, for all $s \in \mathcal{S}$
Output: Optimal policy π_* and optimal state value function V_*
/* Find optimal state value function V_* */
1 **while** not converge **do**
2      $\delta = 0$
3   **for** $s \in \mathcal{S}$ **do**
4         $v = V(s)$
5         $V(s) \leftarrow \max_a \left[R(s,a) + \gamma \sum_{s' \in \mathcal{S}} P(s'|s,a) V(s') \right]$
6         $\delta = \max\left(\delta, |v - V(s)|\right)$
7   **if** $\delta < \epsilon$ **then**
8   Break
/* Compute optimal policy π_* based on V_* */
9 **for** $s \in \mathcal{S}$ **do**
10     $A_* = \arg\max_a \left[R(s,a) + \gamma \sum_{s' \in \mathcal{S}} P(s'|s,a) V(s') \right]$
11   **for** $a \in A(s)$ **do**
12   **if** $a = A_*$ **then**
13           $\pi(a|s) = 1$
14   **else**
15           $\pi(a|s) = 0$

Dynamic programming is an effective approach for solving MDPs when the state and action spaces are small enough to compute the value function or the policy explicitly. However, for larger problems, other methods such as Monte Carlo methods and temporal difference learning are more practical.

3.7 Summary

This chapter provided an overview of dynamic programming (DP), a class of algorithms used to solve Markov decision processes (MDPs) by decomposing complex problems into smaller subproblems that can be solved recursively. We began by discussing how DP can be used to solve Markov reward processes (MRPs), which involve computing the value function representing the expected cumulative reward from a given state.

Next, we focused on using DP to solve MDPs, covering two fundamental steps: policy evaluation and policy improvement. Policy evaluation computes the value function for a given policy, while policy improvement finds a better policy based on the current value function. The policy iteration algorithm alternates between these steps until convergence, but it can be computationally expensive due to multiple iterations of both steps. The general policy iteration algorithm is a more flexible version that allows for varying levels of effort between policy evaluation and policy improvement, making it more practical.

We then introduced the value iteration algorithm, which combines policy evaluation and policy improvement into a single step. Value iteration is computationally efficient and can converge to the optimal policy in a finite number of iterations.

In the next chapter, we will introduce Monte Carlo methods as an alternative approach to solving MDPs, which does not require explicit knowledge of the model (reward function and dynamics function) of the environment.

References

[1] Richard Bellman. *Dynamic Programming*. Princeton University Press, Princeton, NJ, 1957.
[2] Richard S. Sutton and Andrew G. Barto. *Reinforcement Learning: An Introduction*. The MIT Press, second edition, 2018.

Monte Carlo Methods

4

In Chap. 3, we introduced dynamic programming (DP) algorithms [1] that can help the agent find the optimal value functions V_* and Q_* for small MDPs; that's when the state and action spaces are small enough to compute the value function or the policy explicitly. DP algorithms assume the agent has access to the perfect model (reward function and dynamics function) of the environment, making them model-based reinforcement learning methods. However, in most real-world reinforcement learning problems, the true model of the environment is unknown to the agent, and the state or action spaces might be very big. In this chapter, we will introduce Monte Carlo methods [2], which are model-free reinforcement learning algorithms that use the agent's experience to estimate value functions.

The term Monte Carlo refers to the method of using experience to estimate value functions by averaging sample returns. Monte Carlo methods utilize sequences of state-action-reward samples, and the agent often interacts with a simulated environment to generate these samples (or it can use historical samples). The environment only needs to support basic interactions, such as generating successor states and reward signals when the agent takes an action in the environment. Unlike DP methods, Monte Carlo methods do not require the environment to provide all possible successor states and state transition probabilities. However, Monte Carlo methods can only be used to solve episodic reinforcement learning problems since they average returns based on complete episode sequences. The general policy iteration algorithm we introduced in Chap. 3 also applies to Monte Carlo methods.

In summary, this chapter will introduce Monte Carlo methods for model-free reinforcement learning. These methods use the agent's experience to estimate value functions without requiring a model of the environment. While they do not assume the Markov property, Monte Carlo methods can only be used to solve episodic reinforcement learning problems.

4.1 Monte Carlo Policy Evaluation

As a reminder, policy evaluation aims to estimate the state value function for a given policy π. Like in dynamic programming, we start with Monte Carlo policy evaluation (prediction). However, since Monte Carlo is a model-free reinforcement learning method, we cannot use the model of the environment and the Bellman equation as an update rule to iteratively compute the values. So, how can we compute the state values when using the Monte Carlo method?

© The Author(s), under exclusive license to APress Media, LLC, part of Springer Nature 2023
M. Hu, *The Art of Reinforcement Learning*,
https://doi.org/10.1007/978-1-4842-9606-6_4

Let's revisit the definition of the state value function V_π, which measures the expected return starting from state s, then following the policy π, as shown in Eq. (4.1).

$$V_\pi(s) = \mathbb{E}_\pi\left[G_t \mid S_t = s\right]$$

$$= \mathbb{E}_\pi\left[R_t + \gamma R_{t+1} + \gamma^2 R_{t+2} + \cdots \mid S_t = s\right]$$

$$= \mathbb{E}_\pi\left[R_t + \gamma G_{t+1} \mid S_t = s\right] \tag{4.1}$$

Given a sequence of state-action-reward pairs $S_0, A_0, R_0, S_1, A_1, R_1, \cdots, S_{T-1}, A_{T-1}, R_{T-1}, S_T$ generated by an agent following policy π, we can compute the discounted cumulative reward from any time step t to the end of the episode using the formula $G_t = R_t + \gamma G_{t+1}$ as shown in Fig. 4.1. The expected value of this sequence starting from state s can be estimated using Monte Carlo policy evaluation, which involves averaging the returns from a large number of sample episodes. This yields an estimate of the state value function V_π.

In Monte Carlo policy evaluation, the agent learns from its own experience. It collects a large number of sample episodes while following the policy π and averages the returns for the corresponding environment states in these sample episodes. As the number of samples increases, the averaged values will converge to the true state value V_π. However, it's often the case that the agent only visits a few states in a single episode when following the policy π. Therefore, we often need to run the algorithm over a large number of episodes to obtain an accurate estimate of V_π. By the law of large numbers, as the total visits to a particular state s become infinite, the estimated state value $V_\pi(s)$ also converges to the true value when following policy π. But in practice, we often need to stop the evaluation process in a much shorter time frame.

In summary, Monte Carlo policy evaluation is a model-free method that estimates the state value function for a given policy by collecting a large number of sample episodes and averaging the returns for the corresponding environment states. The agent learns from its own experience and does not require a model of the environment or the Bellman equation to estimate the state values. However, the algorithm often requires a large number of episodes to get an accurate estimate of V_π since the agent may only visit a few states in a single episode. Monte Carlo policy evaluation is typically used

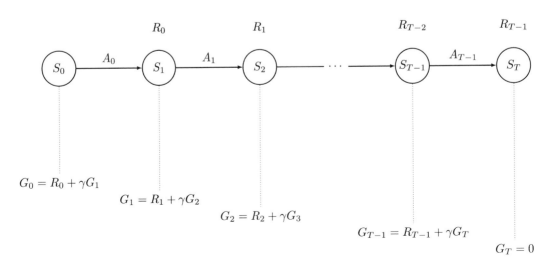

Fig. 4.1 The idea of Monte Carlo policy evaluation that estimates state values for a single episode sequence

to solve episodic reinforcement learning problems, where the process terminates after a finite number of steps.

First-Visit Monte Carlo Policy Evaluation

In the previous section, we discussed the basic idea behind Monte Carlo policy evaluation: we generate a large number of sample episodes by following the policy π and use these episodes to estimate the state value function $V_\pi(s)$.

The first-visit Monte Carlo policy evaluation algorithm, shown in Algorithm 1, is a specific implementation of this idea. For each episode, we compute the returns $G_0, G_1, G_2, \ldots, G_{T-1}$ for every time step $t = 0, 1, 2, \ldots, T - 1$ of the episode, where T is the time step at which the episode terminates. We then go through the states $S_0, S_1, S_2, \ldots, S_{T-1}$ in the episode and update our estimates of the state value function for each state.

To update our estimate for a state S_t, we check whether this is the first time we have visited this state in the current episode (hence the name "first-visit" Monte Carlo). If it is, we update two quantities: the total return $G(S_t)$, which is the sum of the returns G_t for all visits to S_t across episodes, and the visit count $N(S_t)$, which is the number of times we have visited S_t across episodes. We can then compute the average return for state S_t as $\overline{G}(S_t) = G(S_t)/N(S_t)$.

At the end of each episode, we update our estimate of the state value function for each state by taking the average of the returns we have observed in all episodes that have visited that state. This process continues for a fixed number of episodes. One drawback of this method is that we must manually choose the number of episodes to run, and the estimates may not be accurate if the number is too small.

Algorithm 1: First-visit Monte Carlo policy evaluation for V_π

Input: Policy π to be evaluated, discount rate γ, number of episodes K
Initialize: $i = 0$, $V_\pi(s) = 0$, $G(s) = 0$, $N(s) = 0$, for all $s \in \mathcal{S}$
Output: The estimated state value function V_π for the input policy π

1 **while** $i < K$ **do**
2   Generate a sample episode τ by following policy π, where $\tau = s_0, r_0, s_1, r_1, \cdots$
3   Compute the return G_t for every time step $t = 0, 1, 2, \cdots$ in episode τ
4   **for** each time step t till the end of episode τ **do**
5   **if** it's the first time state s_t is visited in episode τ **then**
6   Increment counter of total visits: $N(s_t) \leftarrow N(s_t) + 1$
7   Increment total return $G(s_t) \leftarrow G(s_t) + G_t$
8   Update estimated value $V_\pi(s_t) \leftarrow G(s_t)/N(s_t)$
9      $i = i + 1$

In summary, the first-visit Monte Carlo policy evaluation algorithm provides a way to estimate the state value function for a given policy by generating a large number of sample episodes and using them to compute average returns for each state. While this method has some limitations, it is a useful tool for understanding the behavior of reinforcement learning algorithms and can be applied to a wide range of problems.

First-Visit vs. Every-Visit

Monte Carlo policy evaluation is an algorithm for estimating the state value function for a given policy using simulated experience. One important decision that must be made when implementing this algorithm is whether to use a first-visit or every-visit approach.

In the context of Monte Carlo policy evaluation, the term "visit" refers to a time when the agent enters a state during an episode. Since an episode can involve multiple visits to the same state, it is important to decide how to handle these repeated visits in the evaluation process.

In first-visit Monte Carlo policy evaluation, the algorithm considers only the first time a state is visited during an episode when calculating the state's value. All subsequent visits to the same state in the current episode are ignored. For example, suppose an agent navigates through a maze environment and visits state s multiple times during an episode. The first-visit approach would only consider the return obtained after the first visit to state s while ignoring all subsequent visits to the same state s in the current episode.

On the other hand, every-visit Monte Carlo policy evaluation updates the value of a state every time it is visited during an episode. Continuing with the maze example, if the agent visits state s ten times during an episode, the every-visit approach would update the state's value function with the returns obtained after each of the ten visits.

There are some important differences between the two approaches. The first-visit approach is unbiased, meaning that the expected value of the estimated state value function converges to the true value as the number of episodes approaches infinity. However, the every-visit approach can have lower variance, meaning that the estimated values are less likely to be far from the true values.

In practice, the choice of whether to use a first-visit or every-visit approach will depend on the specific problem being solved and the desired trade-off between bias and variance. In some cases, it may be necessary to use a hybrid approach that combines both methods.

Overall, the choice between first-visit and every-visit Monte Carlo policy evaluation depends on the specific problem and the desired trade-off between bias and variance. The first-visit approach is unbiased but can have higher variance, while the every-visit approach can have lower variance but may be biased. Hybrid approaches that combine both methods can be used to strike a balance between bias and variance.

Worked Example

To better understand the difference between first-visit Monte Carlo policy evaluation and every-visit Monte Carlo policy evaluation, let's look at an example using the service dog MDP we introduced in Chap. 2. Recall the diagram for the service dog MDP shown in Fig. 4.2.

Let's say the agent is following some policy π and generated an episode sequence in the form of $S_0, R_0, S_1, R_1, \cdots, S_{T-1}, R_{T-1}$. For example, we might have an episode as follows:

$$episode = \{\text{Room 1}, -1, \text{Room 2}, -1, \text{Room 3}, -2, \text{Room 2}, -1, \text{Room 3}, 10\}$$

To evaluate the policy for this episode, we need to compute the returns G_t for each time step t in the episode. We use no discount $\gamma = 1.0$. The returns G_t are computed as the sum of the rewards from time step t to the end of the episode, with no discounting applied. For example, for the preceding episode, the returns G_t are

$$returns = \{5.0, 6.0, 7.0, 9.0, 10.0\}$$

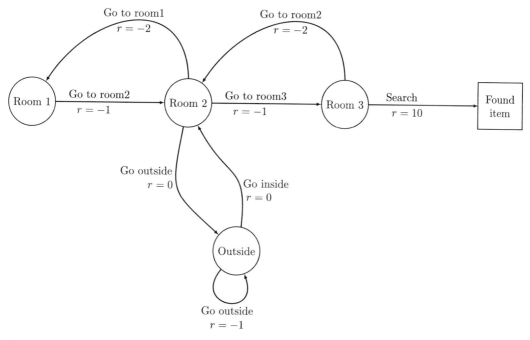

Fig. 4.2 Service dog MDP

Let's focus on state *Room 2*. We can see that it's been visited twice in this single episode, once at $t = 1$ and another at $t = 3$. Suppose our previous estimated value for state *Room 2* is the sample average of returns observed from state *Room 2*, which is $V_\pi(\text{Room 2}) = 0.0$. Additionally, the agent has never visited state *Room 2* before in this policy, so the number of visits to this state is $N(\text{Room 2}) = 0$.

If we use first-visit Monte Carlo policy evaluation, then the new value for state *Room 2* should be $6/1 = 6$. This is because the algorithm will ignore the second time state *Room 2* is visited in this episode and only update the value of the state based on the first visit. In contrast, if we use every-visit Monte Carlo policy evaluation, then the value for state *Room 2* will first be updated to $6/1 = 6$, and then it will be updated to $(6 + 9)/2 = 7.5$. This is because every-visit Monte Carlo policy evaluation updates the value of a state on every visit in an episode.

In summary, first-visit Monte Carlo policy evaluation updates the value of a state only on its first visit in an episode, while every-visit Monte Carlo policy evaluation updates the value of a state on every visit in an episode. The difference between these two approaches can be important for accurately estimating the value of states in an MDP.

Every-Visit Monte Carlo Policy Evaluation

It's fairly easy to extend the first-visit Monte Carlo policy evaluation to the every-visit case; we just need to remove the conditional check inside the for loop, and everything else stays exactly the same, as shown in Algorithm 2.

We used first-visit and every-visit Monte Carlo policy evaluation algorithms to estimate the state values of a random policy π for our service dog MDP. Table 4.1 shows the results. Since Monte Carlo methods involve randomness, we averaged the results over 100 independent runs, with each run

---

Algorithm 2: Every-visit Monte Carlo policy evaluation for V_π

Input: Policy π to be evaluated, discount rate γ, number of episodes K
Initialize: $i = 0$, $V_\pi(s) = 0$, $G(s) = 0$, $N(s) = 0$, for all $s \in S$
Output: The estimated state value function V for the input policy π

1 **while** $i < K$ **do**
2   Generate a sample episode τ by following policy π, where $\tau = s_0, r_0, s_1, r_1, \cdots$
3   Compute the return G_t for every time step $t = 0, 1, 2, \cdots$ in episode τ
4   **for** each time step t till the end of episode τ **do**
5   Increment counter of total visits: $N(s_t) \leftarrow N(s_t) + 1$
6   Increment total return $G(s_t) \leftarrow G(s_t) + G_t$
7   Update estimated value $V_\pi(s_t) \leftarrow G(s_t)/N(s_t)$
8      $i = i + 1$

Table 4.1 State values computed using first-visit and every-visit Monte Carlo policy evaluation, for the same random policy for the service dog MDP. Results were averaged over 100 independent runs; each run is over 2000 episodes; the last row contains true state values computed using DP policy evaluation. We use the same discount $\gamma = 0.9$ in this experiment

	Room 1	Room 2	Room 3	Outside	Found item
First-visit MC	−2.67	−1.85	3.16	−3.32	0
Every-visit MC	−2.63	−1.81	3.2	−3.3	0
DP	−2.66	−1.85	3.17	−3.33	0

consisting of 2000 episodes. For all runs, we set the discount factor γ to 0.9. The values computed using Monte Carlo policy evaluation algorithms are very close to the true values computed using dynamic programming policy evaluation (as shown in the last row of Table 4.1).

4.2 Incremental Update

The concept of incremental update is a widely used technique in reinforcement learning algorithms and is particularly useful in Monte Carlo policy evaluation. The idea behind incremental updates is to update an estimate in small steps, rather than updating it all at once. This has several advantages, including faster convergence, better memory efficiency, and the ability to update estimates in real time as new data becomes available.

The incremental update formula can be expressed as follows:

$$\text{NewEstimate} = \text{OldEstimate} + \text{StepSize}\Big[\text{Target} - \text{OldEstimate}\Big]$$

where OldEstimate is the previous estimate, StepSize is a small positive constant that determines the size of the update, and Target is a new sample value. The term Target − OldEstimate is often referred to as the estimate error and represents the difference between the new sample value and the previous estimate.

In Monte Carlo policy evaluation, incremental updates can be used to estimate the value function for a state. Specifically, we can update the estimated value function for a state S_t using the sample return G_t obtained from a specific episode:

$$V_\pi(S_t) = V_\pi(S_t) + \frac{1}{N(S_t)}\Big(G_t - V_\pi(S_t)\Big) \tag{4.2}$$

where $V_\pi(S_t)$ is the estimated value function for state S_t, and $N(S_t)$ is the number of times the state S_t has been visited. The update formula in Eq. (4.2) represents an incremental update, where the estimated value function is updated based on a small step proportional to the difference between the sample return and the previous estimate.

The incremental Monte Carlo policy evaluation algorithm for the first-visit case is shown in Algorithm 3. In each episode, for each time step t in the episode, we first check if it's the first time the state S_t is visited, we then update the episode returns and the visited number for S_t, the process continues until a terminal state is reached. Finally, we update the estimated value function for each visited state using the incremental update equation in Eq. (4.2). The every-visit case can be easily implemented by removing the condition check inside the for loop.

Algorithm 3: First-visit Monte Carlo policy evaluation with incremental update for V_π

Input: Policy π to be evaluated, discount rate γ, number of episodes K
Initialize: $i = 0$, $V_\pi(s) = 0$, $N(s) = 0$, for all $s \in \mathcal{S}$
Output: The estimated state value function V_π for the input policy π

1 **while** $i < K$ **do**
2   Generate a sample episode τ by following policy π, where $\tau = s_0, r_0, s_1, r_1, \cdots$
3   Compute the return G_t for every time step $t = 0, 1, 2, \cdots$ in episode τ
4   **for** each time step t till the end of episode τ **do**
5   **if** it's the first time state s_t is visited in episode τ **then**
6   Increment counter of total first visits: $N(s_t) \leftarrow N(s_t) + 1$
7   Update estimated value $V_\pi(s_t) \leftarrow V_\pi(s_t) + \frac{1}{N(s_t)}\left(G_t - V_\pi(s_t)\right)$
8      $i = i + 1$

Table 4.2 shows the state values computed using first-visit Monte Carlo policy evaluation and first-visit Monte Carlo policy evaluation with the incremental update method, all for the same random policy for the service dog MDP. Similarly to the experiment shown in Table 4.1, the results were averaged over 100 independent runs, and each run is over 2000 episodes; we use the same discount $\gamma = 0.9$ in this experiment.

Monte Carlo Policy Evaluation for Q_π

We can use Monte Carlo policy evaluation to estimate the state-action value function Q_π for a policy π. The state-action value function $Q_\pi(s, a)$ measures the expected return starting in state s, taking action a, then following policy π.

Monte Carlo policy evaluation estimates Q_π by generating many episodes of the Markov decision process (MDP) under the policy π, then computing the average return for each state-action pair.

Table 4.2 State values computed using first-visit Monte Carlo policy evaluation and first-visit Monte Carlo policy evaluation with incremental update, for the same random policy for the service dog MDP. Results were averaged over 100 independent runs; each run is over 2000 episodes; the last row contains true state values computed using DP policy evaluation. We use the same discount $\gamma = 0.9$ in this experiment

	Room 1	Room 2	Room 3	Outside	Found item
First-visit MC	−2.67	−1.85	3.17	−3.33	0
Incr. first-visit MC	−2.65	−1.84	3.16	−3.34	0
DP	−2.66	−1.85	3.17	−3.33	0

Specifically, for each episode, the agent starts in an initial state s_0 and takes actions according to the policy π until it reaches a terminal state. At each time step t, the agent observes the current state S_t, takes an action A_t according to the policy π, and receives a reward R_t. The agent continues to take actions and receive rewards until it reaches a terminal state, where the episode ends.

After generating multiple episodes, Monte Carlo policy evaluation computes the average return G_t for each state-action pair (s, a) that has been visited during the episodes. This is done by averaging the returns obtained after each visit to the pair (s, a), where the return is the sum of discounted rewards from that time step until the end of the episode:

$$G_t = R_t + \gamma R_{t+1} + \gamma^2 R_{t+2} + \cdots + \gamma^{T-1-t} R_{T-1} \tag{4.3}$$

Here, T is the time step at which the episode ends, and γ is the discount factor that determines the importance of future rewards relative to immediate rewards.

Once we have computed the average return G_t for each state-action pair (s, a), we can update the estimate of $Q_\pi(s, a)$ using the incremental update rule:

$$Q_\pi(S_t, A_t) = Q_\pi(S_t, A_t) + \frac{1}{N(S_t, A_t)}\left(G_t - Q_\pi(S_t, A_t)\right) \tag{4.4}$$

Here, $N(S_t, A_t)$ is the number of times the state-action pair (S_t, A_t) has been visited during the episodes.

The first-visit Monte Carlo policy evaluation is one way to implement Monte Carlo policy evaluation for estimating Q_π, which estimates the value of each state-action pair by averaging the returns obtained after the first visit to that pair. The algorithm is as follows:

Algorithm 4: First-visit Monte Carlo policy evaluation with incremental update for Q_π

Input: Policy π to be evaluated, discount rate γ, number of episodes K
Initialize: $i = 0$, $Q_\pi(s, a) = 0$, $N(s, a) = 0$, for all $s \in \mathcal{S}, a \in A(s)$
Output: The estimated state-action value function Q_π for the input policy π

1 **while** $i < K$ **do**
2   Generate a sample episode τ by following policy π, where $\tau = s_0, a_0, r_0, s_1, a_1, r_1, \cdots$
3   Compute the return G_t for every time step $t = 0, 1, 2, \cdots$ in episode τ
4   **for** each time step t till the end of episode τ **do**
5   **if** it's the first time (s_t, a_t) is visited in episode τ **then**
6   Increment counter of total first visits: $N(s_t, a_t) \leftarrow N(s_t, a_t) + 1$
7   Update estimated value $Q_\pi(s_t, a_t) \leftarrow Q_\pi(s_t, a_t) + \frac{1}{N(s_t, a_t)}\left(G_t - Q_\pi(s_t, a_t)\right)$
8      $i = i + 1$

Figure 4.3 shows the state-action values we've computed using Algorithm 4 for a random policy π for the service dog MDP. As usual, the results were averaged over 100 independent runs; each run is over 2000 episodes. Notice the state values were computed using DP policy evaluation. And we use discount $\gamma = 0.9$ in this experiment.

We can verify the results by computing the state values using these state-action values. Recall that the state value is just the weighted sum over all state-action values for this state.

$$V_\pi(s) = \sum_{a \in A} \pi(a|s) Q_\pi(s, a) \tag{4.5}$$

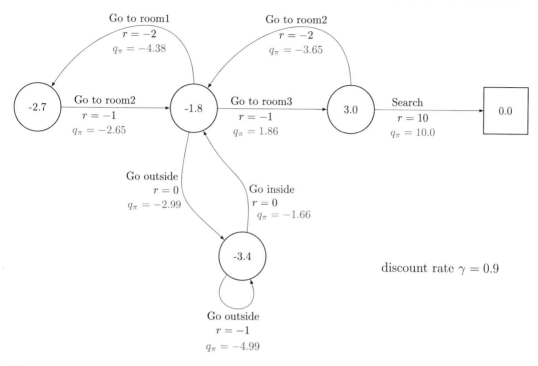

Fig. 4.3 State-action values computed using first-visit Monte Carlo policy evaluation with incremental update of a random policy for the service dog MDP. Results were averaged over 100 runs; each run is over 2000 episodes. Notice the state values were computed using DP policy evaluation. We use the same discount $\gamma = 0.9$ in this experiment

If we look at state *Room 2*, the state-action values for actions *Go to room1, Go to room3, Go outside* are -4.38, 1.86, -2.99. And if we weights each by the probability of selecting the action which is all 0.33, we get $-4.38 * 0.33 + 1.86 * 0.33 - 2.99 * 0.33 = -1.82$, which is very close to the true value of -1.85 computed using DP. The same process could be applied to state *Room 3*, where we get the value $-3.65 * 0.5 + 10 * 0.5 = 3.175$, which is very close to the true value of 3.17.

4.3 Exploration vs. Exploitation

In reinforcement learning, the exploration-exploitation trade-off arises when an agent must decide between taking actions that it already knows to be good (exploitation) and taking actions that it has not yet tried, in the hope of finding even better ones (exploration). The goal is to balance the benefits of exploiting known actions with the potential rewards of exploring new options.

For instance, consider a game of chess. A chess-playing agent must explore new moves in order to improve its chances of winning. However, it must also exploit the strategies it has already learned in order to maximize its chances of success. This balance is critical to the agent's overall performance, as it cannot win by simply repeating the same moves over and over again, nor can it win by making completely random moves.

Another example is a recommender system that suggests products or services to users based on their past behavior. The system must balance exploiting what it already knows about the user's preferences while exploring new products that the user may like. If the system only recommends products that the user has already purchased or viewed, it may miss out on opportunities to suggest new, interesting products that the user may enjoy.

In model-free reinforcement learning methods that rely on value functions (such as Monte Carlo methods), the exploration-exploitation trade-off is particularly important. The agent must learn the value of each action by experiencing the rewards it receives when taking that action in different states. However, it also needs to explore new actions in order to discover whether there are even better actions to take.

In contrast, in dynamic programming (DP), the agent assumes that it has access to the model of the environment (i.e., the dynamics function and reward function) and can therefore consider all possible outcomes when updating its value function. This allows DP to avoid the exploration-exploitation trade-off.

To better understand the issue of exploration vs. exploitation, let's take a look at the Bellman equations for the value function V_π, as shown in Eq. (4.6):

$$V_\pi(s) = \mathbb{E}_\pi\left[R_t + \gamma V_\pi(S_{t+1}) \mid S_t = s\right]$$

$$= \sum_{a \in A} \pi(a|s)\left[R(s, a) + \gamma \sum_{s' \in \mathcal{S}} P(s'|s, a)V_\pi(s')\right], \quad \text{for all } s \in \mathcal{S} \qquad (4.6)$$

The Bellman equation shows how the value of a state is related to the values of its successor states, weighted by the probabilities of transitioning to those states and the rewards received in those states. In DP, this equation can be used to compute the value of each state, assuming that the agent has access to the model of the environment.

```
/* How DP policy improvement computes state-action value function using
      estimated state value function                                       */
1  for s ∈ S do
2  |    for a ∈ A(s) do
3  |    |    Qπ(s, a) ← R(s, a) + γ ∑        P(s'|s, a)Vπ(s')
   |    |                          s'∈S
```

In contrast, in model-free reinforcement learning, the agent must estimate the value function by sampling the environment, which can be difficult when exploring new actions. For example, imagine a robot that is trying to learn how to walk. If it always chooses the same actions, it may not be able to discover the best way to walk, because it has not explored enough of the state space. On the other hand, if it always takes random actions, it may never make any progress, because it is not exploiting the knowledge it has already gained.

To address this issue, various exploration strategies have been proposed, such as ϵ-greedy, where the agent takes a random action with probability ϵ otherwise takes the best known action with probability $1 - \epsilon$. Other strategies include Upper Confidence Bound (UCB), Thompson sampling can also be used to address the issue. These strategies aim to balance exploration and exploitation by encouraging the agent to try new actions while also exploiting the knowledge it has already gained.

In summary, exploration vs. exploitation is a fundamental challenge in reinforcement learning and is particularly important in model-free methods that rely on value functions. Dynamic programming can avoid this challenge by assuming access to the model of the environment, but this is not always possible in practice. Various exploration strategies have been proposed to address this issue, but finding the optimal balance between exploration and exploitation remains an active area of research in reinforcement learning.

Exploration with ϵ-Greedy Policy

One of the fundamental challenges in reinforcement learning is finding a balance between exploration and exploitation. Exploration is the process of trying out new actions to learn more about the environment, while exploitation involves using the current knowledge to take the best possible action in each state. The ϵ-greedy policy is a simple and effective way to encourage exploration while still exploiting the current knowledge.

The ϵ-greedy policy works by selecting a random action with probability ϵ and with probability $1 - \epsilon$, it selects the best known action according to the current state-action value function. The parameter ϵ controls the level of exploration, where $0 \leq \epsilon \leq 1$. When ϵ is high, the agent will be more likely to explore the environment by taking random actions. As ϵ decreases, the agent will rely more on the current knowledge and choose the best known actions more frequently. This gradual shift from exploration to exploitation is often necessary for the agent to achieve good performance in a reinforcement learning task.

Formally, the ϵ-greedy policy is defined as

$$\pi(a|s) = \begin{cases} 1 - \epsilon + \dfrac{\epsilon}{|A(s)|} & \text{if } a = \arg\max_a Q_\pi(s, a) \\ \dfrac{\epsilon}{|A(s)|} & \text{if } a \neq \arg\max_a Q_\pi(s, a) \end{cases} \tag{4.7}$$

where $Q(s, a)$ is the state-action value function, and $|A(s)|$ is the number of legal actions in state s. The policy selects the best known action with probability $1 - \epsilon + \frac{\epsilon}{|A(s)|}$ and a random action with probability $\frac{\epsilon}{|A(s)|}$.

For example, suppose we are training an agent to play a simple game where it needs to navigate a maze to reach a goal. At the start of training, the agent has no knowledge of the environment, so it needs to explore to learn about the maze layout and the effects of different actions. We might set ϵ to a high value like 0.9 so that the agent explores more often. As the agent learns more about the maze and gains more accurate estimates of the state-action values, we can gradually decrease ϵ to encourage more exploitation of the current knowledge.

A common approach for scheduling the decay of ϵ is to use a decreasing function over time. For example, we might use a linear schedule where we start with a high value of ϵ and decrease it by a fixed amount of episodes or time steps until it reaches a minimum value. Alternatively, we might use an exponential decay schedule where ϵ is multiplied by a decay factor after each episode or time step.

In summary, the ϵ-greedy policy is a simple yet effective way to balance exploration and exploitation in reinforcement learning. The parameter ϵ allows us to control the level of exploration, and we can gradually decrease it over time as the agent gains more knowledge about the environment. The choice of decay schedule and minimum value for ϵ depend on the specific application and may require some trial and error to find the optimal values.

4.4   Monte Carlo Control (Policy Improvement)

Now we are ready to start using Monte Carlo methods for policy improvement. In dynamic programming (DP), which is a class of algorithms used for solving Markov decision processes (MDPs), the policy improvement algorithm consists of two steps. The first step is to use the estimated state value function V_π and the model of the environment to compute the state-action value function Q_π. The second step is to compute a better deterministic policy π' based on Q_π.

In Monte Carlo policy evaluation, we can directly estimate the state-action value function Q_π without having to first estimate the state value function. This is shown in Algorithm 4. The algorithm works by simulating complete episodes of interaction with the environment, where an episode is a sequence of state-action-reward transitions that start from an initial state and end in a terminal state. We use the first-visit method to estimate Q_π, which means that we only consider the first occurrence of a state-action pair in each episode when calculating the average return. This ensures that the estimates are unbiased.

Using the estimates of Q_π, we can compute a better policy π'. One way to do this is to use an ϵ-greedy policy, which is a way to balance exploration and exploitation during learning. An ϵ-greedy policy chooses the action with the highest estimated state-action value with probability $1 - \epsilon$ and a random action with probability ϵ. The value of ϵ determines the amount of exploration; a larger value of ϵ leads to more exploration.

The ϵ-greedy policy for Monte Carlo policy improvement works as follows: for each state, we choose the action with the highest estimated state-action value with probability $1 - \epsilon$ and a random action with probability ϵ/K, where K is the number of actions available in that state. This ensures that every action is chosen with nonzero probability, which is necessary for convergence to the optimal policy. The new policy π' is then computed as the greedy policy with respect to Q_π, that is, for each state we choose the action with the highest estimated state-action value.

Algorithm 5: Example of Monte Carlo policy improvement, with ϵ-greedy exploration

---

Input: Estimated state value function Q_π for π, exploration rate ϵ
Output: The improved policy
1 **for** $s \in S$ **do**
2 $\quad$ $A_* = \arg\max_a Q_\pi(s, a)$
3 $\quad$ **for** $a \in A(s)$ **do**
4 $\quad\quad$ **if** $a = A_*$ **then**
5 $\quad\quad\quad$ $\pi(a|s) = 1 - \epsilon + \dfrac{\epsilon}{|A(s)|}$
6 $\quad\quad$ **else**
7 $\quad\quad\quad$ $\pi(a|s) = \dfrac{\epsilon}{|A(s)|}$

In summary, Monte Carlo policy improvement allows us to directly estimate the state-action value function Q_π and use it to improve the policy, without having to first estimate the state value function as in DP. Using an ϵ-greedy policy for exploration, we can balance exploration and exploitation during learning and converge to the optimal policy.

We adapt the general policy iteration template introduced in Chap. 3 for Monte Carlo control, which is to find the optimal policy by learning its value function. But there are some changes we need to make. The first change is, during Monte Carlo policy evaluation, we need to estimate the state-action value function Q_π instead of V_π for the policy π. The second change is when to stop running the algorithm. As in DP, we stop when the policy from the previous iteration is the same as the current iteration, meaning no improvement can be made. But in Monte Carlo, the whole process of policy evaluation is based on the number of sample episodes, so we often define the stop condition based on the number of sample episodes we want the agent to collect. One last optional change, since we're now using ϵ-greedy policy for exploration, we may want to decrease the exploration rate ϵ, for example, after every iteration.

We now introduce the first-visit Monte Carlo control algorithm with the incremental update method, as shown in Algorithm 6. We can easily extend this into the every-visit case by simply removing the condition check during the policy evaluation process.

Algorithm 6: First-visit Monte Carlo control with incremental update and ϵ-greedy exploration

---

Input: Discount rate γ, initial exploration rate ϵ, number of episodes K
Initialize: $i = 0$, $Q_\pi(s, a) = 0$, $N(s, a) = 0$, for all $s \in \mathcal{S}$, $a \in A(s)$, random policy π
Output: Found (optimal) policy π and its state-action value function Q_π

1 **while** $i < K$ **do**
2   Generate a sample episode τ by following policy π, where $\tau = s_0, a_0, r_0, s_1, a_1, r_1, \cdots$, and (optional) decay exploration rate ϵ during the process
3   Compute the return G_t for every time step $t = 0, 1, 2, \cdots$ in episode τ
    /* Policy evaluation   */
4   **for** each time step t till the end of episode τ **do**
5   **if** it's the first time (s_t, a_t) is visited in episode τ **then**
6   Increment counter of total first visits: $N(s_t, a_t) \leftarrow N(s_t, a_t) + 1$
7   Update estimated value $Q_\pi(s_t, a_t) \leftarrow Q_\pi(s_t, a_t) + \dfrac{1}{N(s_t, a_t)}\Big(G_t - Q_\pi(s_t, a_t)\Big)$
    /* Policy improvement   */
8   **for** each time step t till the end of episode τ **do**
9          $a_* = \arg\max_a Q_\pi(s_t, a)$
10   **for** $a \in A(s_t)$ **do**
11   **if** $a = a_*$ **then**
12                 $\pi(a|s_t) = 1 - \epsilon + \dfrac{\epsilon}{|A(s_t)|}$
13   **else**
14                 $\pi(a|s_t) = \dfrac{\epsilon}{|A(s_t)|}$
15     $i = i + 1$

We can run Algorithm 6 to find the optimal policy for our service dog MDP. And amazingly after ten episodes, it found the optimal policy, although the values are not quite close to the true optimal state value function V_*.

We can also see from Table 4.3, as we increase the number of episodes, the values are getting closer to the true optimal state value V_*. One interesting fact is for the state *Outside*, the values are never getting closer to the true value; this might be because the agent has never visited the state that often during the learning process. So the value is not getting updated that often.

Table 4.3 Optimal state value computed using first-visit Monte Carlo control with the incremental update algorithm with different numbers of episodes, for the service dog MDP. Results were averaged over 100 independent runs; the last row contains true state values computed using DP policy iteration. We use initial exploration rate $\epsilon = 1.0$ and discount $\gamma = 0.9$ in this experiment

Number of episodes	Room 1	Room 2	Room 3	Outside	Found item
10	2.26	5.16	10.0	−0.62	0
100	5.71	7.71	10.0	1.52	0
1000	6.14	7.96	10.0	2.54	0
10,000	6.19	8.0	10.0	2.99	0
DP	6.2	8.0	10.0	7.2	0

4.5   Summary

The chapter introduced Monte Carlo methods, a type of reinforcement learning algorithm used to estimate value functions and improve policies based on experience. Unlike dynamic programming, Monte Carlo methods are model-free and do not require knowledge of the environment's reward and dynamics functions.

The chapter provided a clear explanation of Monte Carlo policy evaluation, a method used to estimate the value of states by averaging the observed returns after visiting them. The chapter used concrete examples to illustrate the difference between first-visit Monte Carlo and every-visit Monte Carlo. Additionally, the chapter covered the incremental update technique, which allows for efficient updating of the value function after each episode.

The chapter also highlighted the importance of the exploration-exploitation dilemma in reinforcement learning and how Monte Carlo methods address this issue by balancing the need to explore new states and actions with exploiting existing knowledge. Monte Carlo control was introduced as a method for improving policies by estimating the value of each state under the current policy and updating the policy to be greedy with respect to the estimated values.

References

[1] Richard Bellman. *Dynamic Programming*. Princeton University Press, Princeton, NJ, 1957.
[2] Richard S. Sutton and Andrew G. Barto. *Reinforcement Learning: An Introduction*. The MIT Press, second edition, 2018.

Temporal Difference Learning

5

In the previous chapter (Chap. 4), we introduced Monte Carlo (MC) methods for reinforcement learning, which allow an agent to learn from its own experience without relying on a model of the environment. However, MC methods are limited to episodic reinforcement learning problems, where the agent interacts with the environment in discrete episodes. In this chapter, we introduce temporal difference (TD) learning, a class of algorithms that generalize MC methods to support both episodic and continuing reinforcement learning problems. TD learning is based on the idea of updating value estimates using a combination of current rewards and estimated future rewards, making it a more versatile and efficient approach than MC methods for many types of reinforcement learning problems. We'll explore the theory and implementation of various TD learning algorithms in this chapter.

5.1   Temporal Difference Learning

In Chap. 4, we introduced Monte Carlo methods to solve episodic reinforcement learning problems, where the agent can learn from its own experience without requiring access to the model of the environment. However, Monte Carlo methods have two limitations that make them unsuitable for solving continuing reinforcement learning problems. The first major issue is that they do not support continuing reinforcement learning problems where there is no natural ending of the task, as in episodic problems, there is a terminal state that marks the end of the episode, whereas in continuing problems, the agent interacts with the environment indefinitely. The second minor issue is that the agent has to wait until the end of an episode to compute and update the estimated values, because Monte Carlo methods require the agent to experience a full episode before updating the estimated values. This may not be a problem for small reinforcement learning problems with very short episodes, such as our service dog MDP example. But for problems with long episodes, potentially thousands or millions of time steps, this waiting can slow down the learning process.

Temporal difference (TD) learning proposed by Sutton [1, 2] is a model-free reinforcement learning method that overcomes these limitations. It's called TD learning because the agent updates its estimates based on the difference between the predicted and actual rewards received by the agent in each state. More precisely, the agent updates the estimated value for a particular state s right after it receives the immediate reward r and the successor state s' from the environment, which is a tuple of (s, a, r, s') (often called a transition). This *one-step look-ahead* allows the agent to update its estimates

© The Author(s), under exclusive license to APress Media, LLC, part of Springer Nature 2023
M. Hu, *The Art of Reinforcement Learning*,
https://doi.org/10.1007/978-1-4842-9606-6_5

based on the immediate reward and the estimated value of the next state, without waiting for the end of an episode.

One way to think of TD learning is that the agent is continuously adjusting its estimates of the value of each state based on new information as it becomes available. This is in contrast to Monte Carlo methods, which only update estimates at the end of each episode.

TD learning is widely used in real-world applications and is central to solving reinforcement learning problems. It's an efficient and flexible algorithm that allows agents to learn in a continuous and incremental manner, making it suitable for real-world applications. Other model-free reinforcement learning algorithms, such as Q-learning and SARSA (which we will introduce later in this chapter), build on top of this concept.

To summarize, TD learning is a powerful method for solving reinforcement learning problems, and its "one-step look-ahead" approach allows agents to update their estimates quickly and continuously.

5.2   Temporal Difference Policy Evaluation

As always, we follow the general policy iteration template introduced in Chap. 3. The first step is to do policy evaluation, which involves estimating the state value function V_π (or state-action value function Q_π) for a given (fixed) policy π. Since TD learning is a model-free reinforcement learning method, the agent will use experience generated when following policy π to learn these value functions.

To estimate the state value function using TD learning, we start with the equation for $V_\pi(s)$, which represents the expected return when starting in state s and following policy π. The Monte Carlo method uses a sum of discounted rewards to estimate this return for a complete episode sequence. Alternatively, the equation can be rewritten in a recursive way, as shown in Eq. (5.2), which gives us a clue how the TD learning method works. The TD learning method updates the estimated value for a particular state using a tuple of (s, a, r, s'), which is exactly what we need to complete Eq. (5.2).

$$
\begin{aligned}
V_\pi(s) &= \mathbb{E}_\pi\left[G_t \,\middle|\, S_t = s\right] \\
&= \mathbb{E}_\pi\left[R_t + \gamma R_{t+1} + \gamma^2 R_{t+2} + \cdots \,\middle|\, S_t = s\right] \\
&= \mathbb{E}_\pi\left[R_t + \gamma G_{t+1} \,\middle|\, S_t = s\right] \tag{5.1} \\
&= \mathbb{E}_\pi\left[R_t + \gamma V_\pi(S_{t+1}) \,\middle|\, S_t = s\right] \tag{5.2}
\end{aligned}
$$

As we introduced earlier in this chapter, the TD learning method only requires a tuple of (s, a, r, s') to update the value for the state s. This is exactly what we need to complete Eq. (5.2). The expectation sign reminds us that we need to collect a large amount of samples to compute the expected return when starting in state s and following policy π.

Next, we need to bring in the incremental update method. The general form for the incremental update method is

$$
\text{NewEstimate} = \text{OldEstimate} + \text{StepSize}\left[\text{Target} - \text{OldEstimate}\right]
$$

Here, the "StepSize" represents the learning rate, and "Target" represents the TD target. To update the value estimate for state s, we use the incremental update method with the TD target $R_t + \gamma V(S_{t+1})$, which represents the estimated return from the current state. The difference between the TD target and the current estimated value $V(S_t)$ is called the TD error, which is also used to update the value function estimate. The update rule for TD learning is shown in Eq. (5.3):

$$V_\pi(S_t) = V_\pi(S_t) + \alpha\left(G_t - V_\pi(S_t)\right)$$

$$= V_\pi(S_t) + \alpha\left(\left[R_t + \gamma V_\pi(S_{t+1})\right] - V_\pi(S_t)\right) \tag{5.3}$$

Here, α is the learning rate for TD learning. The TD error is $R_t + \gamma V_\pi(S_{t+1}) - V_\pi(S_t)$, where $R_t + \gamma V_\pi(S_{t+1})$ represents the estimated return for state S_t. By using this update rule, we can iteratively improve our estimates of the state values based on the observed immediate reward and the discounted (estimated) value of successor state.

In TD learning, the agent does not keep a record of the total number of visits for states, so we cannot use $\frac{1}{N(S_t)}$ as the step size. Instead, we need to set the value for step size α manually. The choice of α can have a significant impact on the learning performance and stability, and there are various heuristics and methods to set it appropriately.

Now that we have the equation for the update rule, we can introduce the TD learning algorithm for policy evaluation, as shown in Algorithm 1. Sometimes people also refer to it as the TD(0) policy evaluation algorithm, as it estimates the value using the immediate reward and discounted value of its successor state, without taking additional steps into the future. The general idea is similar to the Monte Carlo policy evaluation, which involves the agent collecting a sample transition while following the policy π. However, instead of a complete episode sequence, here we only need a tuple of (s, a, r, s'). This is why TD methods also work for continuing reinforcement learning problems.

In Monte Carlo, we run the algorithm based on the number of sample episodes the agent needs to complete. But TD learning can also apply to continuing reinforcement learning problems where measuring the number of episodes is impossible. One simple solution is to count the total number of time steps (or environment steps) to run. In general, this solution is also better than simply counting the number of episodes for the episodic environment since the total number of steps on an episode-to-episode basis can vary.

Algorithm 1: Temporal difference [TD(0)] policy evaluation for V_π

Input: Policy π to be evaluated, discount rate γ, step size α, number of steps K
Initialize: $i = 0$, $V_\pi(s) = 0$, for all $s \in S$
Output: The estimated state value function V_π for the input policy π

1 Sample initial state s_0
2 while $i < K$ do
3   Sample action a_t for s_t when following policy π
4   Take action a_t in environment and observe r_t, s_{t+1}
5 $i = i + 1$
6   Compute TD target:
7 $\delta_t = \begin{cases} r_t & \text{if } s_{t+1} \text{ is terminal state} \\ r_t + \gamma V_\pi(s_{t+1}) & \text{otherwise bootstrapping} \end{cases}$
8   Update estimated value $V_\pi(s_t) \leftarrow V_\pi(s_t) + \alpha\left(\delta_t - V_\pi(s_t)\right)$
9 $s_t = s_{t+1}$

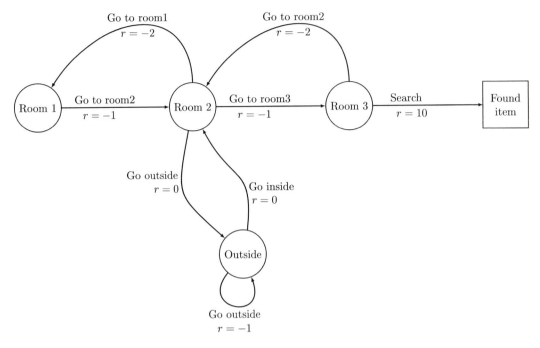

Fig. 5.1 Service dog MDP

Worked Example

To better understand how the TD policy evaluation algorithm works, let's look at an example using the service dog MDP that we introduced in Chap. 2. Recall the diagram for the service dog MDP shown in Fig. 5.1.

Assuming that the agent is following a (fixed) random policy π, we use a step size of $\alpha = 0.01$ and no discount ($\gamma = 1.0$) in this experiment. We start with an initial estimated value of 0 for all states. Since TD is a model-free reinforcement learning method, the agent doesn't know beforehand the reward signal it will receive when taking any action, nor does it know the successor state. We want to see how the algorithm updates the estimated values for the different states.

Suppose we're looking at the first time step $t = 0$ of the first episode, where the current state of the environment is $S_t = Room\ 2$, and the agent chooses action $A_t = Go\ to\ room\ 3$ according to policy π. It receives an immediate reward of $R_t = -1$ and the successor state $S_{t+1} = Room\ 3$ from the environment. We now have the tuple (S_t, A_t, R_t, S_{t+1}), which is what we need to update the estimated value for state $S_t = Room\ 2$. Here's how we update it:

$$V_\pi (Room\ 2) = V_\pi (Room\ 2) + \alpha \left[R_t + \gamma V_\pi (Room\ 3) - V_\pi (Room\ 2) \right]$$

$$= 0 + 0.01 * (-1 + 0 - 0)$$

$$= -0.01$$

After the update, the agent continues to act in the environment following policy π. The current state of the environment becomes $S_t = Room\ 3$, the agent chooses action $A_t = Search$ according to policy π, and it receives an immediate reward of $R_t = 10$. The successor state $S_{t+1} = Found\ item$ is also a terminal state, so the current episode is over. We now have the tuple (S_t, A_t, R_t, S_{t+1}) to update the estimated value for state $S_t = Room\ 3$. Here's how we do it:

$$V_\pi(Room\ 3) = V_\pi(Room\ 3) + \alpha\left[R_t + \gamma V_\pi(Found\ item) - V_\pi(Room\ 3)\right]$$

$$= 0 + 0.01 * (10 + 0 - 0)$$

$$= 0.1$$

In a new episode, if the agent is again in environment state $S_t = Room\ 2$ and chooses action $A_t = Go\ to\ room\ 3$ according to policy π, it receives an immediate reward of R_t and the successor state $S_{t+1} = Room\ 3$ from the environment. We can update the estimated value for $V_\pi(Room\ 2)$ as follows:

$$V_\pi(Room\ 2) = V_\pi(Room\ 2) + \alpha\left[R_t + \gamma V_\pi(Room\ 3) - V_\pi(Room\ 2)\right]$$

$$= -0.01 + 0.01 * \left[-1 + 0.1 - (-0.01)\right]$$

$$= -0.019$$

We can repeat this process over and over. Table 5.1 shows the state values for the random policy after running the TD(0) policy evaluation algorithm with different numbers of updates. The results are averaged over 100 independent runs, and we also include the true state values computed using DP policy evaluation (last row). After 10,000 updates, the estimated state values are getting very close to the true state values.

The key takeaway from these results is that the TD(0) algorithm can learn accurate estimates of state values for some policy without knowing the MDP's transition probabilities and reward function. However, the estimates converge more slowly compared to DP policy evaluation when the number of states is large. Nevertheless, TD(0) is an important algorithm for model-free reinforcement learning and can handle continuous state and action spaces, which makes it useful in many practical applications.

Table 5.1 State value of the same random policy for the service dog MDP, after running the TD(0) policy evaluation algorithm with different numbers of updates. Results were averaged over 100 independent runs; we use discount $\gamma = 0.9$ and step size $\alpha = 0.01$ in this experiment. The last row contains true state values computed using DP policy iteration

Number of updates	Room 1	Room 2	Room 3	Outside	Found item
0	0	0	0	0	0
100	−0.22	−0.32	0.5	−0.26	0
1000	−1.62	−1.18	2.74	−2.0	0
10,000	−2.65	−1.82	3.2	−3.31	0
20,000	−2.65	−1.82	3.2	−3.3	0
50,000	−2.66	−1.87	3.16	−3.31	0
DP	−2.66	−1.85	3.17	−3.33	0

Temporal Difference Policy Evaluation for State-Action Value Function

In this section, we describe how to extend the TD(0) algorithm to estimate the state-action value function Q_π. This is important because we need Q_π to make better decisions using policy improvement, just like we did with the Monte Carlo methods.

Recall the Bellman expectation equation for Q_π:

$$
\begin{aligned}
Q_\pi(s, a) &= \mathbb{E}_\pi\Big[G_t \mid S_t = s, A_t = a \Big] \\
&= \mathbb{E}_\pi\Big[R_t + \gamma R_{t+1} + \gamma^2 R_{t+2} + \cdots \mid S_t = s, A_t = a \Big] \\
&= \mathbb{E}_\pi\Big[R_t + \gamma G_{t+1} \mid S_t = s, A_t = a \Big] \\
&= \mathbb{E}_\pi\Big[R_t + \gamma Q_\pi(S_{t+1}, A_{t+1}) \mid S_t = s, A_t = a \Big]
\end{aligned}
\tag{5.4}
$$

Here, we use $R_t + \gamma Q_\pi(S_{t+1}, A_{t+1})$ to replace the return G_t. To estimate Q_π, we need to collect a tuple of (s, a, r, s', a') and update the value based on the state-action pair (s, a). The equation for updating estimated values is

$$
Q_\pi(S_t, A_t) = Q_\pi(S_t, A_t) + \alpha\left(\Big[R_t + \gamma Q_\pi(S_{t+1}, A_{t+1}) \Big] - Q_\pi(S_t, A_t) \right)
\tag{5.5}
$$

The overall process is very similar to TD(0) policy evaluation for V_π, but there are some key differences:

- We need to collect tuples of (s, a, r, s', a') instead of (s, a, r, s') to estimate Q_π.
- The value update is based on the state-action pair (s, a).

By estimating the state-action value function Q_π, we can make better decisions using policy improvement. Q_π tells us the expected value of taking a particular action in a given state, which is important information for choosing actions that lead to higher rewards. Moreover, Q_π is related to the value function V_π, which tells us the expected value of being in a given state and taking actions according to a given policy π. Specifically, V_π can be expressed as the expected value of Q_π over all possible actions, weighted by the probability of taking each action under the policy π.

The TD(0) policy evaluation algorithm is a way to estimate the state-action value function Q_π for a fixed policy π. Algorithm 2 shows the pseudocode for the algorithm.

In this algorithm, we initialize the state-action value function Q_π to 0 for all state-action pairs. Then, we run episodes using the policy π and update the state-action value function based on the rewards received and the expected future rewards. We repeat this process until the algorithm converges.

Figure 5.2 shows the computed state-action values for a random policy for the service dog MDP. The results were averaged over 100 independent runs; each run is over 20,000 updates. We use discount $\gamma = 0.9$, step size $\alpha = 0.01$. The values are very close to the ones we computed back in Chap. 4 using Monte Carlo policy evaluation as shown in Fig. 4.3.

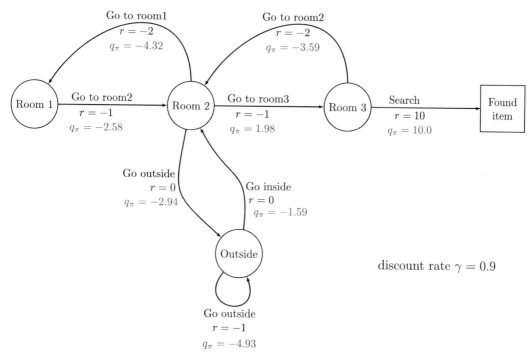

Fig. 5.2 State-action values computed using TD(0) policy evaluation of a random policy for the service dog MDP. Results were averaged over 100 runs; each run is over 2000 updates. We use the discount $\gamma = 0.9$ and step size $\alpha = 0.01$ in this experiment

---

Algorithm 2: Temporal difference [TD(0)] policy evaluation for Q_π

Input: Policy π to be evaluated, discount rate γ, step size α, number of steps K
Initialize: $i = 0$, $Q_\pi(s, a) = 0$, for all $s \in S$, $a \in A(s)$
Output: The estimated state-action value function Q_π for the input policy π

1 Sample initial state s_0
2 **while** $i < K$ **do**
3   Sample action a_t for s_t when following policy π
4   Take action a_t in environment and observe r_t, s_{t+1}
5      $i = i + 1$
6   Sample action a_{t+1} for s_{t+1} when following policy π
7   Compute TD target:

8      $$\delta_t = \begin{cases} r_t & \text{if } s_{t+1} \text{ is terminal state} \\ r_t + \gamma Q_\pi(s_{t+1}, a_{t+1}) & \text{otherwise bootstrapping} \end{cases}$$

9   Update estimated value $Q_\pi(s_t, a_t) \leftarrow Q_\pi(s_t, a_t) + \alpha\left(\delta_t - Q_\pi(s_t, a_t)\right)$
10     $s_t = s_{t+1}$, $a_t = a_{t+1}$

5.3   Simplified ϵ-Greedy Policy for Exploration

After performing TD policy evaluation, the next step is to extend the algorithm to TD control (or policy improvement). TD learning, like Monte Carlo methods, is a model-free reinforcement learning method that faces the challenge of balancing exploration vs. exploitation. In Chap. 4, we introduced the ϵ-greedy policy to help the agent explore when using Monte Carlo methods. We can also use the same ϵ-greedy policy for TD learning methods by updating the policy probability distribution for the different actions using Eq. (5.6).

$$
\pi(a|s) = \begin{cases} 1 - \epsilon + \dfrac{\epsilon}{|A(s)|} & \text{if } a = \arg\max_a Q_\pi(s, a) \\ \dfrac{\epsilon}{|A(s)|} & \text{if } a \neq \arg\max_a Q_\pi(s, a) \end{cases} \tag{5.6}
$$

Recall that the ϵ-greedy policy is derived from the state-action value function Q_π, and it uses Eq. (5.6) to update the policy probability distribution for the different actions. The ϵ-greedy policy is essentially a combination of a random policy and a deterministic policy. With probability ϵ, the policy acts randomly, while with probability $1 - \epsilon$, it acts greedily according to Q_π.

It is worth noting that the values of Q_π are constantly updated in TD policy evaluation, that is, updated every single time step. Therefore, the "deterministic policy" part of the ϵ-greedy policy is also updated from time to time. However, in practice, we often use a simpler and more efficient way to create the ϵ-greedy policy, which does not require updating the policy probability distribution of π.

Instead, we can create a proxy function over Q_π to construct the ϵ-greedy policy. When the agent follows this ϵ-greedy policy and needs to make a decision in the environment, with probability ϵ, this proxy function returns an action that is sampled uniformly for the specific state, and with probability $1 - \epsilon$, it returns the best action according to Q_π for this state. This simplified version of the ϵ-greedy policy is more robust, especially when dealing with large-scale MDPs, as we will discuss in Part II of this book.

To do so, we create a proxy function over Q_π to construct the ϵ-greedy policy, as shown in Algorithm 3. When the agent follows this ϵ-greedy policy and needs to make a decision in the environment, with probability ϵ, this proxy function returns an action that is sampled uniformly for the specific state, and with probability $1 - \epsilon$, it returns the best action according to Q_π for this state.

Algorithm 3: A simplified ϵ-greedy policy over Q_π

---

1 def ϵ-greedy-policy(Q, ϵ, S_t):

    /* Q state-action value function, ϵ exploration rate, S_t environment state   */

2   **if** $probability < \epsilon$ **then**

3   Sample action a randomly for S_t

4   Probability of select a when in S_t is $P(a) = \dfrac{\epsilon}{|A(S_t)|}$

5   **else**

6   Choose action $a = \arg\max_a Q(S_t, a)$

7   Probability of select a when in S_t is $P(a) = 1 - \epsilon + \dfrac{\epsilon}{|A(S_t)|}$

8   **return** $a, P(a)$

Thus, there is no need for the additional step to specifically improve the policy, because there is no such policy anymore, as we can see from the sample code. All we need to do is use this simplified version of the ϵ-greedy policy when the agent needs to choose an action while interacting with the environment.

In summary, the ϵ-greedy policy is a useful tool for balancing exploration vs. exploitation in TD control. While it was originally derived from the state-action value function Q_π, and since Q_π is being constantly updated when using TD methods, we can use a simplified version of the ϵ-greedy policy that does not require updating the policy probability distribution of π.

5.4   TD Control—SARSA

State-Action-Reward-State-Action (SARSA) is the first TD control algorithm we want to introduce. It is an online, on-policy, and model-free reinforcement learning algorithm based on TD learning. SARSA can handle both episodic and continuing reinforcement learning problems and uses the general policy iteration template to find the optimal policy. However, SARSA simplifies the exploration part by using an ϵ-greedy policy. The ϵ-greedy policy selects the action that maximizes the estimated state-action value function Q_π with probability $1-\epsilon$ (exploitation) and selects a random action with probability ϵ (exploration).

In SARSA, we use policy evaluation to estimate the state-action value function Q_π instead of V_π, which is the case for Monte Carlo policy evaluation. The SARSA algorithm updates the estimated state-action value function Q_π using the Bellman equation:

$$Q_\pi(S_t, A_t) \leftarrow Q_\pi(S_t, A_t) + \alpha\big(R_t + \gamma Q_\pi(S_{t+1}, A_{t+1}) - Q_\pi(S_t, A_t)\big) \qquad (5.7)$$

where S_t and A_t denote the current state and action, R_t is the immediate reward received after taking action A_t in state S_t, S_{t+1} and A_{t+1} denote the next state and action, γ is the discount factor, and α is the step size parameter.

The SARSA algorithm uses the estimated state-action value function Q_π to update the policy in an online manner. The policy is updated using a simplified version of ϵ-greedy policy. Specifically, the policy selects the action that maximizes the estimated state-action value function Q_π with probability $1 - \epsilon$ (exploitation) and selects a random action with probability ϵ (exploration). Since the policy is updated in an online manner, we do not need to perform a separate policy improvement step.

SARSA has some advantages over Monte Carlo methods. It can learn online, which means it can update the estimate of the state-action value function Q_π and the policy after every time step, making it more suitable for applications where real-time interaction with the environment is required. SARSA is also an on-policy method, we'll discuss what's on-policy later in this chapter. However, SARSA can suffer from high variance and bias in the estimated state-action value function, especially when the exploration rate is low.

The SARSA algorithm has few parameters that will require additional attention (or as we call it, it's sensitive to these parameters). The first is the step size α. In Monte Carlo methods, we use the $\frac{1}{N(s,a)}$ as the step size, but here in SARSA, we update the state-action values based on the difference between the current estimate and the next estimate. Therefore, we don't keep count of how many times the agent has visited a particular state-action pair (s, a). As a result, we often need to set the step size manually.

The second parameter is the exploration rate ϵ for the ϵ-greedy policy. We can choose to use a fixed value for ϵ, but in practice, a common practice is to use a large initial value for ϵ (like 1.0) and decay ϵ linearly over the first half of the learning process to a small fixed value (like 0.01). Deciding how often to decay the value of ϵ is a difficult question without a clear answer. Decaying it too fast may lead to the agent not exploring enough before ϵ becomes too small, halting exploration altogether. However,

---

Algorithm 4: SARSA

Input: Discount rate γ, initial exploration rate ϵ, step size α, number of steps K
Initialize: $i = 0$, $Q_\pi(s, a) = 0$, for all $s \in S$, $a \in A(s)$
Output: The estimated optimal state-action value function $Q_\pi \approx Q_\star$

1   Sample initial state s_0
2   **while** $i < K$ **do**
3   $\quad$ Sample action a_t for s_t when following ϵ-greedy policy w.r.t. Q_π
4   $\quad$ Take action a_t in environment and observe r_t, s_{t+1}
5   $\quad$ $i = i + 1$
6   $\quad$ (optional) decay ϵ, for example, linearly to a small fixed value
7   $\quad$ Sample action a_{t+1} for s_{t+1} when following ϵ-greedy policy w.r.t. Q_π
8   $\quad$ Compute TD target:
9   $\quad$ $\delta_t = \begin{cases} r_t & \text{if } s_{t+1} \text{ is terminal state} \\ r_t + \gamma Q_\pi(s_{t+1}, a_{t+1}) & \text{otherwise bootstrapping} \end{cases}$
10 $\quad$ Update estimated value $Q_\pi(s_t, a_t) \leftarrow Q_\pi(s_t, a_t) + \alpha\left(\delta_t - Q_\pi(s_t, a_t)\right)$
11 $\quad$ $s_t = s_{t+1}, a_t = a_{t+1}$

Table 5.2 Optimal state value for the service dog MDP after running the SARSA algorithm for different numbers of updates. Results were averaged over 100 independent runs. We use discount $\gamma = 0.9$ and step size $\alpha = 0.01$, use initial exploration rate $\epsilon = 1.0$, and decay it from 1.0 to 0.01. The last row contains true optimal state values V_* computed using DP policy iteration

Number of updates	Room 1	Room 2	Room 3	Outside	Found item
100	−0.12	0.0	0.47	0.0	0
1000	0.28	3.49	8.17	−0.01	0
10,000	5.83	7.85	10.0	4.25	0
20,000	6.02	7.93	10.0	5.53	0
50,000	6.12	7.97	10.0	6.38	0
DP	6.2	8.0	10.0	7.2	0

decaying it too slowly may cause the agent to take longer to learn as it continues to act randomly even when it is no longer necessary. In practice, this parameter requires multiple trials to find a good fit.

Like Monte Carlo policy iteration, SARSA will only converge to the true optimal state-action value in the limit, but in practice, we often terminate the algorithm after a fixed number of iterations or when the change in the state-action values falls below a certain threshold. Therefore, the outcome from the SARSA algorithm may not be the true optimal state-action values if we terminate the learning process too soon.

We ran the SARSA algorithm on our server dog MDP to find the optimal policy and optimal value functions. The optimal policy is the one that maximizes the expected total reward over all time steps, and the optimal value functions are the expected total reward from each state under the optimal policy. We computed the true optimal values using dynamic programming (DP), which is guaranteed to converge to the optimal values in a finite number of iterations. The results shown in Table 5.2 were averaged over 100 independent runs, using a discount factor $\gamma = 0.9$ and a step size $\alpha = 0.01$. We set the initial exploration rate $\epsilon = 1.0$ and decayed it linearly over the first half of the learning process to 0.01. Surprisingly, the algorithm found the optimal policy after 500 updates made to the state-action values, but the estimated state values were not quite close to the true V_* computed using DP (last row).

As we increased the number of updates, the estimated values for the states became very close to the true values.

5.5   On-Policy vs. Off-Policy

Reinforcement learning is a type of machine learning where an agent learns to take actions in an environment to maximize a cumulative reward signal. In this chapter and previous chapters, we have introduced several on-policy reinforcement learning algorithms. However, before we delve further into these algorithms, let's first define what an on-policy algorithm is.

In reinforcement learning, it is important to distinguish between the behavior policy and the target policy. The behavior policy, denoted by μ, is the policy that the agent follows when interacting with the environment to generate sample experience, while the target policy, denoted by π, is the policy that the agent is trying to learn. An on-policy reinforcement learning algorithm is one where the behavior policy and the target policy are the same policy, that is, $\pi = \mu$.

On the other hand, an off-policy reinforcement learning algorithm is one where the behavior policy and the target policy are different, that is, $\pi \neq \mu$. One example of an off-policy algorithm is Q-learning, which we'll cover later in this chapter, which learns the optimal action-value function, Q_*, regardless of the behavior policy.

For example, Monte Carlo and TD(0) learning algorithms such as SARSA learn the state-action value function for the behavior policy, μ, using the experiences generated by following the same policy, μ, to interact with the environment. These kinds of on-policy reinforcement learning algorithms are relatively simple, but they can be less efficient when the behavior policy is less exploratory than the target policy.

Off-policy reinforcement learning algorithms are often more powerful because they can use experience generated by a different behavior policy, which may be more exploratory than the target policy. A commonly used technique in off-policy reinforcement learning is importance sampling.

Importance sampling is a method for computing the expected value of a random variable using samples generated under a different distribution. For example, suppose we have a behavior policy μ and a target policy π, where $\mu \neq \pi$, and we want to estimate the state value function, V_π, for the target policy, π. When the agent is in some state s and following the behavior policy μ, it needs to choose an action, and the probability of selecting some action a when in state s is defined as $\mu(a|s)$. After the agent takes action a in the environment, the environment could transition to some successor state s', and the state transition probability is defined by the dynamics of the environment, denoted by $p(s'|s, a)$. The probability that the agent will end up in a successor state s' while it is currently in state s and following the policy μ is defined as $\mu(a|s)p(s'|s, a)$.

However, if the agent is following the target policy π, the probability of it ending up in a successor state s' while it is currently in state s is defined as $\pi(a|s)p(s'|s, a)$. Since $\mu \neq \pi$, it is likely that these two trajectories will be different. To correct for this difference, we use importance sampling.

Importance sampling involves weighting the return of each sample generated by the behavior policy μ by the ratio of the probabilities of the actions selected by μ and π. This way, we can estimate the expected return for the target policy π using the samples generated by the behavior policy μ.

When using off-policy reinforcement learning and TD policy evaluation for V_π, we need to compute the importance sampling ratio for a single transition tuple (S_t, A_t, R_t, S_{t+1}). The importance sampling ratio is a ratio of the probabilities of taking an action under the target policy π vs. the behavior policy μ. This ratio is used to correct for the fact that the agent's experience may have been generated by a different policy than the one being evaluated (Fig. 5.3).

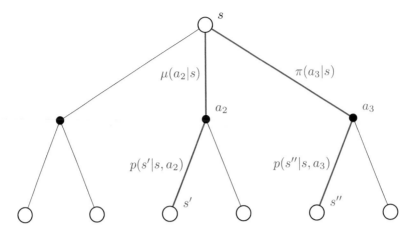

Fig. 5.3 Example of off-policy reinforcement learning to estimate the state value function V_π for target policy π while using data generated from a different behavior policy μ. The red path represents the transition from state s to s' when following μ, and blue path for π

The importance sampling ratio can be computed using the following equations:

$$\rho_t = \frac{\pi(A_t|S_t)p(S_{t+1}|S_t, A_t)}{\mu(A_t|S_t)p(S_{t+1}|S_t, A_t)}$$

$$= \frac{\pi(A_t|S_t)}{\mu(A_t|S_t)} \tag{5.8}$$

where π is the target policy and μ is the behavior policy.

Note that the dynamics of the environment $p(S_{t+1}|S_t, A_t)$ is often unknown to the agent, but it doesn't matter in this case since it appears in both the numerator and the denominator of the ratio and can be canceled out.

In general, off-policy reinforcement learning requires that if $\pi(a|s) > 0$, then $\mu(a|s) > 0$. This means that if the target policy π is to take some actions, we require the behavior policy μ to have at least a non-negative probability of taking these same actions. In most cases, the behavior policy will be a more exploratory policy than the target policy, such as an ϵ-greedy policy with $\epsilon > 0$. This ensures that the probability of each action being selected is always greater than zero.

There is a special case when the denominator in Eq. (5.8) is zero, which occurs when the behavior policy μ selects an action that the target policy π would not have selected. In this case, the importance sampling ratio becomes zero, which means that this transition tuple is ignored when updating the value function.

Off-Policy TD Policy Evaluation

In off-policy reinforcement learning algorithms, the agent learns from experience generated by a behavior policy that is different from the target policy. To account for this, we can use the importance sampling ratio to adjust the TD error during value updates. Specifically, the TD target is weighted by the ratio of the probabilities of taking the action selected by the target policy and the behavior policy in a given state.

The value update rule for off-policy TD(0) policy evaluation for the state value function V_π is shown in Eq. (5.9):

$$V_\pi(S_t) = V_\pi(S_t) + \alpha \frac{\pi(A_t|S_t)}{\mu(A_t|S_t)} \left(\left[R_t + \gamma V_\pi(S_{t+1}) \right] - V_\pi(S_t) \right) \tag{5.9}$$

Here, $\pi(A_t|S_t)$ and $\mu(A_t|S_t)$ represent the probabilities of selecting action A_t in state S_t under the target policy and behavior policy, respectively.

Off-policy TD(0) learning is a popular method for prediction in reinforcement learning. TD(0) is a method for updating the value function using one-step bootstrapping, where the TD target is the immediate reward plus the estimated value of the next state. The overall procedure for off-policy TD(0) is similar to regular TD(0) policy evaluation, but we use a different behavior policy $\mu \neq \pi$ to generate sample experiences. During the value update, the TD target is weighted by the importance sampling ratio, which accounts for the difference between the target policy and the behavior policy. This ratio can help improve the accuracy of the value function estimates by ensuring that the agent assigns more weight to the experience that is relevant to the target policy. For example, if the target policy selects actions that are unlikely under the behavior policy, the importance sampling ratio will be high, and the TD target will be correspondingly weighted more heavily.

The pseudocode for the off-policy TD(0) policy evaluation algorithm for the state value function V_π is shown in Algorithm 5.

Algorithm 5: Off-policy temporal difference [TD(0)] policy evaluation for V_π

Input: Target policy π to be evaluated, discount rate γ, step size α, number of steps K
Initialize: $i = 0$, $V_\pi(s) = 0$, for all $s \in S$, some random behavior policy μ, where $\mu(a|s) > 0$ for all $s \in S, a \in A(s)$
Output: The estimated state value function V_π for the target policy π

1   Sample initial state s_0
2   **while** $i < K$ **do**
3   Sample action a_t for s_t when following policy μ
4   Take action a_t in environment and observe r_t, s_{t+1}
5      $i = i + 1$
6   Compute TD target:
7      $\delta_t = \begin{cases} r_t & \text{if } s_{t+1} \text{ is terminal state} \\ r_t + \gamma V_\pi(s_{t+1}) & \text{otherwise bootstrapping} \end{cases}$
8   Compute importance sampling ratio $\rho_t = \dfrac{\pi(a_t|s_t)}{\mu(a_t|s_t)}$
9   Update estimated value $V_\pi(s_t) \leftarrow V_\pi(s_t) + \alpha \rho_t \left(\delta_t - V_\pi(s_t) \right)$
10     $s_t = s_{t+1}$

However, to extend the off-policy TD(0) policy evaluation algorithm to estimate the state-action value function Q_π for π, it is a little bit different. Since we are estimating the value for a state-action pair (s, a), it does not matter what the probability is that the behavior policy (or target policy) will select action a when in state s, because the state-action value function measures the expected return when in state s, taken action a, and then following the policy π afterwards. We only care about what will happen next. So, to compute the importance sampling for a tuple of $(S_t, A_t, R_t, S_{t+1}, A_{t+1})$, we would start with the second step $t + 1$ for the probability of selecting action A_{t+1} in the successor state S_{t+1} and skip the first step t, which is the probability of selecting action A_t in state S_t.

To better understand this, let's look at an example of how off-policy reinforcement learning estimates the state-action value function Q_π, as shown in Fig. 5.4. Remember, we are estimating

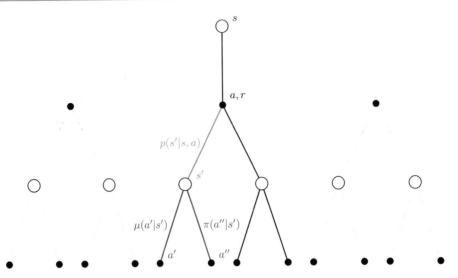

Fig. 5.4 Example of off-policy reinforcement learning to estimate the state-action value function Q_π for target policy π while using data generated from a different behavior policy μ

the value for a state-action pair, not just a state. More specifically, we are only interested in the value for action a in state s. It does not matter what the probability is for selecting a under the behavior policy μ or target policy π, because we will update the value for this specific action a with certainty. If we look at the diagram, part of the path (from s to a) is always the same, so there is no need to include it when we compute the importance sampling ratio. What truly matters are the probabilities of the behavior and target policies selecting the action a' in state s' after the environment has transitioned into a successor state s'.

When using off-policy reinforcement learning and TD policy evaluation for Q_π, we can use the following equation to compute the importance sampling ratio for a single transition tuple $(S_t, A_t, R_t, S_{t+1}, A_{t+1})$:

$$
\rho_{t+1} = \frac{\pi(A_{t+1}|S_{t+1})\, p(S_{t+1}|S_t, A_t)}{\mu(A_{t+1}|S_{t+1})\, p(S_{t+1}|S_t, A_t)}
$$

$$
= \frac{\pi(A_{t+1}|S_{t+1})}{\mu(A_{t+1}|S_{t+1})} \tag{5.10}
$$

Here, ρ_{t+1} represents the importance sampling ratio starting from time step $t + 1$, which corrects for the difference in probability between the behavior policy and the target policy. Specifically, π represents the target policy, while μ represents the behavior policy.

Eq. (5.11) shows the adjusted value update rule for Q_π when using TD(0) and off-policy reinforcement learning:

$$
Q_\pi(S_t, A_t) = Q_\pi(S_t, A_t) + \alpha \frac{\pi(A_{t+1}|S_{t+1})}{\mu(A_{t+1}|S_{t+1})} \left(\left[R_t + \gamma Q_\pi(S_{t+1}, A_{t+1}) \right] \right.
$$

$$
\left. - Q_\pi(S_t, A_t) \right) \tag{5.11}
$$

This equation updates the state-action value function Q_π estimate for the target policy π, using transitions that were generated by a different behavior policy μ. The importance sampling ratio is used to correct for the difference in probability between the behavior policy and the target policy. Specifically, we use the ratio $\frac{\pi(A_{t+1}|S_{t+1})}{\mu(A_{t+1}|S_{t+1})}$ to adjust the update, so that it takes into account the difference between the target policy and the behavior policy. The update is based on the temporal difference error, which is the difference between the estimated Q-value for the current state-action pair and the estimated Q-value for the next state-action pair, plus the reward received at the current state-action pair. The learning rate α controls the size of the update.

Now we introduce the off-policy TD(0) policy evaluation algorithm for Q_π for some given policy π.

Algorithm 6: Off-policy temporal difference [TD(0)] policy evaluation for Q_π

Input: Target policy π to be evaluated, discount rate γ, step size α, number of steps K
Initialize: $i = 0$, $Q_\pi(s, a) = 0$, for all $s \in \mathcal{S}, a \in A(s)$, some random behavior policy μ, where $\mu(a|s) > 0$ for all $s \in \mathcal{S}, a \in A(s)$
Output: The estimated state-action value function Q_π for the target policy π

1   Sample initial state s_0
2   **while** $i < K$ **do**
3   Sample action a_t for s_t when following policy μ
4   Take action a_t in environment and observe r_t, s_{t+1}
5      $i = i + 1$
6   Sample action a_{t+1} for s_{t+1} when following policy μ
7   Compute TD target:
8      $\delta_t = \begin{cases} r_t & \text{if } s_{t+1} \text{ is terminal state} \\ r_t + \gamma Q_\pi(s_{t+1}, a_{t+1}) & \text{otherwise bootstrapping} \end{cases}$
9   Compute importance sampling ratio $\rho_{t+1} = \dfrac{\pi(a_{t+1}|s_{t+1})}{\mu(a_{t+1}|s_{t+1})}$
10   Update estimated value $Q_\pi(s_t, a_t) \leftarrow Q_\pi(s_t, a_t) + \alpha \rho_{t+1} \Big(\delta_t - Q_\pi(s_t, a_t) \Big)$
11     $s_t = s_{t+1}, a_t = a_{t+1}$

In the previous chapter, we introduced Monte Carlo methods for learning action values, and we can extend these methods to off-policy learning. However, the importance sampling ratio becomes more complex, as we need to include the action selection probability and state transition probability until the end of the episode. The equation for computing the importance sampling ratio, $\rho_{t:T-1}$, includes terms such as $\pi(a|s)$ and $\mu(a|s)$, which represent the probability of selecting action a in state s under the target policy and behavior policy, respectively. It's important to note that if either of these probabilities becomes zero at any point in the sequence of transitions, the importance sampling ratio becomes zero as well. This means that the value of the sequence will also become zero and be discarded.

$$
\begin{aligned}
\rho_{t:T-1} &= \frac{\pi(A_t|S_t)p(S_{t+1}|S_t, A_t)\pi(A_{t+1}|S_{t+1}) \cdots p(S_T|S_{T-1}, A_{T-1})}{\mu(A_t|S_t)p(S_{t+1}|S_t, A_t)\mu(A_{t+1}|S_{t+1}) \cdots p(S_T|S_{T-1}, A_{T-1})} \\
&= \frac{\pi(A_t|S_t)\pi(A_{t+1}|S_{t+1}) \cdots \pi(A_{T-1}|S_{T-1})}{\mu(A_t|S_t)\mu(A_{t+1}|S_{t+1}) \cdots \mu(A_{T-1}|S_{T-1})}
\end{aligned} \tag{5.12}
$$

In practice, TD learning with off-policy learning is often used, such as the Q-learning algorithm which we'll introduce in the next section. However, off-policy learning using Monte Carlo methods might be preferred over other methods in situations where

- The behavior policy that generated the data is significantly different from the target policy that the agent is currently following. In this case, the data collected by the behavior policy may not be very useful for learning under the target policy using other methods, such as TD learning. Monte Carlo methods, on the other hand, can still be used to estimate the value of states or actions, as they rely on complete episode sequences rather than individual transitions.
- The agent needs to learn the value of rare or infrequent actions. In some environments, certain actions may be rare or infrequently selected by the behavior policy. TD learning might not provide accurate estimates of the value of these actions, as it relies on individual transitions to update the estimates. In contrast, Monte Carlo methods can provide more accurate estimates of the value of rare or infrequent actions, as they rely on complete episode sequences.
- The agent needs to learn the value of a policy that is different from both the behavior policy and the target policy. In this case, Monte Carlo methods can be used to estimate the value of the desired policy, without requiring the agent to actually follow that policy during data collection.

5.6   Q-Learning

Q-learning is an online, off-policy learning algorithm, developed by Watkins and Peter [3]. It is a model-free reinforcement learning method that also uses TD learning, which means Q-learning supports episodic and continuing reinforcement learning problems. It is probably one of the most used algorithms in the real world to solve reinforcement learning problems. The famous DQN agent built by DeepMind, which can play Atari video games at human level, is based on Q-learning (DQN agent uses deep neural networks to approximate the state-action value function, which we will introduce in Part II of this book).

The SARSA algorithm we introduced before is considered a family of general policy iteration algorithms. It estimates the state-action value function Q_π for a policy π and uses the ϵ-greedy policy to compute a new, better policy π'. Although we use a much simpler version of the ϵ-greedy policy that skips the step to compute a better deterministic policy, the general idea of policy iteration still applies to SARSA.

However, if we remember, there is another DP algorithm we introduced in Chap. 3 called value iteration. Instead of estimating the value function of a policy π like the general policy iteration algorithm does, the value iteration algorithm first tries to estimate the optimal state value function V_*, based on the Bellman optimality equation for V_*:

$$V_*(s) = \max_a \mathbb{E}\left[R_t + \gamma V_*(S_{t+1}) \mid S_t = s, A_t = a \right]$$

$$= \max_a \left[R(s, a) + \gamma \sum_{s' \in \mathcal{S}} P(s'|s, a) V_*(s') \right], \quad \text{for all } s \in \mathcal{S} \tag{5.13}$$

Q-learning adopts the similar idea as value iteration, but instead of estimating the optimal state value function V_*, it estimates the optimal state-action value function Q_*. Recall the Bellman optimality equation for Q_*:

$$Q_*(s, a) = \mathbb{E}\left[R_t + \gamma \max_{a'} Q_*(S_{t+1}, a') \mid S_t = s, A_t = a\right] \tag{5.14}$$

Q-learning uses Eq. (5.14) and the idea of increment update methods for TD learning as the value update rule, which is given by the following update rule:

$$Q(S_t, A_t) = Q(S_t, A_t) + \alpha \left(\left[R_t + \gamma \max_{a'} Q(S_{t+1}, a')\right] - Q(S_t, A_t)\right) \tag{5.15}$$

Q-learning is an off-policy learning algorithm that avoids the need for importance sampling, which is typically used in other off-policy learning algorithms. Importance sampling can be computationally expensive and introduce bias, which makes Q-learning a more efficient and reliable method for estimating the optimal state-action value function.

One advantage of Q-learning is that it always uses the maximum value $Q(S_{t+1}, a')$ across all valid actions for the successor state s', regardless of which action is chosen by the behavior policy. For Q-learning, since the target policy always selects the action with the maximum Q-value for the successor state, the importance sampling ratio simplifies to the probability of selecting the maximum Q-value action under the behavior policy divided by the probability of selecting the maximum Q-value action under the target policy. This ratio is always equal to 1, regardless of the behavior policy, because it does not use the action chose by the behavior policy during the value update. In contrast, other off-policy learning algorithms may require importance sampling to correct for the differences between the behavior policy and the target policy.

Figure 5.5 shows an example of Q-learning estimating the optimal state-action value function Q_* for target policy π. As the diagram shows, the traversed path for value update is always the same, regardless of the action the agent chooses when following the behavior policy μ in the successor state

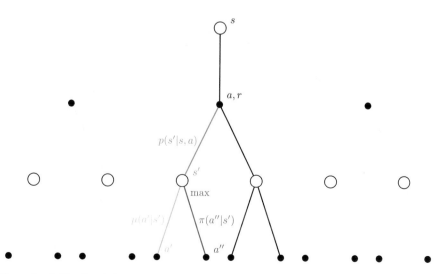

Fig. 5.5 Example of off-policy Q-learning estimating the optimal state-action value function Q_* for target policy π

s'. However, there are special cases where we are required to use importance sampling, such as when using N-step returns, which rely on the behavior policy to select actions.

To update the Q-values in Q-learning, we use the target policy to evaluate the action $a' = \max_{a'} Q(S_{t+1}, a')$ while at the same time using the evaluated value to learn the same target policy. This means that the importance sampling ratio for Q-learning simplifies to 1, as the numerator and denominator are the same. Specifically, the importance sampling ratio can be expressed as

$$\rho_{t+1} = \frac{\pi(A_{t+1}|S_{t+1})p(S_{t+1}|S_t, A_t)}{\pi(A_{t+1}|S_{t+1})p(S_{t+1}|S_t, A_t)}$$

$$= 1 \tag{5.16}$$

By avoiding the need for importance sampling, Q-learning is computationally efficient and can estimate the optimal state-action value function more accurately. For example, suppose we have an agent using Q-learning, and it has already estimated the state-action values for some state s and action a as $Q(s, a) = 10$. Now, the agent takes action a and transitions to a new state s', where the maximum Q-value is $Q(s', a') = 15$. However, the behavior policy used to select the action in state s' is different from the target policy used to update the Q-value. In this case, the importance sampling ratio for updating $Q(s, a)$ involves the ratio of probabilities of selecting a and a' under the behavior policy and the target policy, but because Q-learning always uses the maximum Q-value for the next state, we do not need to consider the probability of selecting the action that led to that maximum value. This simplifies the update process and makes Q-learning more efficient than other off-policy learning algorithms that require importance sampling.

Q-learning is a model-free reinforcement learning algorithm, similar to SARSA, that requires balancing exploration and exploitation to learn an optimal policy. In reinforcement learning, exploration refers to trying out new actions to discover potentially better strategies, while exploitation refers to exploiting known strategies to maximize rewards. To accomplish this, we can employ the ϵ-greedy policy, which selects the best action with probability $1 - \epsilon$ and a random action with probability ϵ.

Q-learning and SARSA share a similar structure and procedure, both using a temporal difference (TD) learning approach. However, during the Q-value update, Q-learning always chooses the action with the highest Q-value for the successor state S_{t+1}, denoted as $\max_{a'} Q(S_{t+1}, a')$, whereas SARSA samples an action from the policy. This is because Q-learning is an off-policy algorithm that updates the Q-values based on the best possible action, whereas SARSA is an on-policy algorithm that updates the Q-values based on the action actually taken.

The pseudocode for the Q-learning algorithm is shown in Algorithm 7. As with SARSA, Q-learning is sensitive to hyperparameters such as the step size α and the exploration rate ϵ of the ϵ-greedy policy. The step size determines how quickly the algorithm updates the Q-values, while the exploration rate determines how often the algorithm explores new actions instead of exploiting the current best action.

Worked Example

Let's look at an example of how Q-learning would update the value for a state-action pair (s, a). Again, we will be using the service dog MDP. But in this case, we use the state-action values for a random policy computed using TD(0) policy evaluation from Fig. 5.2. We only focus on how the algorithm would compute the value for state *Room 1* and action *Go to room2* (Fig. 5.6).

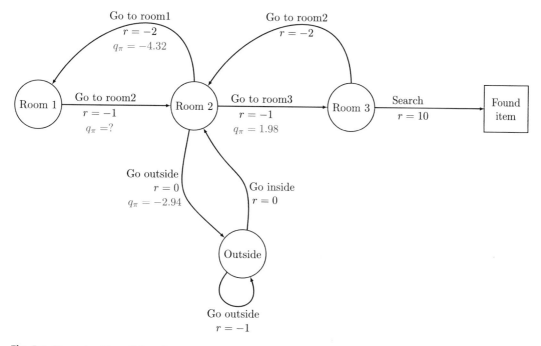

Fig. 5.6 Example of how Q-learning computes state-action values for the service dog MDP. The values for actions in state *Room 2* are borrowed from Fig. 5.2

Algorithm 7: Q-learning with ϵ-greedy exploration

Input: Discount rate γ, initial exploration rate ϵ, step size α, number of steps K
Initialize: $i = 0$, $Q(s, a) = 0$, for all $s \in \mathcal{S}, a \in A(s)$
Output: The estimated optimal state-action value function Q

1 Sample initial state s_0
2 **while** $i < K$ **do**
3   Sample action a_t for s_t from ϵ-greedy policy w.r.t. Q
4   Take action a_t in environment and observe r_t, s_{t+1}
5      $i = i + 1$
6   (optional) decay ϵ, for example, linearly to a small fixed value
7   Compute TD target:
8      $$\delta_t = \begin{cases} r_t & \text{if } s_{t+1} \text{ is terminal state} \\ r_t + \gamma \max_{a'} Q(s_{t+1}, a') & \text{otherwise bootstrapping} \end{cases}$$
9   Update estimated value $Q(s_t, a_t) \leftarrow Q(s_t, a_t) + \alpha\left(\delta_t - Q(s_t, a_t)\right)$
10     $s_t = s_{t+1}$

Suppose the agent took action *Go to room2* when in state *Room 1* and received a reward signal of -1 and successor state *Room 2* from the environment. We use the following variables and parameters:

- $Q(s, a)$: The value of state-action pair (s, a)
- $\alpha = 0.1$: The step size
- $\gamma = 0.9$: The discount factor

The estimated state-action values for state *Room 2* and actions *Go to room1, Go to room3, Go outside* are $-4.32, 1.98, -2.94$. The initial estimated value for state-action pair *Room 1* and *Go to room2* is 0.

How would the Q-learning algorithm estimate the (optimal) value for state-action pair *Room 1* and *Go to room2*? We can plug in those numbers directly into Eq. (5.15). The only thing we need to specify is the value for the successor state-action pair. In Q-learning, we look for the maximum value over all actions for the successor state. In this case, this value is 1.98 for action *Go to room3*. We can compute the value for state-action pair *Room 1* and *Go to room2* as follows:

$$Q(Room\ 1, Go\ to\ room2) = 0 + 0.1 * (-1 + 0.9 * 1.98 - 0)$$

$$= 0.078$$

We conducted experiments to evaluate the performance of the Q-learning algorithm on our service dog MDP, using the same experimental setup as our previous experiments with SARSA. The goal of these experiments was to compare the performance of Q-learning and SARSA on this particular MDP.

We used the following variables and parameters in our experiments:

- $\alpha = 0.01$: The step size used in the Q-learning algorithm
- $\gamma = 0.9$: The discount factor applied to future rewards
- $\epsilon = 1.0$: The exploration rate that determines the probability of choosing a random action

To reduce the exploration rate over the course of the learning process, we applied a linear decay schedule that reduced the rate from its initial value of 1.0 to a final value of 0.01.

As shown in Table 5.3, the Q-learning algorithm found the optimal policy after 500 updates to the value function. It also found the optimal value function V_* after about 20,000 updates, which is much faster than SARSA's 50,000 updates as shown in Table 5.2. However, it is important to note that the optimal value function found by SARSA still did not perfectly match the true optimal value function, despite running more updates. This suggests that Q-learning can converge more quickly than SARSA, and produces better results.

5.7   Double Q-Learning

Double Q-learning is a method that addresses a small issue with Q-learning, where the estimated optimal state-action value Q_* could be biased. This is due to the nature of the value update process, which always uses the action that yields the maximum value for the successor state s'. In

Table 5.3 Optimal state value for the service dog MDP after running the Q-learning algorithm for different numbers of updates, $\gamma = 0.9, \alpha = 0.01$, decay ϵ from 1.0 to 0.01. The last row contains true V_* computed using DP policy iteration

Number of updates	Room 1	Room 2	Room 3	Outside	Found item
100	−0.12	0.0	0.45	0.0	0
1000	0.95	4.09	8.25	0.22	0
10,000	6.2	8.0	10.0	7.05	0
20,000	6.2	8.0	10.0	7.2	0
50,000	6.2	8.0	10.0	7.2	0
DP	6.2	8.0	10.0	7.2	0

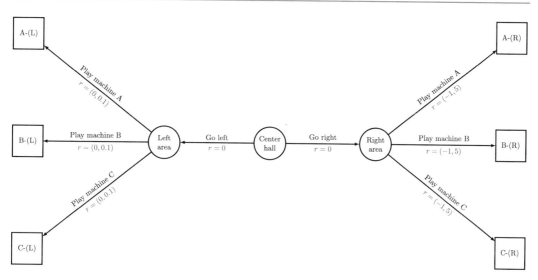

Fig. 5.7 Casino slot machines MDP, only showing three slot machines from each side

some stochastic environments, this approach can lead to problems, as explained in detail by Hado Hasselt [4].

Casino Example

To illustrate the issue of maximization bias, let's consider a simple example derived from Sutton and Barto [2]. Imagine we're at a casino and we want to play some slot machines just for fun. We can model the example as shown in Fig. 5.7, where the center of the casino playground is the center hall, and there are two play areas (left, right) located around the center hall. On each side, there are ten slot machines to choose from, though we've only shown three on each side in the figure.

However, there's a hidden secret about this casino. The slot machines on the left all share the same reward distribution of mean 0 and standard deviation 0.1. This means that if we choose to play any slot machine on the left, we may get a reward between [−0.1, 0.1]. If we play 1000 rounds, we'll break even, since the mean reward is 0.

The slot machines on the right share a totally different reward distribution, with a mean of −1.0 and standard deviation of 5.0. When we play any slot machine on the right, we may get a reward between [−6.0, 4.0]. If we play 1000 rounds, we're guaranteed to lose money (on average, −1 dollar per round). So the obvious choice is to play with the slot machines on the left.

This example demonstrates the potential pitfalls of relying solely on the maximum value estimate in Q-learning and highlights the need for alternative methods like double Q-learning.

We model the problem as an episodic reinforcement learning problem, in which an agent learns to maximize its rewards over time by interacting with a stochastic environment. After each action, the agent receives a reward that is a random variable with an unknown distribution to the agent. The episode always starts with the state *Center Hall*, from which the agent can choose to *Go left* or *Go right* to transition to the corresponding successor state *Left area* or *Right area*. In each area, the agent can choose to play one of several slot machines, each with a different probability distribution for the rewards it can generate. The episode terminates when the agent chooses to play one of the slot machines, regardless of which area it is in. The next episode starts independently of how the previous episode ended, assuming that the agent has an unlimited amount of money to play.

If we use standard Q-learning with no discount factor ($\gamma = 1$), what are the estimated values for the actions *Go left* and *Go right* when the agent is in state *Center Hall*? Let's follow the update rule in Eq. (5.15) to find out. Assuming that the initial Q-values for both actions are 0, the step size is $\alpha = 1$. The update rule can be simplified as $Q(S_t, A_t) = R_t + \max_{a'} Q(S_{t+1}, a')$, where R_t is the immediate reward and $\max_{a'} Q(S_{t+1}, a')$ is the maximum Q-value of the successor state S_{t+1}.

In state *Left area*, the highest reward we can get from any of the slot machines is 0.1 (ignoring outliers), so the estimated Q-value for *Center Hall* and *Go left* is $0 + 0.1 = 0.1$. However, in state *Right area*, the highest reward we can get from any of the slot machines is 4.0, which leads to an estimated Q-value of $0 + 4.0 = 4.0$ for *Center Hall* and *Go right*. Note that this overestimates the true value of the action due to the high variance in the rewards. In practice, the agent would need to try each action enough times to estimate its true value, which is costly in this example.

Double Q-learning is an extension to the standard Q-learning algorithm that mitigates the problem of overestimation that can occur with the latter. In Q-learning, the agent estimates the optimal state-action value function Q and uses it to select actions. However, in certain situations such as the casino machine MDP example, Q-learning can overestimate the value of actions, which can lead to suboptimal policies.

The idea behind double Q-learning is to learn two independent Q-functions, Q_1 and Q_2, and use them to select actions. During the learning process, in half of the time the algorithm updates the values of Q_1, and in the other half, it updates the values of Q_2. The key contribution of double Q-learning is how to choose which Q-value to use when estimating the value of successor state-action pair.

In the standard Q-learning algorithm, we can use the following equation to compute the TD target:

$$TD_{target} = R_t + \gamma \max_{a'} Q(S_{t+1}, a')$$

$$= R_t + \gamma Q\left(S_{t+1}, \arg\max_{a'} Q(S_{t+1}, a')\right)$$

With double Q-learning, we modify the preceding equation for updating Q_1 to

$$TD_{target} = R_t + \gamma \max_{a'} Q(S_{t+1}, a')$$

$$= R_t + \gamma Q_2\left(S_{t+1}, \arg\max_{a'} Q_1(S_{t+1}, a')\right)$$

Suppose the agent is going to update Q_1 and has collected a tuple of (S_t, A_t, R_t, S_{t+1}). The agent uses Q_1 to select the best action $a' = \arg\max_{a'} Q_1(S_{t+1}, a')$, which yields the maximum value in the successor state S_{t+1}. However, instead of using the value $Q_1(S_{t+1}, a')$ directly, the agent uses this best action a' to select the value from Q_2, which is $Q_2(S_{t+1}, a')$. This process helps reduce overestimation and leads to more accurate value estimates.

Similarly, if the agent is going to update Q_2, it uses the same logic, but with Q_1 and Q_2 swapped:

$$Q_2(S_t, A_t) = Q_2(S_t, A_t) + \alpha\left(R_t + \gamma Q_1\left(S_{t+1}, \arg\max_{a'} Q_2(S_{t+1}, a')\right) - Q_2(S_t, A_t)\right)$$

As an example, suppose an agent is learning to play a game where it needs to navigate through a maze to reach a goal. In some situations, the agent may overestimate the value of certain actions, leading to suboptimal policies. With double Q-learning, the agent learns two independent Q-functions, and during the learning process, it alternates between updating the values of Q_1 and Q_2. When selecting actions, the agent uses the best action from Q_1 to select the value from Q_2, or vice versa. This process helps reduce overestimation and can lead to more accurate value estimates, resulting in better policies.

In the context of the casino MDP example, the probability of the same action having the maximum value in both Q_1 and Q_2 is much lower, which mitigates the maximization bias issue.

We now formally introduce the double Q-learning algorithm with ϵ-greedy policy for exploration. The overall procedure is the same compared to standard Q-learning. One small change we need to make is when it comes to the time for the agent to act in the environment, it uses the merged state-action value functions Q_1 and Q_2 as a single state-action value function Q for the ϵ-greedy policy. The ϵ-greedy policy selects the action with the highest value from Q with probability $1 - \epsilon$ and selects a random action with probability ϵ.

Algorithm 8: Double Q-learning with ϵ-greedy exploration

Input: Discount rate γ, initial exploration rate ϵ, step size α, number of steps K
Initialize: $i = 0$, $Q_1(s, a) = 0$, $Q_2(s, a) = 0$, for all $s \in \mathcal{S}, a \in A(s)$
Output: The optimal state-action value function Q

1 Sample initial state s_0
2 **while** $i < K$ **do**
3   Sample action a_t for s_t from ϵ-greedy policy w.r.t. averages (or sum) of Q_1, Q_2
4   Take action a_t in environment and observe r_t, s_{t+1}
5      $i = i + 1$
6   (optional) decay ϵ, for example, linearly to a small fixed value
7   **if** *probability* < 0.5 **then**
        /* Update Q_1 */
8   Compute TD target:
9          $\delta_t = \begin{cases} r_t & \text{if } s_{t+1} \text{ is terminal state} \\ r_t + \gamma Q_2\big(s_{t+1}, \arg\max_{a'} Q_1(s_{t+1}, a')\big) & \text{otherwise bootstrapping} \end{cases}$
10         $Q_1(s_t, a_t) \leftarrow Q_1(s_t, a_t) + \alpha\big(\delta_t - Q_1(s_t, a_t)\big)$
11   **else**
        /* Update Q_2 */
12   Compute TD target:
13         $\delta_t = \begin{cases} r_t & \text{if } s_{t+1} \text{ is terminal state} \\ r_t + \gamma Q_1\big(s_{t+1}, \arg\max_{a'} Q_2(s_{t+1}, a')\big) & \text{otherwise bootstrapping} \end{cases}$
14         $Q_2(s_t, a_t) \leftarrow Q_2(s_t, a_t) + \alpha\big(\delta_t - Q_2(s_t, a_t)\big)$
15     $s_t = s_{t+1}$

We conducted an experiment using standard Q-learning and double Q-learning to find the optimal policy for our casino MDP. We used a fixed exploration rate of $\epsilon = 0.1$, the same step size of $\alpha = 0.1$, and no discount ($\gamma = 1.0$) throughout the experiment. The goal of the experiment was to measure how often the agent chose the action *Go right* when in state *Center Hall*.

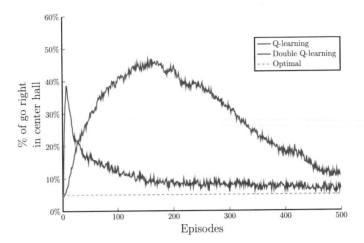

Fig. 5.8 The percentage of choosing *Go right* when in state *Center Hall* on the casino MDP. Results were averaged over 1000 independent runs; each run is over 500 episodes. We use step size $\alpha = 0.02$, no discount $\gamma = 1.0$, and fixed exploration rate $\epsilon = 0.1$ throughout the experiment

The results, shown in Fig. 5.8, were averaged over 1000 independent runs, each consisting of 500 episodes. The optimal policy is to choose *Go right* with a probability of 5% (since we are using a fixed $\epsilon = 0.1$, the agent will act randomly with a probability of 10%, and when acting randomly and choosing between *Go left* and *Go right*, the agent should choose *Go right* with a probability of 0.5 * 10%). We observed that the standard Q-learning agent chose *Go right* more often than the optimal policy during the first 200 episodes. However, the double Q-learning agent quickly discovered the optimal policy after only 50 episodes.

5.8   N-Step Bootstrapping

In reinforcement learning, the goal is for an agent to learn a policy that maximizes its expected cumulative reward over time. One approach to learning such a policy is to estimate the value of each state, which is the expected cumulative reward that an agent can obtain by following its policy from that state onward.

In TD(0) policy evaluation, we estimate the return G_t using the TD target $R_t + \gamma V_\pi(S_{t+1})$, which is the immediate reward R_t plus the discounted value of the successor state $\gamma V_\pi(S_{t+1})$. However, one concern with this method is that if the estimated value for the successor state S_{t+1} is biased, then the updated values $V_\pi(S_t)$ will also be biased. This can lead to suboptimal performance, which means the agent may need to spend more time learning the state values.

Monte Carlo methods don't have this issue because the agent always uses the actual return computed from a complete episode. However, we prefer to use TD methods over Monte Carlo because we want to support continuing reinforcement learning problems, and we want to avoid waiting until the episode is over to update the values. So, is there a way we can estimate the return more accurately while still using TD methods?

To answer this question, let's look at the mathematical equations for the value functions. The state value function V_π can be written recursively as

$$V_\pi(s) = \mathbb{E}_\pi\left[G_t \mid S_t = s\right]$$

$$= \mathbb{E}_\pi\left[R_t + \gamma R_{t+1} + \gamma^2 R_{t+2} + \gamma^3 R_{t+3} + \cdots \mid S_t = s\right]$$

$$= \mathbb{E}_\pi\left[R_t + \gamma\left(R_{t+1} + \gamma R_{t+2} + \gamma^2 R_{t+2} + \cdots\right) \mid S_t = s\right]$$

$$= \mathbb{E}_\pi\left[R_t + \gamma G_{t+1} \mid S_t = s\right]$$

$$= \mathbb{E}_\pi\left[R_t + \gamma V_\pi(S_{t+1}) \mid S_t = s\right] \tag{5.17}$$

In TD(0) policy evaluation, we use $R_t + \gamma V_\pi(S_{t+1})$ (from Eq. 5.17) to replace G_t with the incremental update methods:

$$V_\pi(S_t) = V_\pi(S_t) + \alpha\left(G_t - V_\pi(S_t)\right)$$

$$= V_\pi(S_t) + \alpha\left(\left[R_t + \gamma V_\pi(S_{t+1})\right] - V_\pi(S_t)\right) \tag{5.18}$$

To estimate the return more accurately, we can use a method called N-step bootstrapping, which rearranges the return G_t in a different way that uses multiple rewards received into the future from the current time step t, plus the discounted value of the successor state after the last reward. This is shown in Eqs. (5.19), (5.20), and (5.21):

$$G_t = R_t + \gamma R_{t+1} + \gamma^2 R_{t+2} + \gamma^3 R_{t+3} + \gamma^4 R_{t+4} + \cdots$$

$$= R_t + \gamma R_{t+1} + \gamma^2\left(R_{t+2} + \gamma R_{t+3} + \gamma^2 R_{t+4} + \cdots\right)$$

$$= R_t + \gamma R_{t+1} + \gamma^2 G_{t+2} \tag{5.19}$$

$$G_t = R_t + \gamma R_{t+1} + \gamma^2 R_{t+2} + \gamma^3 R_{t+3} + \gamma^4 R_{t+4} + \cdots$$

$$= R_t + \gamma R_{t+1} + \gamma^2 R_{t+2} + \gamma^3\left(R_{t+3} + \gamma R_{t+4} + \cdots\right)$$

$$= R_t + \gamma R_{t+1} + \gamma^2 R_{t+2} + \gamma^3 G_{t+3} \tag{5.20}$$

$$G_t = R_t + \gamma R_{t+1} + \gamma^2 R_{t+2} + \cdots + \gamma^{n-1} R_{t+n-1} + \gamma^n G_{t+n} \tag{5.21}$$

Inspired by Eq. (5.17), we can rewrite Eq. (5.21) recursively using the state value function V_π, where we use $\gamma^n V_\pi(S_{t+n})$ as an estimator (or bootstrapping) for the missing rewards for time steps $t+n+1, t+n+2, \cdots$. This gives us the *N-step* return or *N-step* bootstrapping, as shown in Eq. (5.22):

$$G_t = R_t + \gamma R_{t+1} + \gamma^2 R_{t+2} + \cdots + \gamma^{n-1} R_{t+n-1} + \gamma^n V_\pi(S_{t+n}) \tag{5.22}$$

We can adapt Eq. (5.22) as part of the incremental value update rule, where we use the N-step returns as the estimated return G_t:

$$V_\pi(S_t) = V_\pi(S_t) + \alpha \left(\left[R_t + \gamma R_{t+1} + \cdots + \gamma^{n-1} R_{t+n-1} \right. \right.$$

$$\left. \left. + \gamma^n V_\pi(S_{t+n}) \right] - V_\pi(S_t) \right) \tag{5.23}$$

Here, α is the step size parameter that determines the magnitude of the update. The update rule is incremental, meaning that the estimated value of a state is updated after each time step based on the observed reward and the estimated value of the next state.

Using N-step bootstrapping is a way to estimate the return more accurately while still using TD methods. By breaking the return down into multiple rewards, we can use the value function to estimate the missing rewards that occur after the N time steps. This approach helps mitigate the bias introduced by using a biased estimate of the successor state value.

One advantage of N-step bootstrapping over Monte Carlo methods is that it can be used for continuing reinforcement learning problems, where episodes do not have a fixed length. Monte Carlo methods can still be used by truncating episodes, but N-step bootstrapping can update the value function after N time steps, making it more efficient.

The N-step bootstrapping method mixes TD(0) and Monte Carlo methods together (for episodic cases). When $N = 1$, we get the standard TD(0) like SARSA. If $N = T$, where T is the length of the episode, we get Monte Carlo methods. This approach is sometimes called "multi-step look-ahead," because the agent uses information from multiple time steps ahead to compute the value estimates (Fig. 5.9).

The choice of N is a trade-off between bias and variance. If N is small, the estimate is more biased because it relies on fewer samples of future rewards, but it is updated more frequently, which can reduce variance. If N is large, the estimate is less biased because it uses more samples, but it is updated less frequently and can have higher variance.

What about the states close to the terminal time step T for the episodic reinforcement learning problems, like $S_{T-3}, S_{T-2}, S_{T-1}$? To compute the N-step return for these time steps, we simply ignore

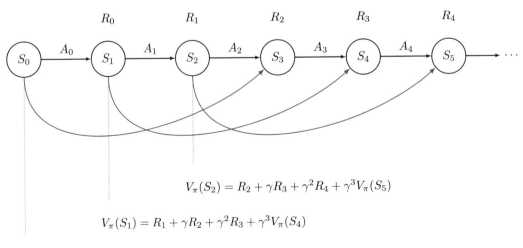

$$V_\pi(S_2) = R_2 + \gamma R_3 + \gamma^2 R_4 + \gamma^3 V_\pi(S_5)$$

$$V_\pi(S_1) = R_1 + \gamma R_2 + \gamma^2 R_3 + \gamma^3 V_\pi(S_4)$$

$$V_\pi(S_0) = R_0 + \gamma R_1 + \gamma^2 R_2 + \gamma^3 V_\pi(S_3)$$

Fig. 5.9 The idea of N-step returns, where $n = 3$

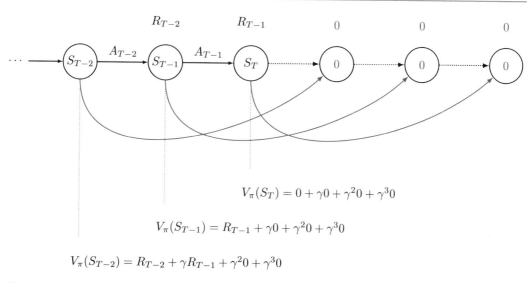

$$V_\pi(S_T) = 0 + \gamma 0 + \gamma^2 0 + \gamma^3 0$$

$$V_\pi(S_{T-1}) = R_{T-1} + \gamma 0 + \gamma^2 0 + \gamma^3 0$$

$$V_\pi(S_{T-2}) = R_{T-2} + \gamma R_{T-1} + \gamma^2 0 + \gamma^3 0$$

Fig. 5.10 How N-step bootstrapping deal with end of episodic sequence, where $n = 3$

the missing time steps. Sometimes, we can also add abstract sequences (both reward and state values are 0) to the end of the episode sequences, which will give us the same results as simply ignoring the missing time steps (Fig. 5.10).

N-step bootstrapping is a powerful technique that can be applied in a variety of settings. Here are two examples of how it can be used in practice:

- In gaming environments, agents often need to make decisions based on a sequence of events that occur over time. For example, in a first-person shooter game, an agent might use N-step bootstrapping to estimate the value of a given state by considering the rewards that will be received over the next N time steps. This can help the agent make more accurate decisions about what actions to take in the game.
- In robotic control, agents must make decisions based on feedback from sensors and other environmental factors. For example, an agent controlling a robot arm might use N-step bootstrapping to estimate the value of a given state by considering the rewards that will be received over the next N time steps. This can help the agent make more accurate decisions about how to move the robot arm in order to achieve a desired outcome.

To apply N-step bootstrapping, there are a few specific steps that should be followed. First, we must select an appropiate value for N that is appropriate for the problem at hand. This will depend on factors such as the length of the time horizon and the rate of change in the environment. Next, the agent uses N-step bootstrapping to estimate the value of each state based on the rewards that will be received over the next N time steps. This can help the agent make more accurate decisions by taking into account the long-term consequences of each action.

However, there are some potential issues with N-step bootstrapping as mentioned in the work by Professor Dimitri P. Bertsekas [5]. For example, in some cases, an incorrect configuration of the algorithm can lead to suboptimal behavior. To illustrate this, consider the following toy example MDP (see Fig. 5.11). In this episodic task, a new episode always starts from the leftmost node (Start), and the episode ends at the rightmost node (End). The agent can only choose one action that keeps it moving forward for all states, except in the initial starting state, where there are two actions that lead the agent to the upper or lower path. The reward signals associated with each state-action pair are

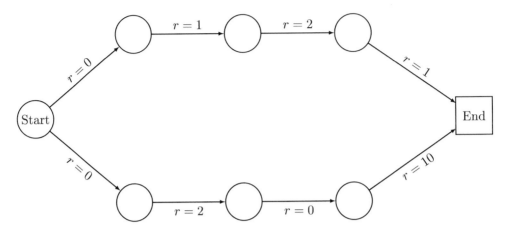

Fig. 5.11 Pitfalls of N-step bootstrapping, where with an incorrect $n = 3$, the algorithm might initially choose the suboptimal path (upper path) instead of the optimal one (lower path)

outlined with the edges between the nodes. The optimal path is the lower path, since there is a high reward signal ($r = 10$) toward the terminal state.

However, with an incorrect configuration, such as $N = 3$, the algorithm would initially choose the upper path, especially when the value estimates are not accurate. To avoid such issues, it is recommended to use the largest value possible for N, although this can be limited by computational constraints.

N-Step SARSA

It's fairly easy to extend the N-step bootstrapping method for state-action value function Q_π. In this case, we update values based on state-action pairs, and for the last successor state S_{t+n}, we also need the action A_{t+n}:

$$Q_\pi(S_t, A_t) = Q_\pi(S_t, A_t) + \alpha\left(\left[R_t + \gamma R_{t+1} + \cdots + \gamma^{n-1} R_{t+n-1}\right.\right.$$

$$\left.\left. + \gamma^n Q_\pi(S_{t+n}, A_{t+n})\right] - Q_\pi(S_t, A_t)\right) \tag{5.24}$$

Here, α is the step size parameter, γ is the discount factor, and t is the current time step.

We now introduce the N-step SARSA on-policy learning algorithm, which builds on the basic structure of SARSA. However, to compute returns for an arbitrary number of N-step, we need to make some changes. This means the values are not updated immediately when the agent makes a decision in the environment. Instead, it has to continue acting in the environment for N steps before it collects the required sample transitions for the N-step return.

To store the necessary sample transitions, we choose to use a list τ that stores tuples of the form (s, a, r, s', a'), where s and s' are the current and next states, respectively; a and a' are the current and next actions, respectively; and r is the reward received for taking action a in state s. The maximum length of τ is N. We use (s, a, r, s', a') instead of (s, a, r, s') because we also need the action A_{t+n} for the last successor state S_{t+n} to complete Eq. (5.24).

The agent only updates the value for the first state-action pair in τ if the length of τ reaches the maximum length, or if it's at the end of an episode, as is the case for episodic reinforcement learning problems. This is because the subsequent state-action pairs in τ are used to compute the N-step return for the first state-action pair, and the value update is only performed after the N-step return has been computed.

Algorithm 9: N-step on-policy SARSA with ϵ-greedy policy for exploration

---

Input: Discount rate γ, initial exploration rate ϵ, step size α, N-step N where $N \geq 1$, number of steps K
Initialize: $i = 0$, $Q_\pi(s, a) = 0$, for all $s \in S$, $a \in A$, empty list τ
Output: The optimal state-action value function Q_π

1   **while** $i < K$ **do**
2   Collect τ, a sequence of N transitions (s_t, a_t, r_t) when following the ϵ-greedy policy w.r.t. Q_π
3   Compute return $G = r_0 + \gamma r_1 + \gamma^2 r_2 + \cdots + \gamma^{N-1} r_{N-1}$ from sequences in τ
4   Compute TD target:
5     $\delta = \begin{cases} G & \text{if } s_N \text{ is terminal state} \\ G + \gamma^N Q_\pi(s_N, a_N) & \text{otherwise bootstrapping} \end{cases}$
6   Update estimated value $Q_\pi(s_0, a_0) \leftarrow Q_\pi(s_0, a_0) + \alpha\left(\delta - Q_\pi(s_0, a_0)\right)$
7   Remove the first transition in τ

We applied the N-step SARSA algorithm to our service dog Markov decision process (MDP) and obtained results that were averaged over 100 independent runs, each consisting of 1000 updates. The algorithm used a step size of $\alpha = 0.01$, a discount rate of $\gamma = 0.9$, and a decayed exploration rate ϵ that started at 1.0 and decreased to 0.01 over the course of 1000 updates. The results are shown in Table 5.4. Increasing the N-step value from 1 to 2 and 3 resulted in state values for *Room 1* and *Room 2* that were closer to the true state values computed using DP policy iteration (as shown in the last row of Table 5.4). However, the values for state *Room 3* and *Outside* did not improve significantly. This is because our service dog MDP is a very simple problem, and under the optimal policy, the length of an episode sequence is only 3. Therefore, using an N-step value greater than 3 does not improve the results.

N-Step Q-Learning

N-step Q-learning is a variant of the traditional Q-learning algorithm that updates the Q-values based on the returns observed over a sequence of N time steps. By incorporating a longer sequence of experiences, N-step Q-learning can lead to more efficient learning and better performance in some

Table 5.4 Optimal state value for the service dog MDP after running the N-step SARSA algorithm for different N. Results were averaged over 100 independent runs; each run is only over 1000 updates. We use discount $\gamma = 0.9$, step size $\alpha = 0.01$, and decay ϵ from 1.0 to 0.01 over the course of 1000 updates. The last row contains true V_* computed using DP policy iteration

N-steps	Room 1	Room 2	Room 3	Outside	Found item
1	0.28	3.49	8.17	0.0	0
2	2.07	6.26	8.58	0.16	0
3	3.86	6.25	8.6	1.06	0
4	3.8	6.45	8.59	0.94	0
5	4.14	6.41	8.59	1.25	0
DP	6.2	8.0	10.0	7.2	0

environments. For example, in environments where rewards are sparse or delayed, using a longer sequence of experiences can help the agent learn faster and make better decisions. Returns refer to the total discounted reward accumulated by the agent over time.

However, N-step Q-learning is more complex than standard Q-learning and requires the use of importance sampling to adjust for differences between the behavior and target policies.

The N-Step Update Rule for Q-Learning

In standard Q-learning, the best action is always used to evaluate the successor state. However, when using N-step returns, the incremental update rule becomes more complex. The update rule involves taking the sum of rewards for the next N steps and then bootstrapping from the value of the state N steps ahead. Bootstrapping refers to using an estimate of the value function to update the estimate of the value function. This allows the agent to learn from a longer sequence of experiences and make better decisions.

Here's the equation for the update rule:

$$Q_\pi(S_t, A_t) = Q_\pi(S_t, A_t) + \alpha\left(\left[R_t + \gamma R_{t+1} + \cdots + \gamma^{n-1} R_{t+n-1}\right.\right.$$

$$\left.\left. + \gamma^n \max_{a'} Q_\pi(S_{t+n}, a')\right] - Q_\pi(S_t, A_t)\right) \tag{5.25}$$

In this equation, $Q(S_t, A_t)$ is the estimated value of taking action A_t in state S_t, α is the step size parameter, R_t is the reward received after taking action A_t in state S_t, γ is the discount factor, and n is the number of steps used for the return. Each variable in the equation has a specific meaning, as explained in the comments following the equation.

N-Step Q-Learning and Importance Sampling

In Q-learning, which is an off-policy reinforcement learning algorithm, it estimates the value of a policy different from the policy that is being followed. When using standard Q-learning with $N = 1$, it is not necessary to use importance sampling to weight the TD target. However, when using N-step with $N \geq 2$, importance sampling is needed because the N-step rewards are generated by the behavior policy μ, which could potentially be different from the target policy π.

For example, if we use $N = 2$ as shown in Fig. 5.12, we can see that the trajectory after the agent takes action a in state s over the course of N-step could be completely different between the behavior and target policies. The behavior policy is more exploratory, meaning it might not choose the best action which is what the target policy would choose. Thus, the path could diverge. This might not be a big issue if we use a relatively small value for N, but the issue might become more significant if the value of N becomes very large.

To compute the importance sampling ratio for N-step Q-learning, we can skip both the first step t and last step $(t + n)$; the reason for skipping the first step is since we're estimating the value for a state-action pair (s, a), it does not matter what the probability is that the behavior policy (or target policy) will select action a when in state s, because it has already happened. The reason for the last step $(t + n)$ is because Q-learning always uses the maximum value $Q(s', a')$ across all valid actions for the successor state S_{t+n}, regardless of which action is chosen by the behavior policy, regardless of the behavior policy, which makes the importance sampling ratio for the last step $t + n$ always equal to 1.

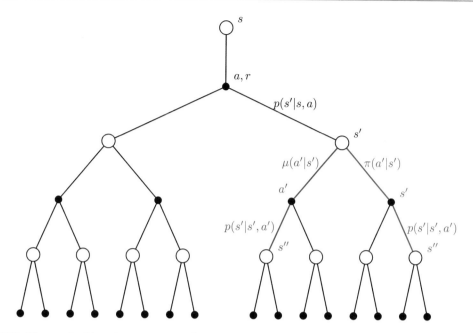

Fig. 5.12 Why we should use importance sampling for N-step Q-learning (for the case where $n = 2$), where π is the target policy and μ is the behavior policy

In other words, we can compute the importance sampling ratio for the remaining steps $t + 1, t + 2, \cdots, t + n - 1$. This simplifies the computation of the importance sampling ratio and reduces the variance of the estimates.

$$
\begin{aligned}
\rho_{t+1:t+n-1} &= \frac{\pi(A_{t+1}|S_{t+1})p(S_{t+2}|S_{t+1}, A_{t+1}) \cdots p(S_{t+n}|S_{t+n-1}, A_{t+n-1})}{\mu(A_{t+1}|S_{t+1})p(S_{t+2}|S_{t+1}, A_{t+1}) \cdots p(S_{t+n}|S_{t+n-1}, A_{t+n-1})} \\
&= \frac{\pi(A_{t+1}|S_{t+1})\pi(A_{t+2}|S_{t+2}) \cdots \pi(A_{t+n-1}|S_{t+n-1})}{\mu(A_{t+1}|S_{t+1})\mu(A_{t+2}|S_{t+2}) \cdots \mu(A_{t+n-1}|S_{t+n-1})}
\end{aligned}
\tag{5.26}
$$

To compute the probabilities for the target policy, we note that the target policy is a deterministic (or greedy) policy that always chooses the action with the highest Q-value for a given state. Thus, the probability of selecting action a in state s is 1 if and only if $a = \arg\max_a Q(s, a)$; otherwise, it is 0. This can be expressed as

$$
\pi(a|s) = \begin{cases} 1, & \text{if} \quad a = \arg\max_a Q(s, a) \\ 0, & \text{if} \quad a \neq \arg\max_a Q(s, a) \end{cases}
\tag{5.27}
$$

Intuitively, importance sampling adjusts the update rule to account for the difference between the behavior and target policies. This is necessary because the behavior policy might choose suboptimal actions, which could lead to biased estimates of the value function. Importance sampling allows the agent to adjust for this bias and learn more efficiently and accurately.

To adapt N-step Q-learning, the agent must store the probabilities of selecting each action $\mu(A_t|S_t)$ choose by the agent over the course of the N-step sequence. During the update step, we compute the probability of the target policy selecting these actions in the corresponding states using Eq. (5.27).

This equation calculates the expected return under the target policy given the current state and is used to update the Q-values for each state-action pair.

The importance sampling ratio is the ratio of the probabilities of selecting the action under the target policy and the behavior policy. It adjusts for the fact that the behavior policy may have selected actions that are suboptimal according to the target policy. By using this ratio, we can update the Q-values to better reflect the optimal policy.

N-step Q-learning is a variant of Q-learning that updates the Q-values based on the returns observed over a sequence of N time steps. By incorporating a longer sequence of experiences, N-step Q-learning can lead to more efficient learning and better performance in some environments. By using importance sampling, we can adjust for the difference between the behavior and target policies, allowing us to learn more efficiently and accurately. The pseudocode is shown in Algorithm 10.

Algorithm 10: N-step Q-learning with importance sampling

Input: Discount rate γ, initial exploration rate ϵ, step size α, N-step N where $N \geq 1$, number of steps K
Initialize: $i = 0$, $Q(s, a) = 0$, for all $s \in \mathcal{S}$, $a \in A(s)$, empty list τ
Output: The optimal state-action value function Q

1 **while** $i < K$ **do**
2   Collect τ, a sequence of N transitions $(s_t, a_t, r_t, \mu(a_t|s_t))$ when following the ϵ-greedy policy w.r.t. μ
3   Compute return $G = r_0 + \gamma r_1 + \gamma^2 r_2 + \cdots + \gamma^{N-1} r_{N-1}$ from sequences in τ
4   Compute TD target:
5      $\delta = \begin{cases} G & \text{if } s_N \text{ is terminal state} \\ G + \gamma^N \max_{a'} Q(s_N, a') & \text{otherwise bootstrapping} \end{cases}$

$/*$ Compute importance sampling ratio, we skip the first time step and last time step. $*/$

6      $\rho_{1:N-1} = \frac{\pi(a_1|s_1)\pi(a_2|s_2)\cdots\pi(a_{N-1}|s_{N-1})}{\mu(a_1|s_1)\mu(a_2|s_2)\cdots\mu(a_{N-1}|s_{N-1})}$
7
8   Update estimated value $Q(s_0, a_0) \leftarrow Q(s_0, a_0) + \alpha\rho_{1:N-1}\left(\delta - Q(s_0, a_0)\right)$
9   Remove the first transition in τ

Adapting N-step Q-learning can be more challenging than standard Q-learning. It involves storing a longer sequence of experiences, which can be more computationally demanding. Additionally, the behavior policy may select suboptimal actions that the target policy would not select. This can cause the N-step rewards to be different between the two policies, which can become a significant issue if the number of steps N is large.

When we use importance sampling for off-policy N-step Q-learning, we need to keep in mind that the values may not get updated that often. It's very likely that in a sequence of N-step transitions, the behavior policy will select some exploratory action that's not the best action according to the target policy. This means the importance sampling ratio will get close to zero, in the extreme case where the policies are deterministic, and the whole TD error term could also becomes zero. Thus, the value update during learning becomes $Q(S_t, A_t) = Q(S_t, A_t)$, which may potentially cause slower convergence in off-policy learning compared to on-policy learning.

In practice, some successful N-step Q-learning algorithms, such as the Rainbow agent proposed by Hessel et al. [6], avoid this issue by using a relatively small N-step ($n = 5$) and not using importance sampling at all to keep things simple. However, it's hard to justify the contribution of N-step bootstrap to the success of the Rainbow algorithm, since it also involves lots other technics and improvements over the standard DQN agent.

5.9   Summary

In the final chapter of Part I, we delved into the powerful framework of temporal difference (TD) learning. TD learning builds upon the ideas from Monte Carlo methods, offering a more efficient approach to reinforcement learning (RL) by updating value estimates based on immediate time step transitions instead of complete episode sequence.

The chapter began with an exploration of TD learning, which forms the foundation of the subsequent sections. TD allows us to update value estimates incrementally, using the difference between successive value estimates as a basis for learning. We then moved on to TD policy evaluation, a technique that enables us to estimate the value function for a given policy. By iteratively updating value estimates using TD updates, we can approximate the true value function without requiring complete episodes.

The chapter continued with a discussion of TD control using the SARSA algorithm. SARSA combines TD learning with an on-policy approach, where the agent's policy is improved while it interacts with the environment. We also explored the differences between on-policy and off-policy learning, highlighting the trade-offs and considerations associated with each approach.

Next, we introduced Q-learning, an off-policy TD control algorithm and core to the famous DQN agent, which we will introduce in Part II of the book. Q-learning maximizes the expected return by updating action-value estimates based on the maximum estimated future rewards. We examined how Q-learning overcomes the limitations of on-policy methods and its impact on RL performance.

In order to mitigate the issue of overoptimistic value estimates, we explored double Q-learning. This technique employs two sets of action-value functions to reduce the overestimation bias, enhancing the stability and accuracy of the learning process.

Finally, the chapter concluded with a discussion on N-step bootstrapping, a method that bridges the gap between TD learning and Monte Carlo methods. N-step bootstrapping enables us to leverage information from multiple steps into the future to update value estimates.

With the completion of this chapter, we wrap up Part I of the book, which focused on the fundamentals of RL and using tabular methods to solve small-scale Markov decision processes (MDPs). In Part II, we will shift our focus to value function approximation, exploring techniques that enable RL algorithms to scale up to larger and more complex problems.

References

[1] Richard S. Sutton. Learning to predict by the methods of temporal differences. *Machine Learning*, 3(1):9–44, Aug 1988.

[2] Richard S. Sutton and Andrew G. Barto. *Reinforcement Learning: An Introduction*. The MIT Press, second edition, 2018.

[3] Christopher J. C. H. Watkins and Peter Dayan. Q-learning. *Machine Learning*, 8(3):279–292, May 1992.

[4] Hado Hasselt. Double q-learning. In J. Lafferty, C. Williams, J. Shawe-Taylor, R. Zemel, and A. Culotta, editors, *Advances in Neural Information Processing Systems*, volume 23. Curran Associates, Inc., 2010.

[5] Dimitri P Bertsekas. *Reinforcement learning and optimal control / by Dimitri P. Bertsekas*. Athena Scientific optimization and computation series. Athena Scientific, Belmont, Massachusetts, 2019.

[6] Matteo Hessel, Joseph Modayil, Hado van Hasselt, Tom Schaul, Georg Ostrovski, Will Dabney, Dan Horgan, Bilal Piot, Mohammad Azar, and David Silver. Rainbow: Combining improvements in deep reinforcement learning, 2017.

Value Function Approximation

Linear Value Function Approximation

6

Part I of this book focuses on using tabular methods to represent the value functions of Markov decision processes (MDPs). However, this approach is limited to small-scale MDPs. In many real-world problems, the state space is too large for a table-based approach to be practical. There are two main reasons for this: first, the sheer number of states may require a large amount of memory to store the values; second, the learned value functions may not generalize well to new states.

To overcome these constraints, Part II of this book shifts its focus toward employing value function approximation (VFA) to estimate the value functions for an MDP. VFA uses a function to estimate the value of a state or state-action pair, rather than storing the value directly in a table. In this chapter, we will focus on the concept of VFA, with emphasis on linear VFA—a popular method that employs a linear combination of features to approximate the value functions. We will explore how to apply linear VFA to estimate the value functions for an MDP.

6.1   The Challenge of Large-Scale MDPs

Using tables to represent value functions can be very effective for small-scale MDPs, where the state space is relatively small and can be fully enumerated. However, there are several challenges and limitations when using tables to represent value functions in larger and more complex MDPs:

- Memory requirements: As the number of states in an MDP increases, the size of the value function table grows exponentially, making it impractical to store the values in memory.
- Generalization: Even if we have a small number of states, it's often difficult to visit every state-action pair during the learning process. This means that we may not have an accurate estimate of the value function for every state-action pair, making generalization to new, unseen states challenging.
- Curse of dimensionality: As the number of state variables (or dimensions) increases, the size of the state space grows exponentially, making it difficult to explore and learn an accurate value function in a reasonable amount of time.
- Discretization: In some cases, continuous state variables may need to be discretized into a finite number of bins in order to represent them in a table. However, this can lead to a loss of information and can make it difficult to learn an accurate value function.

M. Hu, *The Art of Reinforcement Learning*,
https://doi.org/10.1007/978-1-4842-9606-6_6

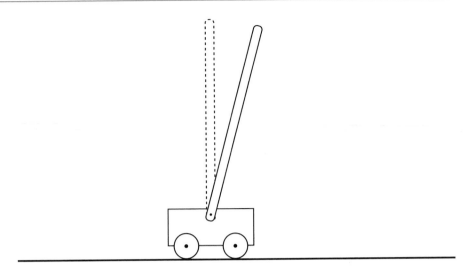

Fig. 6.1 Cart pole; the problem was introduced in the *Reinforcement Learning: An Introduction* book by Richard S. Sutton and Andrew G. Barto

Value function approximation techniques, such as linear function approximation and neural networks, can help overcome these challenges by allowing us to estimate the value function using a smaller number of parameters while also providing better generalization to new states.

To better understand these challenges, let's look at a classic control task called the cart pole task, which was introduced by Richard S. Sutton and Andrew G. Barto.

The Cart Pole Task

The goal of the cart pole task [1] is to keep the pole hinged to the cart from falling over by applying forces to the cart as shown in Fig. 6.1. The task is considered failed if the pole falls to a certain vertical angle or the cart runs off the track, and the pole is reset to the vertical position if the task is failed. The agent can choose to apply forces to the cart to push it to the left or right. The reward signal is +1 for every step when the pole does not fall.

What makes this task different from the examples we've covered so far is the size of its state space. The state of the cart pole task is four-dimensional and contains the numerical values for the position and velocity of the cart, as well as the angle and angular velocity of the pole. All of these dimensions are continuous numerical values, and although some have limits like minimum and maximum values, such as the position being in the range of $[-4.8, 4.8]$ and the pole angle being in the range of $[-0.418, 0.418]$ (in radians), the specific values of the state can be anything in between.

The large state space of the cart pole task presents a significant challenge for reinforcement learning. Because the state space is continuous, the traditional tabular methods for representing value functions are not applicable. Additionally, the agent is unlikely to encounter every possible state during training, which makes it difficult to learn an accurate value function for the task.

In our previous examples, we often used discrete numbers like $0, 1, 2, \cdots$ to represent different states, which allowed us to use a single vector to store the values for these different states. However, in cases where the possible combinations of states are very large, this tabular method is not applicable. For example, in the cart pole example, the state is defined by four continuous values: the position and

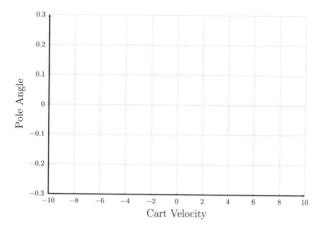

Fig. 6.2 Example of tiling the cart velocity and the pole angle

velocity of the cart and the angle and angular velocity of the pole. This makes the state space infinite and continuous, which cannot be represented by a single vector.

One solution to this problem is to try to make the infinite state finite. For example, if we only look at the cart velocity, we can bin the values into corresponding subsets, which will make the cart velocity dimension discrete. Similarly, we can also extend this method by considering the pole angle, where the bins become tiles. This approach is often called tile coding and works well for small problems. There are open source libraries available that can do this work for us (Fig. 6.2).

However, one major downside of this method is that we have to manually choose the number of bins or tiles to use. For example, if we only use two bins for the cart position, it won't be very useful since the ranges are too wide. On the other hand, if we use too many bins, we might run into the issue of overfitting. Selecting the optimal number of bins requires a lot of human knowledge about the problem at hand.

The other major issue with this method is that it does not scale well for cases where the state of the environment has a very large dimension, such as image-based environments like the Atari video games. In these cases, the state space can contain large number of dimensions, which makes the use of tile coding infeasible. In such cases, other methods like neural networks are often used to address the scalability issue, which we will introduce in the next chapter.

6.2 Value Function Approximation

Function approximation is the process of finding an approximate function that can closely match the behavior of an unknown target function. In reinforcement learning, we can use function approximation to estimate the state value or state-action value functions, which are typically too large to represent explicitly as tables. Instead, we can use a parameterized function that can take in the state or state-action as input and output an estimate of its value.

As an example, consider Fig. 6.3, which illustrates the idea of function approximation. We have a set of sample data generated by a target function g, which we do not know. Our goal is to find an approximate function H from a set of possible functions that can closely approximate the true target function, such that $H(x) \approx g(x)$. In this example, the green dots represent the sample data generated by the target function, and the red line represents a function that approximates the true target function.

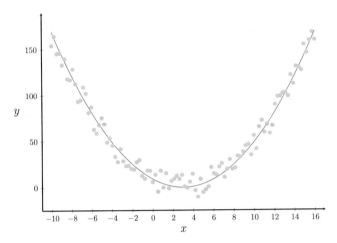

Fig. 6.3 Example to illustrate the concept of function approximation, where the green dots are sample points generated from a target function g, and the red line is the function H to approximate this target function

The use of value function approximation (VFA) in reinforcement learning has several benefits. First, the number of parameters associated with the approximation function is much smaller than the number of states or state-action pairs in the MDP, making it feasible to represent the value functions for large-scale problems. Second, the approximation function can generalize to unseen states or state-action pairs when we only have limited sample transitions, which helps to reduce the amount of data needed for learning. Finally, the space and time required to solve the large-scale MDP can be greatly reduced compared to tabular methods.

However, there are also drawbacks to using VFA. Because we are using a parametric function to estimate the value functions, we may never learn the exact true values for all states or state-action pairs. Instead, we must settle for an approximation that is close enough for practical purposes. This is because any change to the parameters of the function will affect all states or state-action pairs, and it is not possible to optimize the function for one specific state or state-action pair without sacrificing accuracy for others.

To represent the approximate value functions, we use $\hat{V}(s; \boldsymbol{w})$ to denote the function that approximates the true state value function V_π for an arbitrary policy π. Here, $\boldsymbol{w}$ represents the parameters or weights that are used to compute the estimate. Similarly, we use $\hat{Q}(s, a; \boldsymbol{w})$ to denote the function that approximates the true state-action value function Q_π for an arbitrary policy π, and we want to find the weights $\boldsymbol{w}$ that minimize the difference between the approximate value function and the true value function. The process of updating or tuning the parameters $\boldsymbol{w}$ is called training.

In summary, value function approximation is a powerful technique for representing and estimating the value functions in reinforcement learning. It enables us to solve large-scale MDPs within a limited time frame and computational budget. However, it is important to keep in mind the limitations of this approach and the trade-offs involved in finding a good approximation.

Figure 6.4 illustrates the basic idea of value function approximation, which is a way of estimating the value of each state in a reinforcement learning problem. In value function approximation, we use $\hat{V}(s; \boldsymbol{w})$ and $\hat{Q}(s, a; \boldsymbol{w})$ to represent the estimated values for a state or state-action pair, respectively, based on their weights $\boldsymbol{w}$. These functions take in the environment state or state-action pair, and after some computation with respect to their weights $\boldsymbol{w}$, they produce the estimated values as output.

Fig. 6.4 Idea of value
function approximation

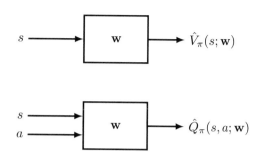

We will discuss in more detail later in this chapter how to use linear methods to compute the values, but for now, we are more concerned with how to update the weights $\mathbf{w}$ of the approximate function. In order to improve an approximation function, we need to be able to measure its performance. However, it may not always be immediately clear how to do this for functions with respect to their parameters.

Let's consider an example of policy evaluation, where we want to estimate the state values for a fixed policy π. We assume we know the true state value function V_π for π. Since we know the true state value, we can measure the difference between the approximated value and true value by using the squared error $(V_\pi(s) - \hat{V}(s; \mathbf{w}))^2$ for a single state s. This also means we can measure the performance of the weights $\mathbf{w}$, because, in this case, the state s and its true state value $V_\pi(s)$ are constant. The only thing that will affect the estimated value from the approximate function is the weights $\mathbf{w}$.

$$\left(V_\pi(s) - \hat{V}(s; \mathbf{w})\right)^2 \tag{6.1}$$

It's worth noting that the agent might encounter a large number of states when following policy π to act in the environment, and often these states are random, especially if the environment or policy is stochastic. Therefore, it is important to measure the accuracy of the approximate function with respect to the weights in a way that is independent of a specific time step. To achieve this, we use the expectation sign $\mathbb{E}_\pi$ in Eq. (6.2). We use S without a subscript t for the state in Eq. (6.2) to emphasize that the measurement of the performance should be independent of a specific time step. We can also think of $\mathcal{S}$ as the entire state space in the equation.

$$\mathbb{E}_\pi\left[\left(V_\pi(S) - \hat{V}(S; \mathbf{w})\right)^2\right] \tag{6.2}$$

We can now define a precise objective for the case where we use $\hat{V}$ to approximate the true state value function V_π, this objective is required in order to find weights $\mathbf{w}$ that minimize the squared error $\left(V_\pi(S) - \hat{V}(S; \mathbf{w})\right)^2$ for all possible states the agent would encounter when it's following policy π to act in the environment. This idea is very similar to a supervised learning problem; in this case, the true label is $V_\pi(S)$, and the predicted label is $\hat{V}(S; \mathbf{w})$. In this book, we often use a shorter term $J(\mathbf{w})$ to denote the objective.

$$J(\mathbf{w}) = \mathbb{E}_\pi\left[\left(V_\pi(S) - \hat{V}(S; \mathbf{w})\right)^2\right] \tag{6.3}$$

Similarly, if we know the true state-action value function Q_π for the policy π, we can also define the objective for approximating $\hat{Q}$. In practice, we're more interested in the state-action value function; after all, we need Q_π to do control (which is to find the optimal policy). The same concept still applies here, where we would like to find the weights $\boldsymbol{w}$ which can minimize the squared error $\Big(Q_\pi(S, A) - \hat{Q}(S, A; \boldsymbol{w}) \Big)^2$.

$$J(\boldsymbol{w}) = \mathbb{E}_\pi \left[\Big(Q_\pi(S, A) - \hat{Q}(S, A; \boldsymbol{w}) \Big)^2 \right] \tag{6.4}$$

We now have defined the objective functions for $\hat{V}$ and $\hat{Q}$; this means we can use any optimization methods that's suitable to minimize the term in Eqs. (6.3) and (6.4) with respect to their weights $\boldsymbol{w}$. In this book, we focus on one particular popular method called stochastic gradient descent, which is widely adapted in machine learning.

6.3   Stochastic Gradient Descent

In optimization, the goal is to find the parameters that minimize or maximize a given mathematical function. The term *gradient descent* refers to using an iterative optimization method to find a *local minimum* of a *differentiable* function. A differentiable function is a mathematical function that has a well-defined derivative at every point in its domain. In other words, the derivative of the function exists and can be calculated for any input value in the function's domain. Geometrically, the derivative of a differentiable function represents the slope of the tangent line to the function's graph at a given point.

A local minimum of a function is a point in the function's domain where the function takes on its smallest value within a small surrounding neighborhood of the point, but not necessarily the smallest value over the entire domain of the function. In other words, it is a point where the function is lower than nearby points, but there may be other points further away that have even lower values. Local minima are important in optimization because they represent candidate solutions to optimization problems, but they are not guaranteed to be the global minimum (i.e., the absolute minimum over the entire domain of the function).

The concept of differentiability is important in optimization because it allows us to calculate the gradient of a function with respect to its inputs, which is essential in many optimization algorithms, such as gradient descent. Given a differentiable function $J(\boldsymbol{w})$ with respect to weights $\boldsymbol{w}$, the goal is to find $\boldsymbol{w}$ so that it minimizes $J(\boldsymbol{w})$.

The algorithm for gradient descent begins by finding the gradients of $J(\boldsymbol{w})$ with respect to $\boldsymbol{w}$. It then updates the weights $\boldsymbol{w}$ by taking a step in the direction of the negative gradient $\nabla_{\boldsymbol{w}} J(\boldsymbol{w})$

$$\boldsymbol{w} = \boldsymbol{w} - \alpha \nabla_{\boldsymbol{w}} J(\boldsymbol{w}) \tag{6.5}$$

The parameter α is a step size (often called the learning rate) that controls how far a step should be when updating the weights $\boldsymbol{w}$. The algorithm repeats this process over a large number of steps in an iterative manner until it finds a good local minimum.

The intuition behind the update rule is that by taking a step in the direction of the negative gradient, we are effectively moving toward the direction of steepest descent of the objective function. This means that at each step, we are moving toward the direction where the function decreases the most. By repeating this process over a large number of steps, we hope to eventually reach a minimum of the function. To see why this makes sense, imagine standing at the top of a hill and wanting to get to the bottom. The fastest way to descend would be to move in the direction of the steepest slope, which is the direction of the direction to the bottom of the hill. Similarly, in optimization, the negative gradient points in the direction of the steepest descent of the objective function, so moving in the direction of the negative gradient will help us reach a minimum more quickly than moving in any other direction.

The learning rate α controls the step size of the update, so it determines how far we move in the direction of the negative gradient at each step. A small learning rate means that we take smaller steps, which may help us avoid overshooting the minimum, while a larger learning rate means that we take larger steps, which may help us converge more quickly. However, if the learning rate is too small, we may converge very slowly, while if it is too large, we may overshoot the minimum and oscillate around it or even diverge from it. Therefore, it is important to choose an appropriate learning rate for the optimization problem at hand.

The gradient for $J(w)$ with respect to weights w is a (column) vector that contains the partial derivatives with respect to each of the weights $w_1, w_2, \cdots, w_n$ in w and has the same dimensionality as the weight vector w. The gradient $\nabla_w J(w)$ is calculated by taking the partial derivative of the function with respect to each weight. It is common to use *stochastic gradient descent* (SGD) in practice, which randomly selects a subset of the data (called a *mini-batch*) at each iteration to calculate an estimate of the gradient.

$$\nabla_w J(w) = \begin{pmatrix} \dfrac{\partial J(w)}{\partial w_1} \\[2mm] \dfrac{\partial J(w)}{\partial w_2} \\[2mm] \vdots \\[2mm] \dfrac{\partial J(w)}{\partial w_n} \end{pmatrix} \tag{6.6}$$

In summary, stochastic gradient descent is an iterative optimization algorithm that uses the gradient of a differentiable function to update its weights in the direction of steepest descent, with the goal of finding a local minimum of the objective function. The learning rate controls the step size of the update, and it is important to choose an appropriate learning rate for the optimization problem at hand. Stochastic gradient descent is commonly used in practice, and it is often implemented using mini-batches to estimate the gradient.

To understand how gradient descent works, let's first look at the simplest case where we assume w only has a single scalar component x, as shown in Fig. 6.5. We have some sample points and the gradient (slope) for each of these sample points. We can see that the point x_3 has a positive slope (gradient), and if we take a step in the direction of the negative gradient by following Eq. (6.1), then it will decrease the value of x. The slope of point x_2 is less steep than that of point x_3, which means the gradient for point x_2 is smaller than the gradient for point x_3. This makes sense since point x_2 is much closer to the minimum of the objective function. For sample point x_1, the slope is negative, which means that if we apply Eq. (6.1), then it will increase the value of x. If we repeat the process over a large number of steps, then, in theory, it will find a local minimum.

In practice, the weight vector w often has multiple scalar components, as shown in Fig. 6.6. This means that when the gradient descent algorithm updates the weights using Eq. (6.1), it will update every component in w with respect to the partial gradients of each component. However, in the case of multiple components, the update rules for w are slightly more complex, as they involve multivariable calculus.

In particular, the partial derivatives used in the update rule must be computed with respect to each component of w. These partial derivatives are combined into a gradient vector, which can be thought of as a generalization of the derivative to the multidimensional case. The gradient vector points in the

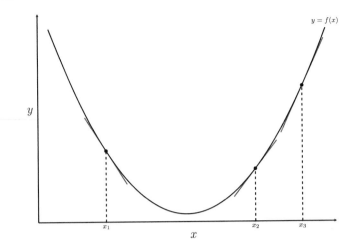

Fig. 6.5 Idea of gradient descent

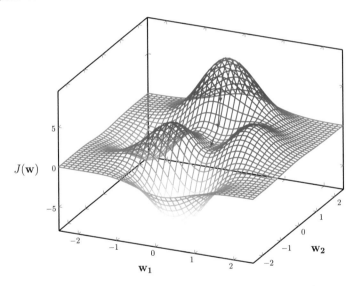

Fig. 6.6 Idea of gradient descent where $\boldsymbol{w}$ is a vector with two scalar components $\boldsymbol{w}_1$ and $\boldsymbol{w}_2$

direction of the steepest increase of the function being optimized, and the update rule modifies $\boldsymbol{w}$ in the direction opposite to the gradient vector to minimize the objective function.

Overall, while the mathematics behind the update rules for multiple components involves multivariable calculus, the underlying intuition of gradient descent remains the same: iteratively updating the weights in the direction of steepest descent to minimize the objective function.

The step size (learning rate) in gradient descent is a crucial parameter. Intuitively, a small step size requires the gradient descent algorithm to take more steps (in terms of weight updates) to find the weights $\boldsymbol{w}$ that can minimize the objective function $J(\boldsymbol{w})$. An appropriate large step size, however, will require fewer steps. However, if the step size is too large, it may cause the algorithm to oscillate

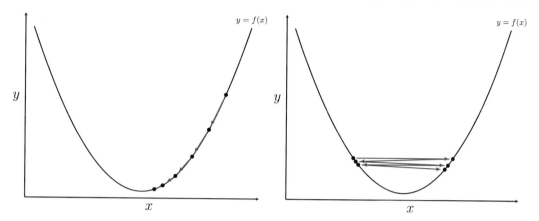

Fig. 6.7 Impact of step size in gradient descent. The left side shows the case where we start with relatively large step size and then decay it over time. The right side shows the case where a too large step size can cause the algorithm to oscillate around the local minimum but never reach it

around the local minimum but never reach it (as shown on the right side of Fig. 6.7). In practice, people often use a relatively large step size at the beginning of the learning process and then diminish it throughout the learning process (as shown on the left side of Fig. 6.7). The exact value for the step size is often a matter of art rather than science because it often requires multiple trials to find a good fit for different problems.

In this chapter, we use a slightly modified equation for gradient descent in the case of VFA, which adds $\frac{1}{2}$ in front of the step size α in Eq. (6.7). This was introduced to simplify the final equation as shown in Eq. (6.8).

$$\boldsymbol{w} = \boldsymbol{w} - \frac{1}{2}\alpha \nabla_{\boldsymbol{w}} J(\boldsymbol{w}) \tag{6.7}$$

$$= \boldsymbol{w} - \frac{1}{2}\alpha \nabla_{\boldsymbol{w}} \mathbb{E}_\pi \left[\left(V_\pi(S) - \hat{V}(S; \boldsymbol{w}) \right)^2 \right]$$

$$= \boldsymbol{w} - \frac{1}{2}\alpha \left(-2\mathbb{E}_\pi \left[\left(V_\pi(S) - \hat{V}(S; \boldsymbol{w}) \right) \nabla_{\boldsymbol{w}} \hat{V}(S; \boldsymbol{w}) \right] \right)$$

$$= \boldsymbol{w} + \alpha \mathbb{E}_\pi \left[\left(V_\pi(S) - \hat{V}(S; \boldsymbol{w}) \right) \nabla_{\boldsymbol{w}} \hat{V}(S; \boldsymbol{w}) \right] \tag{6.8}$$

where $\boldsymbol{w}$ is the weight vector, which is the learnable parameter of the state value function that we want to optimize. α is the step size, which is a hyperparameter that determines the size of the updates to the weight parameters during each iteration of the algorithm. $\nabla_{\boldsymbol{w}}$ represents the vector of partial derivatives of a function with respect to the weights $\boldsymbol{w}$.

To apply gradient descent, we use Eq. (6.3) as the objective function $J(\boldsymbol{w})$. Then, we use the chain rule of calculus to derive the derivatives, which leads to Eq. (6.8). The chain rule of calculus is used to derive the derivatives because the objective function $J(\boldsymbol{w})$ depends on the predicted values of the value function $\hat{V}(S; \boldsymbol{w})$, which in turn depend on the parameters $\boldsymbol{w}$ through the function $\hat{V}(S; \boldsymbol{w})$. This allows us to compute the gradient of the objective function with respect to the parameters $\boldsymbol{w}$.

The intuition behind this equation is to adjust the parameters $\boldsymbol{w}$ in the opposite direction of the squared error between the true value $V_\pi(S)$ and the current predicted value $\hat{V}(S; \boldsymbol{w})$, while the magnitude of change is determined by the partial gradients of each parameter $\nabla_{\boldsymbol{w}} \hat{V}(S; \boldsymbol{w})$ and the step size α. The expectation sign $\mathbb{E}_\pi$ means that we are computing the gradients over expected values. This means that we need to take partial derivatives regarding all states that the agent could encounter when following the policy π in the environment.

When using tabular methods, we would run the algorithm to collect a large amount of transitions and compute the averaging returns. However, when using gradient descent methods, we can use stochastic gradient descent (SGD). SGD is a widely adapted method in machine learning, especially for training deep neural networks. The idea behind SGD is to use one (or a mini-batch of) sample transition to compute the gradients and then update the weights w. This approach is computationally more efficient than using all the transitions due the computation efficient reasons, also because often the cases, it's impossible for the agent to encounter all the states in the entire state space when interacting with the environment.

We can rewrite Eq. (6.8) as Eq. (6.9) to show how to update the weights w using SGD. In this equation, we use one (or a mini-batch of) sample transition to compute the gradients and update the weights. In machine learning, a mini-batch is a subset of the training data that is used to compute the gradient descent update during stochastic gradient descent (SGD) optimization. When training a model on large datasets, it is often computationally infeasible to compute the gradient over the entire dataset in each iteration. Instead, a mini-batch of randomly selected samples from the dataset is used to estimate the gradient, which is then used to update the model's parameters.

$$w = w + \alpha \left[\left(V_\pi(S) - \hat{V}(S; w) \right) \nabla_w \hat{V}(S; w) \right] \tag{6.9}$$

The size of the mini-batch is a hyperparameter that is typically chosen based on the size of the dataset and the available computing resources. A larger mini-batch size can lead to more accurate gradient estimates, but requires more memory and computational resources. A smaller mini-batch size can result in less memory usage, but may cause slower convergence, often with higher variance in the gradient estimates.

Using mini-batches in SGD allows for more efficient training of machine learning models on large datasets, as it reduces the computational burden and memory requirements of computing the gradient over the entire dataset.

Now, let's turn our attention to the methods used to approximate the state (or state-action) values with respect to the weights w. In general, there are linear and nonlinear methods to approximate the value functions. In this chapter, we focus on the simpler linear method. In the next chapter, we will discuss nonlinear methods, specifically using neural networks to approximate the value functions. The linear method has its advantages and disadvantages, which we will explore further in the next section.

6.4   Linear Value Function Approximation

So far, we haven't talked about how exactly the approximate function $\hat{V}$ or $\hat{Q}$, which takes in a raw environment state (or state-action pair) as input, will output the estimated value. We now focus on how to use linear methods to compute these values with respect to the weight vector w.

To get started, it involves the process of mapping raw environment state s (or state-action pair (s, a)) to a feature vector x; this process is often referred to as *feature engineering* or *feature extraction* in traditional machine learning. The number of scalar components in x can be much larger (or smaller) than the dimension of the environment state, depending on the reinforcement learning problem. We can think of the feature vector x as a collection of functions $x_1, x_2, \cdots$ that transform the raw environment state (or state-action pair (s, a)) into corresponding scalar components:

$$x(s) = \begin{pmatrix} x_1(s) \\ x_2(s) \\ \vdots \\ x_n(s) \end{pmatrix} \tag{6.10}$$

The intuition behind the feature vector x is that by carefully crafting these features, we hope to find a weight vector w, so the value for a particular state (or state-action pair) is just the weighted sum over all these feature; this is why we call w the weights, because it weights the corresponding features in x. Mathematically speaking, we take the inner product of x and w (assume both x and w are column vectors, and w, x have the same number of elements), which can be expressed as

$$\hat{V}(s; w) = x(s)^T w = \sum_{j=1}^{n} x_j(s) w_j \tag{6.11}$$

where $\hat{V}(s; w)$ is the estimated value of state s using weight vector w, $x(s)^T$ is the transpose of the feature vector for state s, and $\sum_{j=1}^{n} x_j(s) w_j$ is the weighted sum of the features.

The feature vector x is often handcrafted based on human knowledge or past experience of the reinforcement learning problem. For some simple problems, this process may be as simple as encoding the raw environment state into numerical values, but for more complex problems, it often involves some form of transformation like polynomial functions.

One of the simplest ways to construct the feature vector is to partition the state space into subsets (similar to the bin or tile cases shown in Fig. 6.2) such that any single state s only belongs to one subset. We can then construct the feature vector x using the one-hot encoding method, where the subset to which the state s belongs is represented by a value of one and all other subsets are represented by a value of zero. This is also known as the state aggregation method. Finite MDPs can be treated as a special case of this method, where we have N subsets, here N is the number of sates in the state space, and each subset only has one state.

For example, in the service dog MDP, we can label each state using index values *Room 1 = 0, Room 2 = 1, Room 3 = 2, Outside = 3, Found item = 4*. Then, we can construct the feature vector x for state *Room 2* as a one-hot encoded vector $x = [0, 1, 0, 0, 0]^T$. It's also possible to use neural networks to extract the features automatically, which is a topic we'll cover in the next chapter.

Now that we have a basic idea of how the linear approximation function computes state values, let's finalize how to use stochastic gradient descent to update the weight vector w. We can first define the partial derivatives $\nabla_w \hat{V}(s; w)$ with respect to w for Eq. (6.11). Since it's a linear system and we're taking the derivative with respect to $\boldsymbol{w}$, $\nabla_{\boldsymbol{w}} \hat{V}(s; \boldsymbol{w})$ can be simplified to $\boldsymbol{x}(s)$. This gives us Eq. (6.12) for updating the weights $\boldsymbol{w}$ when using a linear function to approximate the state value function V_π:

$$\boldsymbol{w} = \boldsymbol{w} + \alpha \left[\left(V_\pi(s) - \hat{V}(s; \boldsymbol{w}) \right) \boldsymbol{x}(s) \right]$$

$$= \boldsymbol{w} + \alpha \left[\left(V_\pi(s) - \boldsymbol{x}(s)^T \boldsymbol{w} \right) \boldsymbol{x}(s) \right] \tag{6.12}$$

Here, $\boldsymbol{w}$ represents the weights of the linear approximation function; α represents the step size or learning rate, which determines how much to adjust the weights in each step; $\boldsymbol{x}(s)$ is the feature vector for state s; and $V_\pi(s)$ is the expected value of state s.

Now we know exactly how to update the weights $\boldsymbol{w}$ for a linear function. The only thing that's unknown to us in Eq. (6.12) is the true state value function V_π. We'll start to look at examples of how to use existing methods like Monte Carlo and TD learning to estimate the true state value V_π, which can then be used to update the weights in the preceding equation.

Linear Value Function Approximation for Policy Evaluation

Assuming we have access to the true state value function V_π for the policy π, we can use Eq. (6.12) to update the weights $\boldsymbol{w}$ for a linear approximation function. A linear approximation function is a mathematical function that maps the feature vector for a state to an estimate of its value.

However, in most real-world reinforcement learning problems, we don't know V_π. So, how can we update the weights in this case?

In Monte Carlo policy evaluation, the agent generates a complete sample episode by following the policy π and computes the returns $G_0, G_1, G_2, \cdots$ for each state in the sequence. Here, the returns are the sum of rewards received after visiting a state t until the end of the episode. We can still use this idea and replace $V_\pi(s)$ in Eq. (6.12) with the Monte Carlo sample return G_t as the true target. Therefore, the weights update rule for using linear methods to approximate the state value function with Monte Carlo policy evaluation can be adapted as follows:

$$\boldsymbol{w} = \boldsymbol{w} + \alpha \left[\left(G_t - \hat{V}(S_t; \boldsymbol{w}) \right) \boldsymbol{x}(S_t) \right]$$

$$= \boldsymbol{w} + \alpha \left[\left(G_t - \boldsymbol{x}(S_t)^T \boldsymbol{w} \right) \boldsymbol{x}(S_t) \right] \tag{6.13}$$

Here, $\boldsymbol{w}$ represents the weights of the linear approximation function, α is the learning rate, G_t is the Monte Carlo sample return, $\boldsymbol{x}(S_t)$ is the feature vector for state S_t, and $\hat{V}(S_t; \boldsymbol{w})$ is the predicted value of state S_t using the linear approximation function.

We now introduce the every-visit Monte Carlo policy evaluation algorithm based on linear value function approximation, which is shown in Algorithm 1. In the context of large-scale MDPs where almost all states are unique, the every-visit version is often preferred over the first-visit version. This is because counting how many times a state (or state-action pair) has been visited may not provide meaningful information due to the lack of state or action redundancy. By considering every visit to each state (or state-action pair) in an episode, the every-visit algorithm ensures that all interactions contribute to the value function estimation. Thus, it is better suited for scenarios where state or action redundancy is minimal and counting visits is not informative.

The algorithm takes in an arbitrary policy π and other parameters like discount and learning rate. To get better performance, we generally choose to initialize the weights w randomly in the range of $[-1, 1]$. Random initialization helps to prevent the weights from getting stuck in a local minimum and allows the learning algorithm to explore different weight configurations.

Algorithm 1: Every-visit Monte Carlo linear-VFA policy evaluation

---

Input: Policy π to be evaluated, discount rate γ, learning rate α, number of episodes K
Initialize: $i = 0$, set w randomly in the range of $[-1, 1]$
Output: The approximated state value function $\hat{V}$ parameterized by w for the input policy π

1 **while** $i < K$ **do**
2   Generate a sample episode τ by following policy π, where $\tau = s_0, r_0, s_1, r_1, \cdots$
3   Compute the return G_t for every time step $t = 0, 1, 2, \cdots$ in episode τ
4   **for** $t = 0, 1, \ldots, T - 1$ **do**
5   Update weights $w \leftarrow w + \alpha \left(G_t - x(s_t)^T w \right) x(s_t)$
6       $i = i + 1$

The algorithm generates a sample episode by following the input policy, computes the return for every time step, and updates the weights of the linear function which we use to approximate the true state value function using the Monte Carlo update rule. This process is repeated for a fixed number of episodes.

The Monte Carlo update rule is guaranteed to converge to the true value function as the number of episodes approaches infinity, regardless of the initial values of the weight parameters.

Similarly, we can adapt TD learning to use linear value function approximation. In TD(0) policy evaluation, it estimates the return as the sum of the immediate reward R_t and the discounted value of successor state $\gamma V_\pi(S_{t+1})$. Here, we can use the same idea and replace $V_\pi(S_{t+1})$ with the predicted value of the successor state from the linear function, which is $x(S_{t+1})^T w$. So the weights update rule for TD(0) policy evaluation becomes

$$w = w + \alpha \left[\left(R_t + \gamma \hat{V}(S_{t+1}; w) - \hat{V}(S_t; w) \right) x(S_t) \right]$$

$$= w + \alpha \left[\left(R_t + \gamma x(S_{t+1})^T w - x(S_t)^T w \right) x(S_t) \right] \tag{6.14}$$

The complete algorithm for TD(0) policy evaluation with linear value function approximation is shown in Algorithm 2. Similarly to the Monte Carlo policy evaluation algorithm, here we also initialize the weights w randomly in the range of $[-1, 1]$, which allows the algorithm to explore different weight configurations and avoid getting stuck in local minima. We did not specify exactly how to construct the feature vector x because the optimal features vary depending on the specific reinforcement learning problem.

So far, we've only been using the linear function to approximate the state value function V_π, but in practice, in order to do control (find the optimal policy), we need to learn the state-action value function Q_π. The overall concept and process to use $\hat{Q}$ to approximate Q_π is the same compared to using $\hat{V}$ to approximate V_π. The major difference is that now we need to consider the specific action a in the process. This includes when we construct the feature vector x, which needs to be done with respect to the state-action pair (s, a).

Algorithm 2: TD(0) linear-VFA policy evaluation

Input: Policy π to be evaluated, discount rate γ, learning rate α, number of steps K
Initialize: $i = 0$, set $\boldsymbol{w}$ randomly in the range of $[-1, 1]$
Output: The approximated state value function $\hat{V}$ parameterized by $\boldsymbol{w}$ for the input policy π

1 **while** $i < K$ **do**
2 | Sample action a_t for s_t when following policy π
3 | Take action a_t in environment and observe r_t, s_{t+1}
4 | $i = i + 1$
5 | Update weights $\boldsymbol{w} \leftarrow \boldsymbol{w} + \alpha\left(r_t + \gamma \boldsymbol{x}(s_{t+1})^T \boldsymbol{w} - \boldsymbol{x}(s_t)^T \boldsymbol{w}\right)\boldsymbol{x}(s_t)$

$$\boldsymbol{x}(s, a) = \begin{pmatrix} \boldsymbol{x}_1(s, a) \\ \boldsymbol{x}_2(s, a) \\ \vdots \\ \boldsymbol{x}_n(s, a) \end{pmatrix} \tag{6.15}$$

For the simplest case where we have a finite MDP with a small state space and action space, we can first create the one-hot vectors for both state s and action a, then concatenate these two one-hot vectors to construct the feature vector $\boldsymbol{x}$. For the same service dog MDP example, if we label each state using index values *Room 1 = 0, Room 2 = 1, Room 3 = 2, Outside = 3, Found item = 4*, and each action with index values *Go to room1 = 0, Go to room2 = 1, Go to room3 = 2, Search = 3, Go outside = 4, Go inside = 5*, we can construct the one-hot encoded vector for state *Room 2* as $[0, 1, 0, 0, 0]$ and the one-hot encoded vector for action *Go to room3* as $[0, 0, 1, 0, 0, 0]$. We can then concatenate these two one-hot encoded vectors together, which results in the feature vector $\boldsymbol{x} = [0, 1, 0, 0, 0, 0, 0, 1, 0, 0, 0]^T$.

We also need to make a small change to the weights update rules to incorporate the action chosen by the agent. Eq. (6.16) shows the weights update rules for the Monte Carlo policy evaluation when using linear function $\hat{Q}$ to approximate the state-action value function Q_π:

$$\boldsymbol{w} = \boldsymbol{w} + \alpha\left[\left(G_t - \hat{Q}(S_t, A_t; \boldsymbol{w})\right)\boldsymbol{x}(S_t, A_t)\right]$$

$$= \boldsymbol{w} + \alpha\left[\left(G_t - \boldsymbol{x}(S_t, A_t)^T \boldsymbol{w}\right)\boldsymbol{x}(S_t, A_t)\right] \tag{6.16}$$

Here, $\boldsymbol{x}$ is the feature vector constructed based on the action A_t taken by the agent when in state S_t, G_t is the Monte Carlo sample return, and α is the step size.

Similarly, the weights update rule of the TD(0) policy evaluation when using linear function $\hat{Q}$ to approximate the state-action value function Q_π becomes

$$\boldsymbol{w} = \boldsymbol{w} + \alpha\left[\left(R_t + \gamma \hat{Q}(S_{t+1}, A_{t+1}; \boldsymbol{w}) - \hat{Q}(S_t, A_t; \boldsymbol{w})\right)\boldsymbol{x}(S_t)\right]$$

$$= \boldsymbol{w} + \alpha\left[\left(R_t + \gamma \boldsymbol{x}(S_{t+1}, A_{t+1})^T \boldsymbol{w} - \boldsymbol{x}(S_t, A_t)^T \boldsymbol{w}\right)\boldsymbol{x}(S_t, A_t)\right] \tag{6.17}$$

After we have the weights update rules, Eqs. (6.16) and (6.17), we can extend Algorithms 1 and 2 to approximate the state-action value function. Which are outlined in Algorithm 3 and Algorithm 4 respectively. The overall structure and procedure stay the same; all we need to do is replace the relevant weights update rule.

Algorithm 3: Every-visit Monte Carlo linear-VFA policy evaluation for Q_π

Input: Policy π to be evaluated, discount rate γ, learning rate α, number of episodes K
Initialize: $i = 0$, set $\boldsymbol{w}$ randomly in the range of $[-1, 1]$
Output: The approximated state-action value function $\hat{Q}$ parameterized by $\boldsymbol{w}$ for the input policy π
1 **while** $i < K$ **do**
2   Generate a sample episode τ by following policy π, where $\tau = s_0, a_0, r_0, s_1, a_1, r_1, \cdots$
3   Compute the return G_t for every time step $t = 0, 1, 2, \cdots$ in episode τ
4   **for** $t = 0, 1, \ldots, T$ **do**
5   Update weights $\boldsymbol{w} \leftarrow \boldsymbol{w} + \alpha\Big(G_t - \boldsymbol{x}(s_t, a_t)^T \boldsymbol{w}\Big)\boldsymbol{x}(s_t, a_t)$
6      $i = i + 1$

Algorithm 4: TD(0) linear-VFA policy evaluation for Q_π

Input: Policy π to be evaluated, discount rate γ, learning rate α, number of steps K
Initialize: $i = 0$, set $\boldsymbol{w}$ randomly in the range of $[-1, 1]$
Output: The approximated state-action value function $\hat{Q}$ parameterized by $\boldsymbol{w}$ for the input policy π
1 **while** $i < K$ **do**
2   Sample action a_t for s_t when following policy π
3   Take action a_t in environment and observe r_t, s_{t+1}
4      $i = i + 1$
5   Sample action a_{t+1} for s_{t+1} when following policy π
6   Update weights $\boldsymbol{w} \leftarrow \boldsymbol{w} + \alpha\Big(r_t + \gamma \boldsymbol{x}(s_{t+1}, a_{t+1})^T \boldsymbol{w} - \boldsymbol{x}(s_t, a_t)^T \boldsymbol{w}\Big)\boldsymbol{x}(s_t, a_t)$
7      $s_t = s_{t+1}, a_t = a_{t+1}$

Linear Value Function Approximation for Control

Once we know how to estimate the state-action values by using the policy evaluation algorithm introduced earlier using linear function approximation, we can start to improve the policy (do control). To do this, we need to deal with the exploration problem, which we can use the ϵ-greedy policy introduced in Chap. 5. Briefly, the ϵ-greedy policy selects the action with the highest estimated value with probability $1 - \epsilon$ and a random action with probability ϵ. This allows the agent to explore less-visited states with some probability while still exploiting its current knowledge of the value function.

We can use the Monte Carlo policy iteration algorithm to improve the policy, which uses a linear function $\hat{Q}$ to approximate the true state-action value function Q_π.

Algorithm 5 is very similar to Algorithm 3; the difference between these two is that the agent is acting according to the ϵ-greedy policy, which was build upon the linear function $\hat{Q}$, where $\hat{Q}$ is parameterized by $\boldsymbol{w}$.

We first run the every-visit Monte Carlo policy iteration algorithm with linear value function approximation on the service dog MDP. We construct the feature vector $\boldsymbol{x}$ by concatenating the one-hot encoded vectors for both state s and action a as discussed earlier in this chapter. The results

---

Algorithm 5: Every-visit Monte Carlo control with linear-VFA ϵ-greedy policy for Exploration

---
Input: Discount rate γ, learning rate α, initial exploration rate ϵ, number of episodes K
Initialize: $i = 0$, set $\boldsymbol{w}$ randomly in the range of $[-1, 1]$
Output: The approximated optimal state-action value function $\hat{Q}$ parameterized by $\boldsymbol{w}$

1 **while** $i < K$ **do**
2   Generate a sample episode τ by following ϵ-greedy policy w.r.t. $\boldsymbol{w}$, where $\tau = s_0, a_0, r_0, s_1, a_1, r_1, \cdots$
3   Compute the return G_t for every time step $t = 0, 1, 2, \cdots$ in episode τ
4   **for** each time step $t = 0, 1, \ldots, T$ **do**
5   Update weights $\boldsymbol{w} \leftarrow \boldsymbol{w} + \alpha\Big(G_t - \boldsymbol{x}(s_t, a_t)^T \boldsymbol{w}\Big)\boldsymbol{x}(s_t, a_t)$
6      $i = i + 1$
7   (optional) decay exploration rate $\epsilon = 1/i$

Table 6.1 Optimal state value computed using the every-visit Monte Carlo policy iteration algorithm with linear value function approximation for the service dog MDP. Results were averaged over 100 independent runs; the last row contains true state values computed using DP policy iteration. We use learning rate $\alpha = 0.01$, discount $\gamma = 0.9$, and initial exploration rate $\epsilon = 1.0$ and decay it after every episode

Number of episodes	Room 1	Room 2	Room 3	Outside	Found item
100	5.23	6.85	8.68	0.08	1.56
1000	6.19	8.0	10.0	0.15	0
10,000	6.2	8.0	10.0	0.31	0
DP	6.2	8.0	10.0	7.2	0

shown in Table 6.1 were averaged over 100 independent runs. We can see that after 1000 episodes, the estimated state value from the linear approximation function is very close to the true state values computed using DP policy iteration (last row), except for state *Outside*, which we guess it's because the agent did not visit that state too often.

To this point, it's fairly easy to adapt TD learning algorithms to use linear function $\hat{Q}$ to approximate the true state-action value function Q_π. Algorithm 6 shows the SARSA algorithm based on such linear methods.

Algorithm 6: SARSA linear-VFA with ϵ-greedy policy for Exploration

---
Input: Discount rate γ, learning rate α, initial exploration rate ϵ, number of steps K
Initialize: $i = 0$, set $\boldsymbol{w}$ randomly in the range of $[-1, 1]$
Output: The approximated optimal state-action value function $\hat{Q}$ parameterized by $\boldsymbol{w}$

1 **while** $i < K$ **do**
2   Sample action a_t for s_t when following ϵ-greedy policy w.r.t. $\boldsymbol{w}$
3   Take action a_t in environment and observe r_t, s_{t+1}
4      $i = i + 1$
5   (optional) decay ϵ, for example, linearly to a small fixed value
6   Sample action a_{t+1} for s_{t+1} when following ϵ-greedy policy w.r.t. $\boldsymbol{w}$
7   Update weights $\boldsymbol{w} \leftarrow \boldsymbol{w} + \alpha\Big(r_t + \gamma\boldsymbol{x}(s_{t+1}, a_{t+1})^T \boldsymbol{w} - \boldsymbol{x}(s_t, a_t)^T \boldsymbol{w}\Big)\boldsymbol{x}(s_t, a_t)$
8      $s_t = s_{t+1}, a_t = a_{t+1}$

Last but not least, we'll cover how to use linear function $\hat{Q}$ to approximate the true optimal state-action value function Q_* for Q-learning. Recall that Q-learning learns the optimal state-action value function directly; it does this by using the immediate reward and discounted value of the best successor state-action pair as the TD target:

$$TD_{target} = R_t + \gamma \max_{a'} Q(S_{t+1}, a') \tag{6.18}$$

We can then adapt the weights update rule when using linear value function approximation for Q-learning:

$$\boldsymbol{w} = \boldsymbol{w} + \alpha \left(R_t + \gamma \max_{a'} \boldsymbol{x}(S_{t+1}, a')^T \boldsymbol{w} - \boldsymbol{x}(S_t, A_t)^T \boldsymbol{w} \right) \boldsymbol{x}(S_t, A_t) \tag{6.19}$$

We then formally introduce the Q-learning algorithm which uses linear function $\hat{Q}$ to approximate the true optimal state-action value function Q_*, based on ϵ-greedy policy for exploration, as shown in Algorithm 7.

Algorithm 7: Q-learning linear-VFA with ϵ-greedy policy for Exploration

---

Input: Discount rate γ, learning rate α, initial exploration rate ϵ, number of steps K
Initialize: $i = 0$, set $\boldsymbol{w}$ randomly in the range of $[-1, 1]$
Output: The approximated optimal state-action value function $\hat{Q}$ parameterized by $\boldsymbol{w}$

1   **while** $i < K$ **do**
2   Sample action a_t for s_t from ϵ-greedy policy w.r.t. $\boldsymbol{w}$
3   Take action a_t in environment and observe r_t, s_{t+1}
4      $i = i + 1$
5   (optional) decay ϵ, for example, linearly to a small fixed value
6   Update weights $\boldsymbol{w} \leftarrow \boldsymbol{w} + \alpha \left(r_t + \gamma \max_{a'} \boldsymbol{x}(s_{t+1}, a')^T \boldsymbol{w} - \boldsymbol{x}(s_t, a_t)^T \boldsymbol{w} \right) \boldsymbol{x}(s_t, a_t)$
7      $s_t = s_{t+1}$

It's also possible to adapt the N-step bootstrapping method which we introduced in Chap. 5 together with value function approximation. The overall structure stays the same; we only need to change the weights update rule. The following shows an example for the N-step Q-learning weights update rule (without using importance sampling), where N=3:

$$\boldsymbol{w} = \boldsymbol{w} + \alpha \Big(R_t + \gamma R_{t+1} + \gamma^2 R_{t+2}$$

$$+ \gamma^3 \max_{a'} \boldsymbol{x}(S_{t+3}, a')^T \boldsymbol{w} - \boldsymbol{x}(S_t, A_t)^T \boldsymbol{w} \Big) \boldsymbol{x}(S_t, A_t) \tag{6.20}$$

For large-scale MDPs like the cart pole, it's impossible to measure the optimal state values for all the states since the state space is too large. How can we assess the learning progress or performance of the agent? One simple method for the episodic problems is to measure the total rewards (often nondiscounted) of an episode when the agent is interacting with the environment to generate the sample transitions. In practice, we often use a separate simulation environment to run the evaluation process, often with different exploration rate ϵ and seed. Another method we can use to measure the learning progress is the objective function $J(\boldsymbol{w})$ or loss function; this is suitable for both episodic and continuing problems.

However, it is important not to rely solely on the training loss as a metric for measuring the performance of an RL agent. Unlike supervised learning settings where the training samples are fixed, RL agents generate their own training samples through interactions with the environment. As a result,

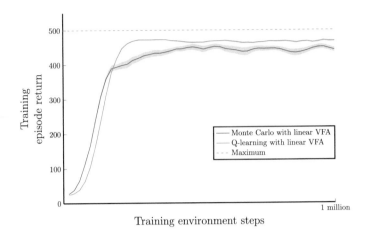

Fig. 6.8 Mean episode returns and 95% confidence interval for the every-visit Monte Carlo control agent and Q-learning agent with linear value function approximation on the cart pole task. The results were averaged over five independent runs and then smoothed using a moving average with a window size of five. We use tile coding to construct the feature vector x (8 tilings and 16 tiles for each tiling), discount rate $\gamma = 0.99$, learning rate $\alpha = 0.0025$, and initial exploration rate $\epsilon = 1.0$ and decay to 0.05 over 100,000 environment steps

the behavior of the agent may differ significantly from what is indicated by the training loss. For example, the loss function may diverge at a certain level, but the RL agent could still be performing well in terms of its overall task. Therefore, it is necessary to consider other evaluation metrics such as total rewards, success rate, or specific domain-specific metrics to get a comprehensive understanding of the agent's performance.

To solve the cart pole reinforcement learning problem, we use an open source tile coding library developed by Sutton and Barto [1][1] to construct the feature vector x. We use linear value function approximation for both every-visit Monte Carlo and Q-learning in the experiment. The results shown in Fig. 6.8 were averaged over five independent runs. We use the same learning rate, discount, and exploration rate for this experiment. The plot has been smoothed using a moving average with a window size of five. We can see that the Monte Carlo agent tends to have high variances compared to the Q-learning agent on this particular task.

We use a prebuilt cart pole environment from the OpenAI Gym [2] for this experiment.[2]

6.5   Summary

In this chapter, we delved into the concept of value function approximation and explored the use of linear methods to approximate the value functions for MDPs. We started by addressing the challenge of dealing with large-scale MDPs, where using the tabular methods to represent the value functions becomes computationally infeasible.

To overcome this challenge, we introduced the notion of value function approximation, which enables us to estimate the value function based on a set of features. By representing the value function

---

[1] Official link for the tile coding library developed by Sutton and Barto: http://incompleteideas.net/tiles/tiles3.html.

[2] Official link for the cart pole environment from OpenAI Gym: www.gymlibrary.dev/environments/classic_control/cart_pole/.

as a linear combination of these features, we can effectively approximate the value function without explicitly storing values for every state.

The chapter proceeded to discuss the technique of stochastic gradient descent, which plays a crucial role in updating the weights of the linear value function approximation. Stochastic gradient descent allows us to iteratively adjust the weights based on the observed rewards and the differences between predicted and actual values.

Furthermore, we delved into the core topic of linear value function approximation. By using a linear combination of features, we constructed a linear function that can estimate the value function. This approach offers simplicity and interpretability, making it a popular choice in various simple reinforcement learning applications.

While linear value function approximation is a valuable technique, it has inherent limitations when it comes to approximating complex value functions accurately. Additionally, it requires the manual crafting of feature vectors, which can pose challenges for certain problems. To address this issue, the subsequent chapter will explore nonlinear methods, particularly neural networks, for approximating value functions. These nonlinear methods can capture more intricate relationships and enable more precise value function approximation in complex domains.

References

[1] Richard S. Sutton and Andrew G. Barto. *Reinforcement Learning: An Introduction*. The MIT Press, second edition, 2018.
[2] Greg Brockman, Vicki Cheung, Ludwig Pettersson, Jonas Schneider, John Schulman, Jie Tang, and Wojciech Zaremba. Openai gym, 2016.

Nonlinear Value Function Approximation

7

In this chapter, we will explore the use of nonlinear methods to approximate value functions in reinforcement learning. Value functions play a crucial role in reinforcement learning as they estimate the expected reward an agent can obtain in a given state and follow the policy afterwards. In the previous chapter, we discussed linear methods that rely on constructing a feature vector and computing a weighted combination of features as the state or state-action value. However, these methods have limitations when applied to complex and challenging domains. In such cases, a simple linear combination may fail to capture all the relevant features and information. Additionally, determining the appropriate features that capture the necessary information for the task at hand requires substantial domain knowledge and experience.

To overcome these limitations, we can leverage nonlinear methods, specifically neural networks, to approximate value functions in reinforcement learning. Neural networks are a type of machine learning model that can learn from data to make predictions or decisions. In reinforcement learning, neural networks can learn to approximate value functions by taking the environment state as input data and predicting the expected return from that state when following some policy.

The power of neural networks lies in their ability to automatically learn complex and nonlinear relationships between inputs and outputs. In other words, instead of manually specifying a set of features to represent the state, we can directly feed the raw state information into the neural network, which will then learn to extract the relevant information by adjusting the weights and biases of its layers of neurons. This process is known as feature learning, and it enables the neural network to learn representations that are more suited to the task at hand.

Neural networks have several benefits over linear methods for approximating value functions in reinforcement learning:

- Nonlinearity: Neural networks are nonlinear function approximators, which means they can model complex, nonlinear relationships between inputs and outputs. This is important because many real-world problems involve nonlinearities, and linear methods may not be able to capture these complex relationships. In contrast, neural networks can learn to approximate complex value functions with high accuracy.
- Generalization: Neural networks can generalize well to new, unseen data. This means that they can learn to approximate value functions based on a limited amount of data and then apply this knowledge to new situations. This is important in reinforcement learning because the agent needs

to be able to generalize its knowledge to new states and actions in order to make effective decisions in the environment.

- Feature learning: Neural networks can learn to extract relevant features from the raw input data, which can improve their accuracy in approximating value functions. In contrast, linear methods require hand-engineered features, which can be time-consuming and may not capture all relevant information.
- Flexibility: Neural networks are highly flexible and can be easily adapted to different types of value functions and reinforcement learning problems. They can be used to approximate both state value functions and action-value functions and can be used with a variety of reinforcement learning algorithms.

Overall, the benefits of using neural networks over linear methods for approximating value functions in reinforcement learning include their ability to capture nonlinear relationships, generalize to new situations, learn relevant features from raw input data, and adapt to different types of reinforcement learning problems. Readers that are familiar with neural networks and have experience on how to train a deep neural network can skip 7.1 and 7.2.

7.1   Neural Networks

A neural network is a type of machine learning model that is inspired by the structure and function of biological neurons in the human brain. It is composed of layers of interconnected nodes, or artificial neurons, that process information through a series of mathematical operations.

The basic building block of a neural network is the artificial neuron, or node. Each node receives one or more inputs, performs a simple computation on those inputs, and produces an output. The output of each node is then passed on as input to the next layer of nodes.

In mathematics's term, a neural network is a mathematical construct that consists of a series of linear and nonlinear functions. At an abstract level, we can represent it as a chain of (parameterized) linear functions and (differentiable) nonlinear functions. For example, we can represent a simple neural network as shown in Eq. (7.1), where f_1 and f_2 are linear functions with their own sets of parameters or weights. Similarly, σ_1 and σ_2 are nonlinear functions that are differentiable. The output from the first linear function $f_1(x)$ becomes the input to the first nonlinear function σ_1. The output of σ_1 is then fed into the second linear function f_2, and so on.

$$H(x) = \sigma_2\left(f_2\left(\sigma_1\left(f_1(x)\right)\right)\right) \tag{7.1}$$

More precisely, we can represent a linear function f as a function of its input data x and its weights W, as shown in Eq. (7.2). Here, W is a weight matrix with dimensions m x n, and $\mathbf{b}$ is a bias vector with m components. The output of the linear function is a vector with m scalar components. After we apply the nonlinear function σ to this output vector, we then feed it to the next linear function, and so on.

$$f(x) = x^T W + \mathbf{b} \tag{7.2}$$

In the context of reinforcement learning, neural networks are a powerful tool for approximating value functions or even the policy, as we'll cover in Part III of the book. By training the weights of the neural network using experience data, we can learn to generalize from observed states and actions to unseen ones. The abstract representation of a neural network as a series of linear and nonlinear functions provides a flexible framework for approximating complex functions that relate states and actions to values.

Feedforward Neural Networks

A feedforward neural network, also known as a multilayer perceptron (MLP), is a type of neural network in which the information flows in one direction, from the input layer, through one or more hidden layers, to the output layer. In a feedforward neural network, the output of one layer serves as the input to the next layer, with no loops between the layers.

The architecture of a feedforward neural network typically consists of an input layer, one or more hidden layers, and an output layer. The input layer receives the raw input data, which is then passed through the hidden layers, where the information is transformed and processed, before being processed by the output layer. Each layer in the network consists of a set of neurons, which are connected to neurons in the adjacent layers. Each neuron in a layer takes the weighted sum of the inputs from the previous layer, applies an activation function to the result, and outputs the result to the next layer.

Fully Connected Neural Networks

A fully connected neural network, also known as a dense neural network, is a type of neural network in which each neuron in one layer is connected to every neuron in the next layer. This means that all the neurons in each layer are connected to all the neurons in the adjacent layer. A fully connected neural network can have any number of layers, but typically has one or more hidden layers between the input and output layers.

In practice, a fully connected neural network is often a specific implementation of a feedforward neural network, where each layer is fully connected to the adjacent layers. The terms feedforward neural network and fully connected neural network are often used interchangeably. However, the term feedforward neural network is often used more generally to refer to any neural network in which the information flows in one direction, regardless of the type of connections between the layers.

Let's illustrate the idea of fully connected neural networks graphically, as shown in Fig. 7.1. Each layer in this example network is a combination of a linear transformation and a pointwise nonlinear activation function. The number of computation units in each layer can vary, as illustrated in Fig. 7.1. The input layer is an abstraction layer that transforms the raw input data into a format that can be processed by the subsequent layers. It has the same number of units as the dimension of the input data, and no weights or activation functions are associated with its units. The output layer has the same number of units as the desired output, and its computation units are responsible for producing the final output of the network.

The layers between the input and output layers are called hidden layers. In this example, two hidden layers are shown, but in practice neural networks can have a large number of hidden layers, leading to what are called deep neural networks. The term deep learning often refers to the use of deep neural networks to solve specific task, such as classification or prediction. The idea behind hidden layers is to allow the network to learn complex representations of the input data by gradually combining simpler features learned by the previous layers. Each hidden layer consists of a set of computation units that apply a nonlinear transformation to the output of the previous layer, followed by a linear transformation to produce the input to the next layer. The nonlinear transformation is usually

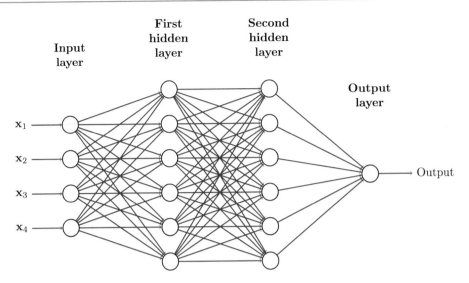

Fig. 7.1 Example of a simple fully connected neural network with two hidden layers, each consisting of six units, and an output layer, consisting of a single unit

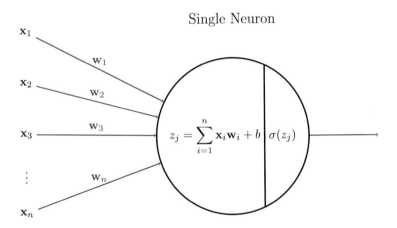

Fig. 7.2 Example of how a single neuron works

a pointwise function such as the sigmoid, ReLU, or tanh function, which introduces nonlinearity and enables the network to model more complex input-output relationships.

We can see in Fig. 7.1 that each unit in the previous layer is connected to all the units in the successor layer (except the input layer). This type of architecture is called a *fully connected layer*, and if a neural network only consists of fully connected layers, we call it a fully connected network. Each layer (except the input layer, which has no activation function) can have its own differentiable nonlinear function σ, which is often referred to as an activation function.

Each unit (or neuron) in a fully connected layer has its own set of parameters (except units in the input layer). Figure 7.2 shows the computation of the jth neuron in a fully connected layer. The neuron takes in a vector x as input data, where x can be the output from the previous layer or the input data if this neuron is in the first hidden layer. It first computes the inner product between the input vector x and the weight vector w (which represents the strength of the connections between the neurons), denoted as $z_j = x^T w$. The neuron then adds a scalar parameter b, known as the bias, to the results z_j. The bias is an additional parameter that can shift the results z_j. After that, the neuron applies an activation function σ to z_j, and the output of the activation function is passed as input to the next layer (or the output layer if the neuron is in the last layer).

It's worth noting that the weight vector w is learned through the training process. During training, the neural network adjusts the values of w and b so that the network can better approximate the desired output.

Activation functions play a critical role in neural networks by adding nonlinearity to the system. Without them, a neural network would be reduced to a linear function with many parameters, which would be unable to capture complex relationships between inputs and outputs.

The idea behind deep neural networks is that by applying a series of nonlinear transformations, we can extract a rich set of features from the input data. These features often have a hidden hierarchy, where lower-level features are combined to form higher-level ones. Activation functions help to create this hierarchy by enabling each layer to extract different features from the input data.

Some of the most common activation functions include ReLU (rectified linear unit), sigmoid, and tanh as shown in Fig. 7.3. ReLU is probably the most popular activation function due to its simplicity and effectiveness, but other functions such as LeakyReLU and ELU are also worth considering. The choice of activation function can have a significant impact on the performance of a neural network, and it is often a matter of trial and error to determine the best one for a particular task.

For example, ReLU is known to work well for most applications, but it can cause the "dying ReLU" problem, where some neurons get stuck in a state of zero activation and stop learning. LeakyReLU and its variants can help to mitigate this problem by allowing a small, nonzero gradient for negative inputs. Sigmoid and tanh functions are also useful for specific applications, such as binary classification or image generation.

In summary, choosing the right activation function is an important step in designing a neural network. It requires understanding the strengths and weaknesses of different functions and experimenting with different combinations to find the one that works best for a particular task.

Convolutional Neural Networks

A convolutional neural network (CNN) is a type of neural network that is specifically designed for processing and analyzing data with a grid-like structure, such as images or videos. CNNs are inspired by the organization and function of the visual cortex in animals, which has specialized neurons that respond to specific features in the visual field.

In a CNN, the input data is typically an image, which is represented as a grid of pixels. The network consists of multiple layers, with each layer performing a different type of processing on the input data. The first layer is typically a convolutional layer, which applies a set of filters to the input image, each filter looking for a specific pattern or feature in the image. These filters move across the image, computing a dot product between the filter and the image, producing a feature map.

The output of the first layer is then passed to the next layer, which may be another convolutional layer, a pooling layer, or a fully connected layer. The pooling layer downsamples the output of the previous layer by computing the maximum or average value of adjacent regions of the feature map. The fully connected layer takes the flattened output of the previous layer and applies a set of weights to it, producing a prediction or classification of the input image.

Fully connected neural networks are effective for processing numerical data, but they are not well suited for image data, which typically has three dimensions: width, height, and color channels. When

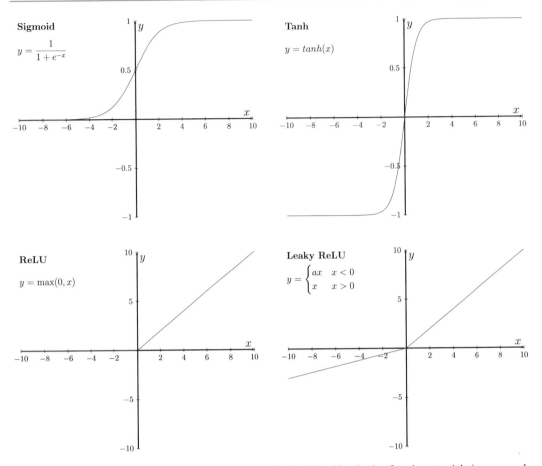

Fig. 7.3 Different activation functions. Top left is an example of a sigmoid activation function; top right is an example of a Tanh activation function; bottom left is an example of a ReLU activation function; and bottom right is an example of a Leaky ReLU activation function

using fully connected neural networks, we would need to flatten the image pixels to create a vector; flattening the image pixels into a vector causes us to lose a region of interest and the correlations between pixels, making it difficult to recognize objects, as shown in Fig. 7.4. Additionally, the number of parameters in a fully connected neural network can become very large when processing images.

There are two main issues with this approach. First, objects in an image often span multiple pixels, and each pixel is correlated with its surroundings. Flatting the image into a vector would make it more difficult to recognize objects; imagine we are training a reinforcement learning agent to do autonomous driving, the failure to successfully recognize pedestrians could mean fatal error and tragedy. Second, the number of parameters in a fully connected neural network can become very large when processing images. For example, a medium-sized image with dimensions of $600 \times 400 \times 3$ results in a vector with 7.2×10^5 elements. When fed into a fully connected neural network, the input data for the first hidden layer has 7.2×10^5 dimensions. If we only have 10 hidden units in the first layer, the network still requires 7.2×10^6 parameters for that layer alone. This number grows even larger when we have more hidden layers or units, making it difficult to scale up to larger images.

To address these challenges, we often use convolutional neural networks (CNNs) to process images. A convolution operation is a mathematical process that involves applying a filter (or kernel) to an input

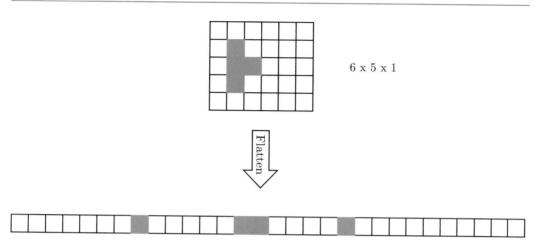

Fig. 7.4 Example of flattenning a grayscale image to a vector

image or feature map to extract relevant features. In deep learning, convolution is used in convolutional neural networks (CNNs) to analyze visual imagery, and it is a key component in many state-of-the-art computer vision models.

Convolution works by taking a small matrix of values (called filter or kernel) and sliding it across the input image or feature map, computing the dot product between the filter and the local patch of the image or feature map at each position. The result of this operation is a new matrix called the output feature map, which captures the most important features of the input.

The use of convolution in deep learning has many advantages, including the ability to capture spatial dependencies in visual data, reduce the number of parameters required for a model, and enable the reuse of learned features across different parts of an image. These properties make convolution an effective tool for analyzing visual data and have led to its widespread use in a variety of computer vision applications.

CNNs consist of one or more convolutional layers, which take the image data directly without flattening it. This preserves the region of interest and pixel correlations in the original image. Additionally, convolutional layers have a fixed number of parameters that is independent of the input image size, which makes it easy to scale up to larger images.

At the core of a convolutional layer is a filter that performs convolution on the input data. A convolutional layer usually consists of several filters, each with its own set of parameters: weights W and bias b. The weights W is a matrix with dimensions of kernel size x input channels. The kernel size is a configurable variable that determines the size of the working area, or receptive field, when the filter is convolving the image. The filter moves across the image with a configurable variable called stride (default 1) to perform the convolution operation. The stride controls how far the filter moves for each convolution operation, such as moving one pixel at a time, until it reaches the right edge of the image.

Figure 7.5 illustrates the operation of a single filter in a convolutional layer of a convolutional neural network. The input data on the left has one channel (e.g., a grayscale image), and the single filter in the middle has a 3×3 kernel size. The weight matrix W for this filter has the dimensions of $3 \times 3 \times 1$, since the input data has only one channel.

To extract features from the receptive field of the input data that is being convolved with the filter, the dot product is computed between the receptive field X and the weight matrix W using the dot

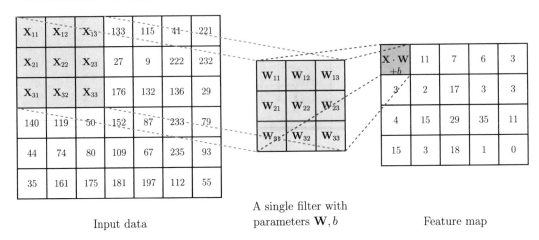

Fig. 7.5 Example of how a single convolutional filter works. The numbers are generated randomly

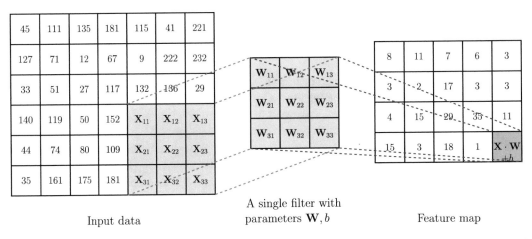

Fig. 7.6 Example of how a single convolutional filter works for the last receptive field. The numbers are generated randomly

product notation $X \cdot W$. The receptive field is the region of the input data being convolved with the filter, and its width and height are identical to the kernel size, as shown by the orange color in Fig. 7.5.

The dot product result is added to the bias b, which is a scalar parameter, using the addition operator, resulting in a scalar value that becomes part of the output feature map. The output feature map is computed using the equation $y = X \cdot W + b$.

Since a filter's kernel size is often much smaller than the input image or feature maps from previous convolutional layers, it needs to perform the same process repeatedly by swiping over the entire image. Each time, the filter reuses its parameters, which include weights and bias. Figure 7.6 shows how the filter works for the last receptive field.

In practice, a single convolutional layer will have multiple filters, and each filter is responsible for detecting specific features. One interesting fact about convolutional layers is that the number of channels in the output feature maps is determined solely by the number of filters in the layer and is independent of the channels in the input data. For example, as shown in Fig. 7.7, if the input data has a single channel but the convolutional layer has four filters (indicated by different colors), the output feature maps will have four channels.

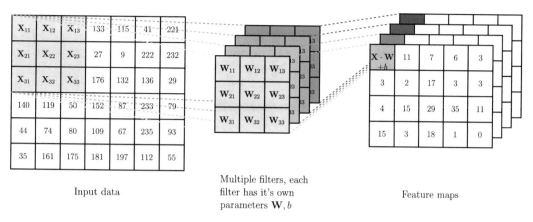

Fig. 7.7 Example of how multiple convolutional filters work; the number of channels in the output feature maps is only dependent on the number of filters. The numbers are generated randomly

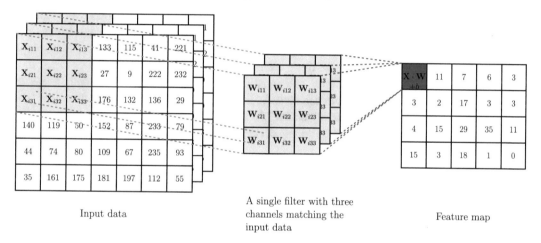

Fig. 7.8 Example of how a single convolutional filter works, where the input data have three channels and we have a single filter. The numbers are generated randomly

The preceding example only considers input data with one channel, but in practice, the input data often has multiple channels. For instance, an RGB image has three color channels, or the feature maps coming from previous hidden layers may have multiple channels. In these cases, the weights W for each filter in the convolutional layer will have the same number of channels as the input data. For example, as shown in Fig. 7.8, if the input data has three channels, the weights W of each filter in the layer will also have three channels. The output feature maps will just be the combined dot products between the corresponding channels: $\sum_i X_i \cdot W_i$. However, since we are talking about a single filter, the output feature map will have only one channel.

There are several important details worth noting about the architecture of convolutional neural networks (CNNs), such as controlling the stride and padding and using pooling layers to downscale images. However, this book primarily focuses on the theory and mathematics of reinforcement learning. Interested readers can find additional resources elsewhere to learn more about these neural network architecture.

When it comes to processing images, CNNs offer a significant advantage over fully connected neural networks. For example, consider an input image with dimensions of $600 \times 400 \times 3$. Using

100 filters with a kernel size of 3×3 for the first hidden layer yields only 2700 parameters (not including bias terms). This is much smaller than the 7.2×10^6 parameters required for ten units in a fully connected layer. Furthermore, CNNs are designed to be independent of the input image size, making them more versatile for image processing tasks.

7.2   Training Neural Networks

Training neural networks refers to the process of teaching a neural network to learn a mapping between input data and output data. During training, the neural network is presented with a set of input-output pairs, and the network adjusts its parameters, such as weights and biases, to minimize the difference between the predicted output and the actual target, such as in the case of supervised learning.

The idea of using stochastic gradient descent (SGD) to update the parameters of a neural network using sample gradients still applies. However, the challenge lies in how to compute the gradients with respect to the different weights and biases in a neural network. Since neural networks often have multiple layers, and each layer can have different activation functions, the derivatives with respect to the weights and biases can differ from layer to layer. This means we can't use a single update rule to update the parameters for these different layers for these different layers, such as the cases we've introduced in the previous chapter for the linear methods.

It is still possible to manually derive the equations for each layer, but this method can be challenging for people not familiar with calculus. Additionally, deep neural networks can often have dozens or even hundreds of layers, making manual derivation impractical. If changes are made to the architecture of the neural network, such as adding or removing layers or changing activation functions, most of the work will need to be redone.

Luckily, the backpropagation algorithm offers a simple but elegant solution to the challenges of computing gradients in neural networks. The algorithm uses a computation graph to represent a complex differentiable function and then uses the chain rule of calculus to compute the gradients for each node in the graph. Each node in the computation graph performs only very basic operations, such as addition, multiplication, division, or grouped operations like an activation function. Since we already know the derivatives of these basic operations, it is relatively easy to derive the derivatives with respect to the variables in those nodes.

Figure 7.9 shows a simplified version of a computation graph for a fully connected layer. Here, each node in the graph performs a specific computation such as the dot product between two variables, addition, or grouped operations like an activation function σ. Note that node $\mathbf{z}_1$ represents the dot product between the input data $\mathbf{x}$ and each weight vector $\mathbf{w}$ for all units in this linear layer. Node $\mathbf{z}_2$ is similar, but adds the bias terms from all units to the results coming from $\mathbf{z}_1$.

Using the backpropagation algorithm not only simplifies the computation of gradients but also makes neural networks more flexible. If changes are made to the architecture of the neural network, the algorithm can be easily adjusted to accommodate those changes, without requiring a complete redo of the work.

After we use the objective function to compute the loss with respect to the predicted outcome y from the neural network and true target y^*, we can use the backpropagation algorithm to compute the gradients with respect to the parameters of the network, and then using these gradients to update the network's parameters. The backpropagation algorithm works backward through the computation graph, computing the gradients (sometimes also called local gradients) with respect to the local parameters at each node and then propagating these gradients back to the node prior to the last node. It does this using the chain rule of calculus, which allows it to compute the gradients with respect to the local parameters and the upstream gradients coming from the previous node. This works not only for

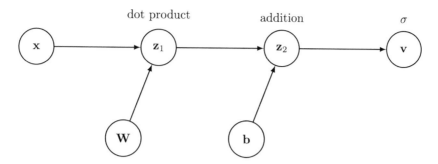

Fig. 7.9 A simplified example of a computation graph for a fully connected layer

fully connected neural networks but any other type of neural network, including convolutional neural networks (CNNs) and recurrent neural networks (RNNs).

Let's look at a very simple fully connected neural network with one hidden layer and one output layer. We use the sigmoid activation function for the hidden layer and no activation function for the output layer. We use the squared error loss as our objective function, which measures the difference between the predicted value and the true target value. Figure 7.10 shows an example of how backpropagation computes gradients for each layer. After the forward pass through the neural network to compute the predicted value y, we use the true target y^* to compute the loss. Then the backpropagation algorithm works backward, starting with the last node l in the graph, which is where the loss comes out. For each node, it uses the upstream gradients and the local gradients to compute the gradients for the specific computation node, until it reaches the first hidden layer. In practice, we are only interested in the gradients with respect to the weights W and bias b for these different layers.

After we know the gradients with respect to the weights and bias for these different layers of a neural network, we can start to update the parameters using the stochastic gradient descent algorithm.

As the number of layers in a neural network increases, it's hard to keep track of the weights and bias for each layer. So we often use θ to represent all the parameters of the neural network, where θ is a vector that contains all the parameters (weights and biases) for a single neural network. To sum it up, in general, training a neural network consists of these major steps:

- Feed input data x to the neural network parameterized by θ to get predicted value y.
- Compute the loss using the objective function with respect to y from the preceding step and the true target y^*.
- Use backpropagation to compute gradients for θ with respect to the loss from the preceding step.
- Use the stochastic gradient descent algorithm to update θ with respect to the gradients from the preceding step.

Luckily, modern deep learning frameworks, like PyTorch and TensorFlow, have built-in automatic differentiation capabilities that can perform backpropagation for us automatically. These tools also have built-in optimizers that can update the parameters of a neural network with respect to the gradients of these parameters. Automatic differentiation and backpropagation help us to compute the gradients with respect to the parameters of a neural network. Optimizers are algorithms that use the gradients to update the parameters of the network. So we only need to focus on the architecture of the neural network and define an objective function (loss function), and these tools will handle the rest of the work for us automatically.

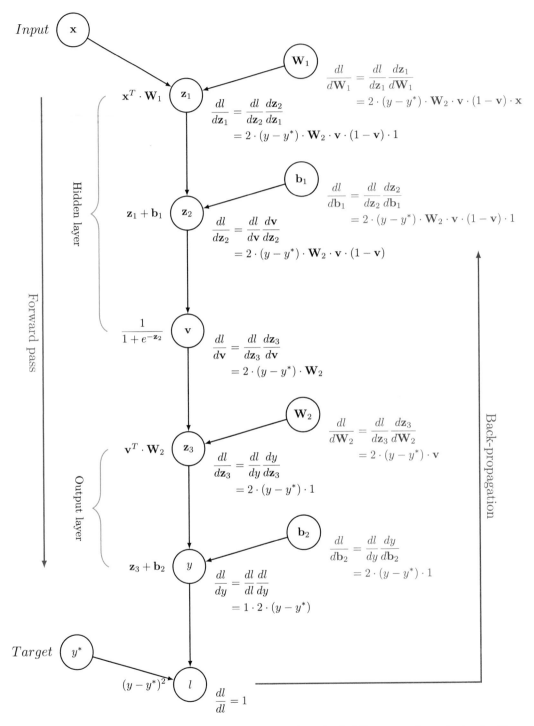

Fig. 7.10 Example of how backpropagation computes gradients for a simple fully connected neural network with one hidden layer and one output layer, where we use a sigmoid activation function for the hidden layer and no activation function for the output layer

In the following example, we demonstrate how to use the PyTorch deep learning framework to train a neural network. First, we create an instance of a neural network and an optimizer (in this case, stochastic gradient descent, or SGD) for updating the network parameters during training. We pass in the network parameters and a learning rate as arguments to the optimizer, so it knows which parameters to update and how big of an update step to take.

During training, we feed input data from the training set into the network to compute a predicted output. We then compute the loss, which is a measure of how far off the predicted output is from the true target value in the training set. In this example, we use mean squared error (MSE) loss as our objective function.

Once we've computed the loss, we use backpropagation to automatically compute the gradients for the different parameters of the network. These gradients tell us how much each parameter contributes to the loss, and we can use them to perform a stochastic gradient descent step to update the network parameters. We repeat this process for each batch of training data until the network has learned to accurately predict the target values.

Here is the code with additional comments that explain what each line does:

```
# create instance of the neural network
network = create_network_instance()

# create instance of optimizer (SGD) for updating the network parameters
optimizer = optim.SGD(network.parameters(), lr=0.0001)

...

# step 1: forward pass
optimizer.zero_grad()    # Clear gradients from previous steps
output = network(input_data)    # compute predicted output from input data

# step 2: compute loss
loss = loss_fn(output, target)    # compute loss (MSE)

# step 3: compute gradients using back-propagation
loss.backward()    # use backpropagation to compute gradients of the loss with
    respect to the network's parameters

# step 4: update network parameters
optimizer.step()    # use the gradients to update the network parameters
```

It's worth noting that some of the terms and concepts used in this example may be unfamiliar to those who are new to PyTorch or deep learning. For example, a neural network is a series of interconnected nodes that can learn to map input data to output data, and an optimizer is an algorithm that updates the network's parameters to minimize the loss function. Backpropagation is a method for computing the gradients of the loss function with respect to the network's parameters, and gradients tell us how much each parameter contributes to the loss. By understanding these concepts, we can create and train neural networks to solve a wide range of problems.

While the framework and tools have made training deep neural networks easier, there are also other challenges that can arise when training deep neural networks. One such challenge is the problem of gradient vanishing or explosion. This issue occurs when the gradients become too small or too large for layers close to the input layer, such as the first hidden layer. This problem often arises with very deep neural networks. To address this challenge, researchers have proposed solutions such as using skip-connection architectures, like ResNet architecture proposed by He et al. [1], to build the neural network. These architectures allow us to train deep neural networks with hundreds of layers, while avoiding the gradient vanishing problem.

Another challenge is the problem of overfitting, which occurs when the number of parameters in a neural network is much larger than the number of training data. This can lead to a neural network that is overly complex and has learned to fit the training data too closely, resulting in poor performance on new, unseen data. Regularization methods, such as L2 and L1 regularization, as well as using dropout, can be used to address the overfitting issue. L2 and L1 regularization add a penalty term to the loss function, which helps to prevent the model from overfitting by reducing the magnitude of the weights. Dropout, on the other hand, randomly drops out a percentage of neurons during training, which helps to prevent the network from relying too heavily on any one neuron or set of neurons.

In the context of training deep neural networks, weight initialization plays a crucial role. Typical initialization methods include random initialization, Xavier initialization [2], and He initialization [3]. Random initialization lacks guidance, while Xavier initialization takes into account layer connections, and He initialization adjusts for ReLU activation. The choice of initialization method depends on the network architecture and the specific activation functions being used. Poor initialization can hinder convergence and lead to suboptimal solutions, whereas appropriate initialization accelerates learning and improves performance.

While deep learning frameworks have made it easier to train neural networks, it is important for users to be aware of the challenges that can arise when working with these models. By understanding these issues, readers can better evaluate the performance of their models and make more informed decisions about how to address these challenges.

To learn more about these topics in greater detail, readers may want to explore the many resources available online, including academic papers, tutorials, and discussion forums. By combining theory with hands-on practice, we can develop a deeper understanding of how to train neural networks and its underlying mathematical principles.

7.3   Policy Evaluation with Neural Networks

In this section, we explore the use of neural networks for policy evaluation in reinforcement learning. The goal is to approximate the state value function V_π for an arbitrary policy π using a neural network with parameters θ. We assume that we know the true state value function V_π of the policy, and our objective is to find the parameters θ that minimize the squared error between the true values and the neural network predictions.

We start with the objective function:

$$J(\theta) = \mathbb{E}_\pi\left[\left(V_\pi(S) - \hat{V}(S; \theta)\right)^2\right] \tag{7.3}$$

where $\hat{V}(S; \theta)$ is the neural network's predicted value for state S. We can use stochastic gradient descent to update the parameters θ using the sample gradients computed using mini-batch of experience data, which gives us a simplified objective function for the given mni-batch:

$$J(\theta) = \left(V_\pi(S) - \hat{V}(S; \theta)\right)^2 \tag{7.4}$$

To use this approach for policy evaluation in practice, we relax the assumption that we know the true state value function V_π and use sample returns from Monte Carlo or TD learning as the target. For Monte Carlo policy evaluation, the objective function becomes

$$J(\theta) = \left(G_t - \hat{V}(s_t; \theta)\right)^2 \tag{7.5}$$

where G_t is the total return obtained from state s_t.

For TD(0) policy evaluation, we use the immediate reward r_t and the discounted value of the successor state $\gamma \hat{V}(s_{t+1}; \theta)$ as the target, and the objective function becomes

$$J(\theta) = \left(r_t + \gamma \hat{V}(s_{t+1}; \theta) - \hat{V}(s_t; \theta)\right)^2 \tag{7.6}$$

By using these objective functions and applying stochastic gradient descent with backpropagation, we can update the neural network parameters θ to better approximate the state value function and improve the policy evaluation in our reinforcement learning algorithm.

Reinforcement learning distinguishes itself from supervised learning by using Monte Carlo or TD learning to estimate the target in the objective function. Monte Carlo and TD learning use reward signals and the value of successor states to estimate the target, which changes over time, unlike the fixed target in supervised learning, this also makes training neural network to approximate value functions more challenging.

We can also use neural works to approximate the state-action value function for some policy π, denoted as $\hat{Q}$. The objective function for TD(0) policy evaluation for $\hat{Q}$ with respect to the parameters θ becomes

$$J(\theta) = \left(r_t + \gamma \hat{Q}(s_{t+1}, a_{t+1}; \theta) - \hat{Q}(s_t, a_t; \theta)\right)^2 \tag{7.7}$$

where r_t represents the reward signal, γ is a discount factor that determines the relative importance of future rewards, and θ represents the parameters of the neural network. By minimizing this objective function, we can update the neural network's parameters to better approximate the true state-action value function Q_π for some policy π.

7.4   Naive Deep Q-Learning

In reinforcement learning, Q-learning is an off-policy learning algorithm that directly learns the optimal state-action value function (Q_*). To compute the estimated target, Q-learning uses the immediate reward and discounted value of the best successor state-action pair as the TD target. When

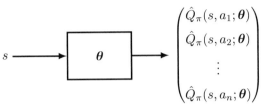

Fig. 7.11 The concept of using neural networks to approximating state-action value function Q_π, where we only feed state s to the neural network, and the output from the neural network is a vector that contains values for every action for the given state

using neural networks to approximate the optimal state-action value function, our objective function for Q-learning can be expressed as

$$J(\boldsymbol{\theta}) = \left(r_t + \gamma \max_{a'} \hat{Q}(s_{t+1}, a'; \boldsymbol{\theta}) - \hat{Q}(s_t, a_t; \boldsymbol{\theta})\right)^2 \tag{7.8}$$

Here, $\boldsymbol{\theta}$ represents the parameters of the neural network.

In the previous chapters, we often computed the value for a specific state-action pair (s, a) using tabular and linear methods. However, if we want to find out which action yields the maximum value $\max_a Q_\pi(s, a)$ for an arbitrary state s, we have to enumerate all the (legal) actions one by one. This means that we have to feed the same state s into a neural network multiple times, which can hurt performance if the action space is quite large. When working with neural networks, we want to avoid these sequential operations and take advantage of parallel computing capabilities especially when working with accelerated hardware like GPUs.

To address this issue, we can make a small change to how we compute the action values for an arbitrary state when using neural networks. Instead of feeding a state-action pair (s, a) into the neural network, we only use state s as input, and we compute the values for each action in one go, as shown in Fig. 7.11. The output of the neural network is a vector, where each scalar component corresponds to the estimated value for the corresponding action. This approach avoids repeatedly feeding the same state into the neural network, which can be computationally expensive. It also allows us to easily find which action has the maximum value for a given state.

We now introduce the naive deep Q-learning algorithm, which uses a (deep) neural network to approximate the optimal state-action value function Q_*. The overall procedure is the same compared to using linear methods to approximate Q_*. The difference is how to compute the gradients and update the parameters. If we use some deep learning framework software like PyTorch, we don't need to manually compute the gradients and then update the weights or bias of different layers of the neural network. Instead, we rely on these tools (like using backpropagation and optimizer) to do the work automatically for us. We only need to compute the loss using the predicted values and the TD (temporal difference) target, which is a measure of the difference between the predicted value and the actual reward received by the agent. It is worth noting that this is not the same DQN agent that DeepMind used to play different Atari games. To develop the same DQN agent, we would need to make some improvements to the naive deep Q-learning algorithm.

We evaluated the performance of the naive deep Q-learning agent on the classic cart pole task. The results are displayed in Fig. 7.12. To evaluate the performance of the agent, we ran 20,000 evaluation environment steps (approximately 40 episodes) on a separate evaluation environment with an exploration epsilon of 0.05 after every 20,000 training environment steps. The results were averaged over five independent runs and then smoothed using a moving average with a window size of five.

Algorithm 1: Naive deep Q-learning with ϵ-greedy policy for Exploration

---

Input: Discount rate γ, learning rate α, initial exploration rate ϵ, number of steps K
Initialize: $i = 0$, initialize the parameters $\boldsymbol{\theta}$ of the neural network
Output: The approximated optimal state-action value function $\hat{Q}$

1 **while** $i < K$ **do**
2   Sample action a_t for s_t from ϵ-greedy policy w.r.t. $\boldsymbol{\theta}$
3   Take action a_t in environment and observe r_t, s_{t+1}
4      $i = i + 1$
5   (optional) decay ϵ, for example, linearly to a small fixed value
6   Compute TD target:
7      $\delta_t = \begin{cases} r_t & \text{if } s_{t+1} \text{ is terminal state} \\ r_t + \gamma \max_{a'} \hat{Q}(s_{t+1}, a'; \boldsymbol{\theta}) & \text{otherwise bootstrapping} \end{cases}$
8   Compute loss $l = \left(\delta_t - \hat{Q}(s_t, a_t; \boldsymbol{\theta})\right)^2$
9   Do stochastic gradient descent step based on loss l w.r.t. $\boldsymbol{\theta}$
10     $s_t = s_{t+1}$

Fig. 7.12 Performance of the naive deep Q-learning agent on cart pole. To evaluate the performance of the agents, we run 20,000 evaluation environment steps (about 40 episodes) on a separate evaluation environment with exploration epsilon 0.05 after every 20,000 training environment steps. The results were averaged over five independent runs and then smoothed using a moving average with a window size of five. We use discount $\gamma = 0.99$, learning rate $\alpha = 0.00025$, and decay exploration rate ϵ from 1.0 to 0.1 over 100,000 environment steps during training

The results indicate that the overall performance of the naive deep Q-learning agent is worse than using linear methods to approximate the value functions. Furthermore, the agent failed to reach the maximum episode returns during the evaluations. One might expect better results if we try to fine-tune the hyper-parameters like learning rate and exploration rate, but there are some underlying issues with the naive deep Q-learning agent which we need to address in order to make it a strong agent.

7.5 Deep Q-Learning with Experience Replay and Target Network

In machine learning, particularly in deep learning, we often rely on training data that is independent and identically distributed (IID). However, in reinforcement learning, consecutive states and reward signals in a sequence are often correlated. This is because an agent's choice of action can affect future states and rewards, resulting in a non-IID dataset. Additionally, the naive deep Q-learning algorithm only uses each transition (s, a, r, s') once to update the network parameters, leading to poor data efficiency.

To address these issues, we can use a technique called experience replay, where we store the most recent N transitions (s, a, r, s') in a ring buffer structure and randomly sample a mini-batch of transitions from the buffer during training. This allows the algorithm to reuse the same transitions multiple times, improving data efficiency. Moreover, by randomly sampling from a large pool of transitions, we break the correlations between consecutive states and rewards.

However, using experience replay introduces a new challenge. Since the data in the buffer may come from a long time ago, the network parameters may have changed significantly since the data was collected. This can lead to training on samples generated under a different behavior policy, which is not ideal for on-policy learning algorithms like SARSA. Therefore, it is crucial to use an off-policy learning algorithm like Q-learning with experience replay.

Implementing experience replay is straightforward. We need to allocate space for N transitions and add methods to add new transitions and randomly sample mini-batches from the buffer.

Algorithm 2: Uniform random experience replay

Input: Capacity of replay N
Initialize: $c = 0$, initialize replay buffer D to capacity N
1 **def** *add(d)*:
     /* Stores a single transition d into buffer D */
2   Compute insertion index $i = c\%N$ for transition d
3   Store transition d at index i in buffer D (with replacement)
4   Increase counter $c = c + 1$
5
6 **def** *sample(n)*:
     /* Samples a batch of n transitions from buffer D randomly   */
7   **if** $c \geq n$ **then**
8   Initialize L, T to store sampled indices and transitions
9   Fill in L with n randomly selected indices in the range of $[0, \min(c, N))$
10   **for** i in L **do**
11   Pick i_{th} transition from D and put it into T
12   **return** T

However, experience replay is not the only technique used to improve deep Q-learning. Another key technique is the use of a target network. The target network is a separate copy of the Q-network that is periodically updated to match the Q-network's parameters. This is done to prevent the algorithm from getting stuck in a feedback loop, where the Q-network's estimates are used to update itself, leading to instability. By using a separate network for target values, we can reduce the variance in the target estimates and stabilize the learning process.

In summary, experience replay and target network are two key techniques used to improve the performance and stability of deep Q-learning. By storing and reusing transitions in a buffer and using a separate network for target values, we can achieve better data efficiency and stability in our training process.

To ensure the stability of training deep neural networks for Q-learning, a separate target network can be used to compute the TD target. The target network calculates the value of the next state-action pair and updates its parameters after a certain number of updates to the standard Q-network. This helps avoid the problem of using values from the standard Q-network as the TD target that are getting updated very frequently.

The objective function for deep Q-learning with a target network is

$$J(\theta) = \left(r_t + \gamma \max_{a'} \hat{Q}(s_{t+1}, a'; \theta^-) - \hat{Q}(s_t, a_t; \theta)\right)^2 \tag{7.9}$$

Here, θ represents the parameters for the standard Q-network, and θ^- represents the parameters for the target network. Initially, the two networks have the same parameters. After every C number of updates to θ, θ^- is updated to θ.

The DQN agent is a modified deep Q-learning algorithm with experience replay and target network. It starts by initializing the standard Q-network θ and the target network θ^- with the same parameters. A replay buffer D with a capacity of N is also initialized to store most recent N transitions.

During training, the agent interacts with the environment using an ϵ-greedy policy with respect to the standard Q-network θ. Instead of computing the loss and performing stochastic gradient descent on the transition data, the agent stores the data in the replay buffer D. When the number of samples in the buffer D reaches a certain threshold m (where $m \geq b$), the agent samples a mini-batch of transitions from D and uses them to compute the TD target and squared error loss. The loss is then optimized using stochastic gradient descent to update the parameters θ of the standard Q-network. The agent continues to interact with the environment using the updated θ, generating better quality training data. The agent then samples another mini-batch of transitions from D to continue learning. After every C updates to θ, θ^- is updated to θ.

Although both experience replay and target network improve the performance of Q-learning, experience replay is generally more important. However, in practice, both techniques are usually used together when training deep Q-learning models.

Algorithm 3: [DQN] Deep Q-learning with experience replay and target network, using ϵ-greedy policy for Exploration

---

Input: Discount rate γ, learning rate α, initial exploration rate ϵ, batch size b, capacity of replay N, minimum replay size to start learning m where $m \geq b$, interval to update target network C, number of steps K
Initialize: $i = 0$, initialize the parameters θ, θ^- of the neural networks, set $\theta^- = \theta$, initialize replay buffer D to capacity N
Output: The approximated optimal state-action value function $\hat{Q}$

1 **while** $i < K$ **do**
2   Sample action a_t for s_t from ϵ-greedy policy w.r.t. θ
3   Take action a_t in environment and observe r_t, s_{t+1}
4       $i = i + 1$
5   (optional) decay ϵ, for example, linearly to a small fixed value
6   Store transition (s_t, a_t, r_t, s_{t+1}) in D
7       $s_t = s_{t+1}$
8   **if** length of $D > m$ and length of $D > b$ **then**
9   Sample a mini-batch of b transitions (s_j, a_j, r_j, s_{j+1}) randomly from D
10   Compute TD target for mini-batch $\delta_j = \begin{cases} r_j & \text{if } s_{j+1} \text{ is terminal state} \\ r_t + \gamma \max_{a'} \hat{Q}(s_{j+1}, a'; \theta^-) & \text{otherwise bootstrapping} \end{cases}$
11   Compute loss for mini-batch $l = \left(\delta_j - \hat{Q}(s_j, a_j; \theta)\right)^2$
12   Do stochastic gradient descent step based on loss l w.r.t. θ
13   Update target network $\theta^- = \theta$ after every C updates to θ

In episodic reinforcement learning problems, such as playing Atari games, terminal states represent the end of an episode sequence, after which the environment is reset to the initial state. One key

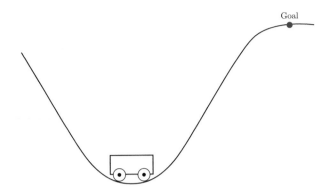

Fig. 7.13 Mountain car; the goal is to drive an underpowered car to reach the goal position. The problem was originally introduced by Sutton and Barto [4]

consideration is how to handle mini-batch transitions that may contain one or more terminal states, since the value of a terminal state is alway zero. When randomly sampling from the replay buffer during training, it is possible to select a mini-batch that includes one or more terminal states. To address this, there are different strategies that can be used.

One possible solution is to store a discount rate γ together with each transition in the replay buffer, such as $(s_t, a_t, r_t, s_{t+1}, \gamma)$. In this case, when s_{t+1} is a terminal state, γ is set to zero. During learning, the stored discount rate is used to compute the temporal difference (TD) target for the transitions.

Another solution is to include a boolean indicator along with each transition in the replay buffer, such as $(s_t, a_t, r_t, s_{t+1}, done)$, where the indicator is set to true when s_{t+1} represents a terminal state. This way, during learning, the discount can be set to zero when "done" is true, which in turn sets the value of the successor state to zero.

By handling terminal states appropriately, the agent can learn more efficiently and effectively. For example, when the discount rate is set to zero for a terminal state, the neural network will not try to learn anything about the value of that state. Providing proper treatment of terminal states is important in ensuring the robustness of a reinforcement learning algorithm.

Mountain Car Task

We now use another classic control task called *mountain car* to demonstrate the performance improvement of using experience replay and target network. The goal of the mountain car task is to drive an underpowered car to reach a predefined goal position as shown in Fig. 7.13. What makes this task difficult is the mountain is too steep to climb for the underpowered car. The only solution is to first drive the car in the opposite direction of the goal and using the gravity as an additional force to build enough velocity to eventually reach the goal.

The environment state contains only two continuing variables: the position $x \in [-1.2, 0.5]$ and velocity $\dot{x} \in [-0.07, 0.07]$ of the car. The reward for the task is a -1 for every step until the car reaches the goal position. There are three actions the agent can choose from: accelerate forward (to the right), accelerate reverse (to the left), and no acceleration. The dynamics of the environment is defined by a simplified physics model, where the position of the car is updated by the following equation:

$$x_{t+1} = \text{bound}(x_t + \dot{x}_{t+1})$$

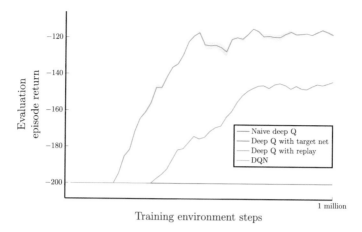

Fig. 7.14 Evaluation episode return for different Q-learning agents on the mountain car task. The results were averaged over five independent runs and then smoothed using a moving average with a window size of five

where bound(x) is a function that returns x if it is within the bounds $[-1.2, 0.5]$ and the nearest bound if it is outside this range. The velocity of the car is updated by the following equation:

$$\dot{x}_{t+1} = \text{bound}\,(\dot{x}_t + 0.001 \cdot a_t - 0.0025 \cdot \cos(3x_t))$$

where a_t is the chosen acceleration at time t.

If the car reaches the left top mountain (position left bound), the velocity of the car is reset to zero, so it does not run out of the valley. Each episode starts from a random position $x_0 \in [-0.6, -0.4]$ and zero velocity. To avoid running the task in an infinite loop, we can limit the maximum number of steps per episode.

The following experiment evaluates the performance of different Q-learning agents on the mountain car task, using a prebuilt mountain car environment[1] from the OpenAI Gym [5]. The experiment settings are

- Maximum episode step: 200
- Learning rate (α): 0.0005
- Discount rate (γ): 0.99
- Batch size: 32
- Replay capacity: 50,000
- Exploration rate (ϵ) decay: From 1.0 to 0.05 over 100,000 updates
- Target network update interval (C): 200

Figure 7.14 shows the performance for naive deep Q-learning, DQN with only target network, and DQN with experience replay and target network. To evaluate the performance of the agent, we ran 20,000 evaluation environment steps (approximately 40 episodes) on a separate evaluation environment with an exploration epsilon of 0.05 after every 20,000 training environment steps. The results were averaged over five independent runs and then smoothed using a moving average with a window size of five.

---

[1] Official URL for mountain car environment from OpenAI Gym: www.gymlibrary.dev/environments/classic_control/mountain_car/.

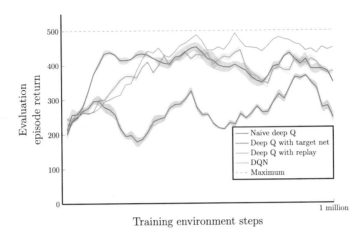

Fig. 7.15 Evaluation episode return for different Q-learning agents on the cart pole task. The results were averaged over five independent runs and then smoothed using a moving average with a window size of five

We use learning rate $(\alpha) = 0.00025$, discount rate $(\gamma) = 0.99$, batch size $= 32$, and replay capacity $= 50,000$ in this experiment. We decay exploration rate (ϵ) from 1.0 to 0.1 over 100,000 environment steps and update the target network (if any) every 200 updates.

This experiment highlights the crucial role of experience replay in the success of Deep Q-Network (DQN) algorithms developed by Mnih et al. [6]. The results indicate that naive deep Q-learning and deep Q-learning with a target network perform similarly to a random agent. In contrast, deep Q-learning with experience replay exhibits a smoother learning curve. While using an additional target network may slow down the learning speed depending on the complexity of the problem and the frequency of target network updates, it leads to smoother learning progress overall.

We can also observe similar results from another experiment that evaluates the performance of different Q-learning agents on the cart pole task, as shown in Fig. 7.15.

7.6   DQN for Atari Games

So far, we've only considered toy problems like cart pole and mountain car, which we can use linear methods instead of neural networks to approximate the value functions. However, for more complex and challenging reinforcement learning problems, neural networks are essential. One example is the Atari video game test bed, which is a collection of games owned by Atari Interactive, Inc. that require advanced AI techniques to play well.

In this section, we'll cover the details of how to train a DQN agent to play some of these Atari games. The DQN agent is a neural network–based approach for solving reinforcement learning problems, as we mentioned in Chap. 1. To adapt the DQN algorithm to the Atari games, we need to make some changes to the training pipeline. Specifically, we'll need to prepare the environment state, define the neural network architecture, and modify the training details. These changes are based on the work of Mnih et al. [7] and [6], which we'll follow closely.

Limitations and Challenges of Using the Atari Game Environment

Although the Atari game environment has proven to be a valuable tool for testing and developing deep reinforcement learning algorithms, it is not without its limitations and challenges. One of the main limitations of the Atari environment is that it is a simplified simulation of the real world, which means that the skills and strategies learned by an agent in the Atari environment may not necessarily transfer to real-world applications. In other words, the Atari environment may be too artificial to capture the full complexity and variability of real-world environments.

The second limitation of the Atari environment is that it is a relatively low-dimensional and discrete environment, which means that it may not be suitable for testing and developing deep reinforcement learning algorithms that require high-dimensional and continuous state and action spaces. Some real-world applications, such as robotics and autonomous driving, have state and action spaces that are much more complex and continuous than those in the Atari environment.

Another limitation of the Atari environment is that it is a static environment, which means that it does not change over time. Real-world environments, on the other hand, are dynamic and constantly changing, which poses additional challenges for deep reinforcement learning algorithms. For example, an autonomous driving agent must be able to adapt to changing traffic conditions and road conditions in real time, which is much more challenging than playing a game of Space Invaders.

Despite these limitations, researchers are actively working on developing more realistic and complex environments for testing and developing deep reinforcement learning algorithms. Some of these environments include simulated robotics tasks, traffic simulations, and video game environments with more realistic physics and graphics. These environments can provide a more challenging and realistic test bed for deep reinforcement learning algorithms and help to bridge the gap between simulated environments and real-world applications.

Environment Preprocessing

Atari games are a type of deterministic, episodic reinforcement learning problem. The screen image, or frame, is used as the environment state, with the time step of an episode sequence measured by the different frames of the game. However, there is one problem with this approach: a single frame does not have the Markov property. To understand this problem, let's look at a single frame from the Breakout video game, as shown in Fig. 7.16. From this single frame, it is hard to tell which direction the ball (small red block in the middle left of the screen) is moving. It could be moving up because the agent just hit the ball with the paddle, or it could be falling from the right corner after hitting the wall on the right side. The problem is that this single frame does not contain enough information to fully describe the direction or movement of the ball.

The Markov property (introduced in Chap. 2) says that the successor state s_{t+1} only depends on the current state s_t and action a_t, not all previous states and actions $(s_0, a_0, s_1, a_1, \cdots, s_t, a_t)$. However, in this case (as shown in Fig. 7.16), the successor state s_{t+1} depends on the past. For example, if the ball is moving down from the right corner and the agent chooses to move the paddle to the right, it might miss the ball, which will have a negative consequence (loss of a life). However, if the ball is moving up, and the agent also moves the paddle to the right, the agent may get some rewards since it is likely the ball will hit the top wall. For the same single frame and action, the agent may get different successor states and rewards.

One solution to this problem is to add the past frames together. Figure 7.17 shows an example of this by stacking the past 28 frames to construct a single image. Now, we can tell that the ball is moving toward the bottom of the screen after hitting the right-side wall.

Fig. 7.16 Example of a
single frame of the
Breakout video game

However, stacking all the historical frames is not good for space and computation efficiency since most Atari games can last for millions of frames. Most of these games generate 30–60 frames per second, and the movements in these games are very simple, so the changes over two consecutive frames are so small that we can barely tell if they are different or not. DQN uses a technique called "action repeat" (sometimes called skip frame) to address this issue. With this technique, the agent repeats the same action k times (default $k = 4$) when interacting with the environment. At the beginning of any episode, the agent selects some action a by following the ϵ-greedy policy with respect to the network θ, then repeats the same action a over k times while interacting with the environment (except for the terminal state, which does not require the action to be repeated). At the $k + 1$ frame, the agent selects another action a' and repeats a' over k times. This process is repeated until the episode is terminated. As a result, the agent only sees every kth frame from the environment, and it efficiently skips processing $(k - 1)/k$ percent of the frames. This simple technique alone can save a lot of space and computation time for the agent.

Figure 7.18 shows an example of action repeat with $k = 4$, where the grayed-out frames represent the skipped frames by repeating the same action.

Finally, to ensure that the environment state has the Markov property while at the same time save computation and space, DQN uses a history of the most recent four frames (after applying the action repeat and frame skip). These four frames are stacked together, as shown in Fig. 7.19, to construct a single image that represents the environment state. Before being fed into the neural network, the image is downsized from $160 \times 210 \times 3$ (the default size of Atari frames) to $84 \times 84 \times 3$ and then converted to grayscale to save additional computation. This downsizing and grayscale conversion helps to speed up the training process by reducing the size of the input data. Finally, the input data is normalized by

Fig. 7.17 Example of stacking the past 28 frames together to construct a single image for the Breakout video game

dividing by 255, so that the data is in the range of [0, 1], before being passed to the neural network for processing.

In summary, the preprocessing steps used by DQN for Atari games involve action repeat, frame skipping, frame stacking, downsizing, grayscale conversion, and normalization. These steps help to reduce the computational burden of processing millions of frames and enable the neural network to learn efficiently from the input data.

Figure 7.19 shows an example of stacking the last four frames, after using action repeat (frame skip) of four in the Breakout video game.

There are various modifications we can apply to the standard Atari game environment to improve the learning progress of the DQN agent. For example, we can treat the loss of a life as a soft-terminal state, which means that we only use the reward as the TD target, without bootstrapping by using the discounted value from the successor state. In a soft-terminal state, the game environment does not reset. Another method is to randomizing the initial state the initial state of each episode by applying some no-ops action for some steps can also be effective in improving the agent's learning progress. Additionally, clipping the rewards in the range of [−1, 1] is often necessary, especially for games with very large reward signals. For instance, some games may have rewards in the hundreds or thousands, which can overwhelm the learning process and make it difficult for the agent to converge to an optimal policy.

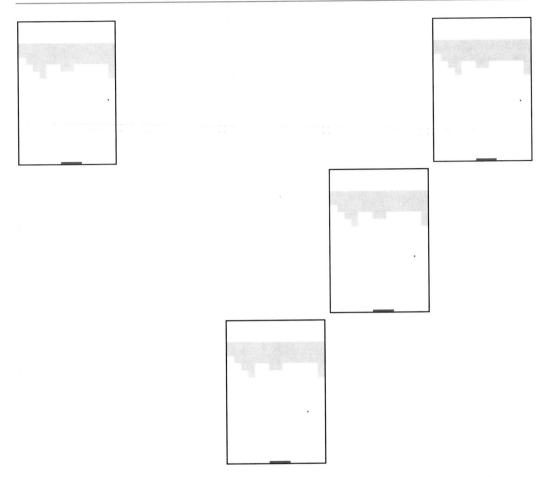

Fig. 7.18 Example of action repeat $k = 4$ for the Breakout video game; grayed-out frames represent the frames that are skipped by the agent

Convolutional Neural Network

The DQN agent uses a very simple convolutional neural network. It consists of three convolutional layers; the input data to the neural network is an $84 \times 84 \times 4$ image. The first hidden convolutional layer has 32 filters with kernel size 8×8 and stride 4. This means the output feature maps will have 32 channels for the first hidden layer. The second hidden convolutional layer has 64 filters with kernel size 4×4 and stride 2. And the third hidden convolutional layer has 64 filters with kernel size 3×3 and stride 1. We apply a ReLU activation function after each of these three convolutional layers. The feature maps from the last convolutional layer are then flattened and fed into a fully connected layer with 512 units; after that, we apply a ReLU activation function. The output from the fully connected layer is then fed into the output layer which has N units and no activation function. Here, N is the number of valid actions for the individual Atari game, which can vary between 4 and 18 depending on the game. Figure 7.20 illustrates the network architecture. Notice we only take the environment state s as input to the neural network, which is a $84 \times 84 \times 4$ image and output a vector contains state-action values for each valid action, for the input state s.

Why don't we use an activation function for the final output layer? Most activation functions will transform the value in some form. But the value for an arbitrary state-action pair could be any real

Fig. 7.19 Example of
stacking last four frames
after action repeat $k = 4$
for the Breakout video
game

number $(s, a) \in \mathbb{R}$. And the limits for the minimum or maximum value for these different Atari games are not the same. To reuse the same neural network for these games, it's best to not transform the values from the output layer.

Other Changes to the Training Procedure

There are some small changes made to the training procedure when we train the DQN agent to play Atari games. One particular change is how often the agent should sample a mini-batch of transitions from the experience replay to do a learning step (where we count one update to the parameters θ as a learning step). In Algorithm 3, the agent would sample a mini-batch at every time step as long as the buffers have enough samples. However, since Atari games can last over millions of steps (frames), learning at every step might not be a good idea from a computation efficiency point of view; it might also affect the performance of the agent since since it has lesser time to generate more transitions, this might also cause the neural network to overfitting existing experience data. So the DQN agent actually adapts a slightly different learning schedule, where it will only do one learning step after the agent selects every four actions. That's every 16 frames, because one action is repeated over four times.

Figure 7.21 shows the performance of the DQN agent on the Atari game Pong, which is one of the easiest games in the Atari test bed. The results display the average episode return (total undiscounted rewards) and a 95% confidence interval. To evaluate the agent's performance, we ran 200,000 evaluation steps on a separate testing environment with an ϵ-greedy policy and a fixed exploration epsilon ($\epsilon = 0.05$) at the end of each training iteration, which consisted of 250,000 training steps or 1 million frames, where no reward clipping or soft termination on loss of life is

Fig. 7.20 Convolutional
neural network architecture
for the DQN agent

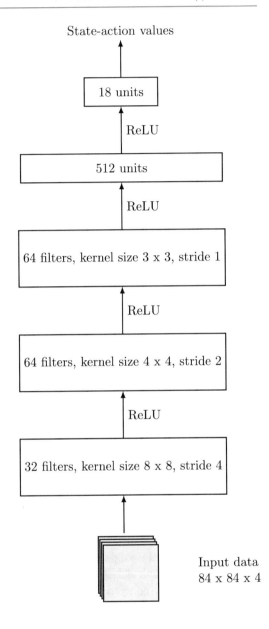

applied to the evaluation environment. The results were averaged over five independent runs and smoothed using a moving average with a window size of five.

Note that for the game of Pong, the maximum episode return is 21. We can observe that the agent converges to an optimal policy after 10 million training environment frames (corresponding to 2.5 million frames that the agent has actually seen after frame skipping).

Figure 7.22 shows the performance of the DQN agent on the Atari game Breakout, which is a slightly more challenging game than Pong. We use the same settings and hyperparameters as for the experiment on Pong.

Figure 7.23 shows the performance of the DQN agent on the Atari game River Raid, which is a much challenging game than Pong and Breakout. We use the same settings and hyperparameters as

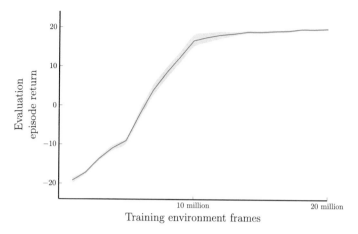

Fig. 7.21 DQN agent on Atari Pong. The results display the average episode return (total undiscounted rewards) and a 95% confidence interval. The results were averaged over five independent runs and then smoothed using a moving average with a window size of five

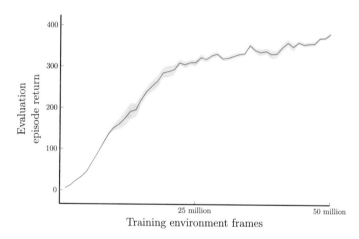

Fig. 7.22 DQN agent on Atari Breakout. The results display the average episode return (total undiscounted rewards) and a 95% confidence interval. The results were averaged over five independent runs and then smoothed using a moving average with a window size of five

for the experiment on Pong. We can see the performance is improving for the first 20 million frames and then plateaus at 20–30 million frames.

We use standard settings from the DQN Nature paper as listed in Table 7.1 for the preceding experiments.

7.7 Summary

In this chapter, we delved into the application of neural networks as nonlinear value function approximators in reinforcement learning (RL), building upon the concepts discussed in the previous chapter on linear value function approximation. Our focus was on utilizing more expressive function approximators that can effectively capture complex relationships and handle high-dimensional state spaces.

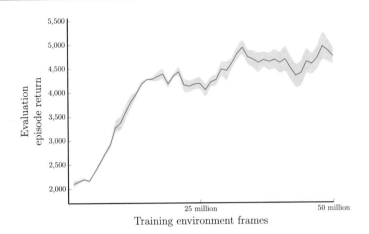

Fig. 7.23 DQN agent on Atari River Raid. The results display the average episode return (total undiscounted rewards) and a 95% confidence interval. The results were averaged over five independent runs and then smoothed using a moving average with a window size of five

Table 7.1 Parameters for training DQN agent

Hyperparameter	Value	Description
Frame width	84	Game environment frame width
Frame height	84	Game environment frame height
Grayscale	True	Game environment frame grayscale instead of RGB
Clip reward	True	Clip the reward from environment to $[-1, 1]$
Terminal on loss life	True	Treat loss of a life as a soft-terminal state; notice this does not actually reset the environment
Max noop actions	30	Maximum noop (do nothing) actions taken at the beginning of every episode
Frame skip	4	We make sure the agent only sees every k-th frame from the game by repeating the same action k times
Frame stack	4	Stack the most recent k frames
Update frequency	4	How often to sample transitions from replay and update network parameters, measured in the number of action selection done by the agent (or every 4×4 frames if we don't count frame skip)
Target network update frequency	2500	How often to update target network parameters, measured in the number of parameter updates to the standard network (the parameter C in the algorithm)
Learning rate	0.00025	The learning rate for the Adam neural network optimizer
Discount	0.99	The discount rate
Initial exploration rate	1	The exploration rate ϵ at the beginning of the training
Minimum exploration rate	0.1	The final (minimum) exploration rate ϵ after decaying over some steps; this final value is kept over the course of the training session
ϵ decay steps	1,000,000	Decay exploration rate ϵ linearly from the initial value to minimum value over the number of steps or frames
Batch size	32	Number of transitions sampled from the experience replay to update the network parameters during training
Replay buffer capacity	1,000,000	Store most recent N transitions into the experience replay

We began by exploring the architecture and components of neural networks, highlighting their remarkable ability to model nonlinear relationships. Neural networks, with their capacity to learn from data, present an exciting avenue for RL algorithms. We then delved into the training process of neural networks, covering topics such as gradient descent and backpropagation. Additionally, we provided a minimal example of training a neural network using the PyTorch deep learning framework. Understanding these training mechanisms is crucial for effectively training neural networks in RL settings.

Next, we shifted our focus to policy evaluation using neural networks. We investigated how neural networks can be utilized to estimate the value function when a fixed policy is given. Additionally, we introduced the concept of naive deep Q-learning, discussing how neural networks can be integrated into the Q-learning algorithm to approximate the action-value function. This approach equips RL agents to handle complex environments and acquire more sophisticated strategies.

Furthermore, we presented an important enhancement to the naive deep Q-learning algorithm. We examined the incorporation of experience replay and target networks, which serve to stabilize and improve learning. Experience replay allows the agent to learn from past experiences, reducing the correlation between consecutive updates. On the other hand, target networks provide a fixed target for Q-value estimation, resulting in more stable and reliable learning.

Finally, we explored the application of Deep Q-Networks (DQN) to Atari games. We showcased the impressive success achieved by combining neural networks with reinforcement learning in solving visually rich and complex environments. Particularly, we highlighted the remarkable accomplishments of the DQN algorithm in surpassing human-level performance in numerous Atari games.

In conclusion, this chapter extensively explored the powerful capabilities of neural networks as nonlinear value function approximators in reinforcement learning. From understanding their architecture and training procedures to their applications in value function approximation, neural networks offer an exciting pathway for advancing RL algorithms. The subsequent chapter will further delve into enhancements to the standard DQN algorithm, paving the way for even more effective and efficient RL agents.

References

[1] Kaiming He, Xiangyu Zhang, Shaoqing Ren, and Jian Sun. Deep residual learning for image recognition, 2015.

[2] Xavier Glorot and Yoshua Bengio. Understanding the difficulty of training deep feedforward neural networks. In *International Conference on Artificial Intelligence and Statistics*, 2010.

[3] Kaiming He, Xiangyu Zhang, Shaoqing Ren, and Jian Sun. Delving deep into rectifiers: Surpassing human-level performance on imagenet classification, 2015.

[4] Richard S. Sutton and Andrew G. Barto. *Reinforcement Learning: An Introduction*. The MIT Press, second edition, 2018.

[5] Greg Brockman, Vicki Cheung, Ludwig Pettersson, Jonas Schneider, John Schulman, Jie Tang, and Wojciech Zaremba. Openai gym, 2016.

[6] Volodymyr Mnih, Koray Kavukcuoglu, David Silver, Andrei A. Rusu, Joel Veness, Marc G. Bellemare, Alex Graves, Martin Riedmiller, Andreas K. Fidjeland, Georg Ostrovski, Stig Petersen, Charles Beattie, Amir Sadik, Ioannis Antonoglou, Helen King, Dharshan Kumaran, Daan Wierstra, Shane Legg, and Demis Hassabis. Human-level control through deep reinforcement learning. *Nature*, 518(7540):529–533, Feb 2015.

[7] Volodymyr Mnih, Koray Kavukcuoglu, David Silver, Alex Graves, Ioannis Antonoglou, Daan Wierstra, and Martin Riedmiller. Playing atari with deep reinforcement learning, 2013.

Improvements to DQN

8

The success story of the DQN agent, which achieved human-level performance in playing Atari games, marked a major breakthrough in the field of artificial intelligence and reinforcement learning. Since then, extensive research and development efforts have been dedicated to building upon the foundations of DQN. Notably, these endeavors have focused on enhancing the network architecture and refining techniques for optimal utilization of experience replay samples.

This chapter delves into three classic improvements to the standard DQN algorithm that have significantly advanced the field of reinforcement learning. These enhancements have played a pivotal role in elevating the performance and robustness of the DQN algorithm. By thoroughly examining these improvements, our goal is to gain a profound understanding of the fundamental principles that have rendered the DQN agent such a formidable asset in the realm of reinforcement learning.

8.1   DQN with Double Q-Learning

DQN (Deep Q-Network) is a type of deep reinforcement learning algorithm that has shown great success in solving complex decision-making problems. One of the improvements that can be made to the standard DQN agent is to adapt the double Q-learning method, which has been proven to be effective by the work of Hado Hasselt (2010) [1]

Double Q-learning is a method that involves learning two state-action value functions, Q_1 and Q_2, and using a slightly different notation to compute the TD target. The agent learns these two functions equally over time and uses the best action derived from one Q-network (for example Q_1) to select the value for the successor state-action pair from the other Q-network (for example Q_2). Equation (8.1) shows how double Q-learning would compute the TD target for the tabular case when we want to update Q_1, where it's using the best action a' as a selector to pick the value for successor state s_{t+1} from Q_2.

$$TD_{target} = r_t + \gamma Q_2\left(s_{t+1}, \arg\max_{a'} Q_1(s_{t+1}, a')\right) \tag{8.1}$$

However, in the case of DQN, we already have a standard Q-network with parameters θ and a target network with parameters θ^-. We can use this target network θ^- as the second state-action value function Q_2 and make some changes to the TD target computation equation to get the value update rule when using double Q-learning with DQN.

© The Author(s), under exclusive license to APress Media, LLC, part of Springer Nature 2023
M. Hu, *The Art of Reinforcement Learning*,
https://doi.org/10.1007/978-1-4842-9606-6_8

We can then make some changes to Eq. (8.1) to get the equation for how to compute the TD target when using neural networks θ, θ^- for Q_1, Q_2:

$$TD_{target} = r_t + \gamma \hat{Q}\left(s_{t+1}, \arg\max_{a'} \hat{Q}(s_{t+1}, a'; \theta); \theta^-\right) \tag{8.2}$$

We can define the objective function for using double Q-learning with neural networks θ and θ^- as a squared difference between the TD target and the estimated state-action value.

$$J(\theta) = \left(r_t + \gamma \hat{Q}\left(s_{t+1}, \arg\max_{a'} \hat{Q}(s_{t+1}, a'; \theta); \theta^-\right) - \hat{Q}(s_t, a_t; \theta)\right)^2 \tag{8.3}$$

Since the standard DQN agent will periodically update the parameters of the target network from the standard Q-network, so that $\theta^- = \theta$. We can actually skip the process of spending 50% of the time to learn the second state-action value function Q_2 as done in the tabular case. We can now introduce the DQN with double Q-learning algorithm developed by Hasselt et al. [2]. The overall is the same when compared to the standard DQN algorithm; the only difference is how we use the target network θ^- to compute the TD target. Depending on the reinforcement learning problem, we might also want to increase C which is the interval to update the target network (e.g., 2x–3x of standard DQN settings).

Algorithm 1: DQN with double Q-learning and experience replay and ϵ-greedy policy for Exploration

---

Input: Discount rate γ, learning rate α, initial exploration rate ϵ, batch size b, capacity of replay N, minimum replay size to start learning m where $m \geq b$, interval to update target network C, number of steps K

Initialize: $i = 0$, initialize the parameters θ, θ^- of the neural networks, set $\theta^- = \theta$, initialize replay buffer D to capacity N

Output: The approximated optimal state-action value function $\hat{Q}$

1 **while** $i < K$ **do**
2   Sample action a_t for s_t from ϵ-greedy policy w.r.t. θ
3   Take action a_t in environment and observe r_t, s_{t+1}
4       $i = i + 1$
5   (optional) decay ϵ, for example, linearly to a small fixed value
6   Store experience (s_t, a_t, r_t, s_{t+1}) in D
7       $s_t = s_{t+1}$
8   **if** length of $D > m$ and length of $D > b$ **then**
9   Sample a mini-batch of b transitions (s_j, a_j, r_j, s_{j+1}) randomly from D
10   Compute TD target for mini-batch

$$\delta_j = \begin{cases} r_j & \text{if } s_{j+1} \text{ is terminal state} \\ r_t + \gamma \hat{Q}\left(s_{j+1}, \arg\max_{a'} \hat{Q}(s_{j+1}, a'; \theta); \theta^-\right) & \text{otherwise bootstrapping} \end{cases}$$

11   Compute loss for mini-batch $l = \left(\delta_j - \hat{Q}(s_j, a_j; \theta)\right)^2$
12   Do stochastic gradient descent step based on loss l w.r.t. θ
13   Update target network $\theta^- = \theta$ after every C updates to θ

Figure 8.1 shows the performance of DQN with double Q-learning and standard DQN agent on the Atari River Raid game. The results display the average episode return (total undiscounted rewards) and a 95% confidence interval. To evaluate the agent's performance, we ran 200,000 evaluation steps

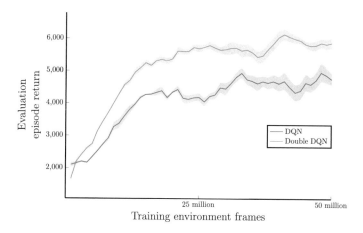

Fig. 8.1 DQN with double Q-learning vs. standard DQN on the Atari game River Raid. The results display the average episode return (total undiscounted rewards) and a 95% confidence interval. The results were averaged over five independent runs, then further smoothed using a moving average with a window size of five

on a separate testing environment with an ϵ-greedy policy and a fixed exploration epsilon ($\epsilon = 0.05$) at the end of each training iteration, which consisted of 250,000 training steps or 1 million frames, where no reward clipping or soft termination on loss of life is applied to the evaluation environment. The results were averaged over five independent runs and smoothed using a moving average with a window size of five.

We use learning rate 0.00025, discount rate 0.99, batch size 32, and replay capacity 100,000 for this experiment. We set C to 5000 which is the interval to update the target network. Figure 8.1 shows the episode return (averaged over five independent runs) for these two agents. We can see that using DQN with double Q-learning DQN gains better performance in terms of speed of convergence and training stability.

8.2   Prioritized Experience Replay

The next improvement to the standard DQN agent comes from how to sample transitions from the experience replay buffer. The current practice is for the DQN agent to sample a mini-batch of transitions randomly from the N most recent transitions. However, it's reasonable to think that some transitions in the replay buffer are more valuable than others. For example, a transition could contain a state that the agent has never encountered before, or it could have a higher reward signal. In both cases, we want the agent to spend more time learning those valuable transitions or prioritize them over others. But how do we define a specific transition as more valuable than others?

One approach to assigning priority is to use the sample frequency of a transition. Transitions that have been sampled less will have a higher priority compared to others, and those that have never been sampled will have the highest priority. The intuition behind this approach is that we should pay more attention to transitions that have not been used to update the network parameters. However, this method has its limitations since simply counting the usage of these transitions does not always reflect their importance or priority. For example, the agent may overuse newly generated transitions, even if it has already learned enough knowledge regarding similar or identical situations. We want the agent to spend less time learning familiar transitions and instead focus on those that are unfamiliar and more interesting.

A systematic way to determine the priority of a transition is by using the TD error δ, as proposed by Schaul et al. [3]. The goal of training a neural network is to minimize these errors, which are typically squared. Transitions with larger TD errors should have higher priorities than those with smaller TD errors. As a reminder, the TD error is defined as the difference between the TD target and the current estimate, as discussed in Chap. 5. For DQN with a target network θ^-, we can define δ for a transition tuple (s_t, a_t, r_t, s_{t+1}) as follows:

$$\delta = TD_{target} - \hat{Q}(s_t, a_t; \boldsymbol{\theta})$$
$$= r_t + \gamma \max_{a'} \hat{Q}(s_{t+1}, a'; \boldsymbol{\theta}^-) - \hat{Q}(s_t, a_t; \boldsymbol{\theta}) \tag{8.4}$$

If we use DQN with double Q-learning, we can then define the TD error δ as

$$\delta = TD_{target} - \hat{Q}(s_t, a_t; \boldsymbol{\theta})$$
$$= r_t + \gamma \hat{Q}\left(s_{t+1}, \arg\max_{a'} \hat{Q}(s_{t+1}, a'; \boldsymbol{\theta}); \boldsymbol{\theta}^-\right) - \hat{Q}(s_t, a_t; \boldsymbol{\theta}) \tag{8.5}$$

When the agent samples a mini-batch of transitions from the replay buffer to update the parameters $\boldsymbol{\theta}$, it computes the TD errors for each transition in the mini-batch and updates their priorities in the replay buffer accordingly.

However, using this method with experience replay for learning raises an important issue. Some transitions in the replay buffer may have been generated by a worse performing agent a long time ago, since DQN is off-policy learning and we're saving a large amount of historical transitions in the replay buffer, resulting in larger TD errors due to the influence of the behavior policy at that time. Meanwhile, newly generated transitions tend to have smaller TD errors as the agent becomes better at predicting the values for different state-action pairs. If the agent samples transitions based solely on their priority, it may favor using the older transitions with larger TD errors, which could lead to overfitting.

To address this issue, Schaul et al. [3] suggest to introduce some noise into the sampling process. One approach is to use proportional prioritization, where the probability of sampling the ith transition is based on the priority of the transition, denoted by p_i, raised to the power of a variable β. Setting $\beta = 0$ corresponds to uniform random sampling, where all transitions have an equal chance of being sampled. The probability of sampling the ith transition is given by

$$P(i) = \frac{p_i^\beta}{\sum_k p_i^\beta} \tag{8.6}$$

However, since we have introduced noise into the sampling process, we need to correct for it when updating the network parameters using the sampled mini-batch. One way to do this is to use some method to re-weight the TD error term. For prioritized replay, the sampling weight for the ith transition is computed using the following equation:

$$w_i = \left(\frac{1}{N} \cdot \frac{1}{P(i)}\right)^\eta \tag{8.7}$$

Here, N is the total number of transitions in the replay buffer, and η is a variable that controls how much we want to correct the TD error term. A value of $\eta = 1$ means we want to fully correct it, while $\eta = 0$ means no correction at all. We can use some linear method to anneal η as the learning progresses, such as linear annealing from 0.4 to 1.0 as used in the original paper.

And the sampling weight vector $\boldsymbol{w}$ is just a column vector with n entries, where n is the mini-batch size.

$$
\boldsymbol{w} = \begin{pmatrix} \left(\dfrac{1}{N} \cdot \dfrac{1}{P(0)} \right)^{\eta} \\ \left(\dfrac{1}{N} \cdot \dfrac{1}{P(1)} \right)^{\eta} \\ \vdots \\ \left(\dfrac{1}{N} \cdot \dfrac{1}{P(n)} \right)^{\eta} \end{pmatrix} \tag{8.8}
$$

In practice, normalizing the sampling weights $\boldsymbol{w}$ is often necessary to achieve stable training progress. This is achieved by dividing the weights by the maximum weight value among all samples. This normalization is represented in Eq. (8.9)

$$
\frac{\boldsymbol{w}}{\max_i w_i} \tag{8.9}
$$

When a new transition is generated, we need to determine its priority. While we could compute the priority using the same approach as before, it's more common to assign a predefined priority, such as 1.0, to every new transition. Alternatively, we could assign the maximum priority the agent has encountered so far during training. We do this because we are still biased toward newly generated transitions and want the agent to use them for training. However, if a transition has a low priority (i.e., TD error), it will be sampled less frequently in the future. Additionally, assigning a maximum priority to all new transitions can save computational resources since agents typically generate millions of transitions.

We now introduce the proportional prioritization for the DQN with double Q-learning algorithm, originally developed by Schaul et al. [3]. This algorithm builds upon the DQN with double Q-learning algorithm that we introduced earlier in this chapter. The main difference is that when the agent stores a single transition in the replay buffer, it assigns a priority value to it. By default, this priority value is set to the maximum value of the priorities seen so far.

Figure 8.2 shows the performance of the prioritized double DQN and the standard DQN agent on the Atari River Raid game. The results display the average episode return (total undiscounted rewards) and a 95% confidence interval. To evaluate the agent's performance, we ran 200,000 evaluation steps on a separate testing environment with an ϵ-greedy policy and a fixed exploration epsilon ($\epsilon = 0.05$) at the end of each training iteration, which consisted of 250,000 training steps or 1 million frames, where no reward clipping or soft termination on loss of life is applied to the evaluation environment. The results were averaged over five independent runs and smoothed using a moving average with a window size of five.

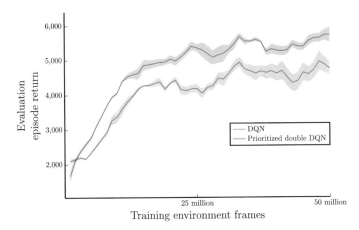

Fig. 8.2 Prioritized Double DQN vs. DQN on the Atari game River Raid. The results display the average episode return (total undiscounted rewards) and a 95% confidence interval. The results were averaged over five independent runs, then further smoothed using a moving average with a window size of five

---

Algorithm 2: Double DQN with prioritized experience replay, using ϵ-greedy policy for Exploration

---

Input: Discount rate γ, learning rate α, priority exponent β, sampling exponent η, initial exploration rate ϵ, batch size b, capacity of replay N, minimum replay size to start learning m where $m \geq b$, interval to update target network C, number of steps K

Initialize: $i = 0$, initialize the parameters θ, θ^- of the neural networks, set $\theta^- = \theta$, initialize replay buffer D to capacity N

Output: The approximated optimal state-action value function $\hat{Q}$

1 **while** $i < K$ **do**
2   Sample action a_t for s_t from ϵ-greedy policy w.r.t. θ
3   Take action a_t in environment and observe r_t, s_{t+1}
4      $i = i + 1$
5   (optional) decay ϵ, for example, linearly to a small fixed value
6   Store experience (s_t, a_t, r_t, s_{t+1}) in D with maximum priority p_{max}
7      $s_t = s_{t+1}$
8   **if** length of $D > m$ and length of $D > b$ **then**
9   Sample a mini-batch of b transitions (s_j, a_j, r_j, s_{j+1}) from D according to the priority $P(j) = \dfrac{p_j^\beta}{\sum_k p_j^\beta}$
10   Compute TD target for mini-batch

$$\delta_j = \begin{cases} r_j & \text{if } s_{j+1} \text{ is terminal state} \\ r_t + \gamma \hat{Q}\left(s_{j+1}, \arg\max_{a'} \hat{Q}(s_{j+1}, a'; \theta); \theta^-\right) & \text{otherwise bootstrapping} \end{cases}$$

11   Compute TD errors for the mini-batch $\delta = \delta_j - \hat{Q}(s_j, a_j; \theta)$
12   Compute the sampling weights $w = \dfrac{(N \cdot P(j))^{-\eta}}{\max_i w_i}$
13   Re-weight TD errors by sampling weights and compute loss $l = \left(w \cdot \delta\right)^2$
14   Do stochastic gradient descent step based on loss l w.r.t. θ
15   Update priorities for mini-batch $p = |\delta|$
16   Update target network $\theta^- = \theta$ after every C updates to θ

Similar to the double DQN experiment, we use learning rate 0.00025, discount rate 0.99, batch size 32, and replay capacity 100,000 in this experiment. We update the target network every 5000 updates.

8.3 Advantage function and Dueling Network Architecture

The advantage function and dueling network architecture proposed by Wang et al. [4] is another improvement to the standard DQN algorithm.

In reinforcement learning, the state-action value function, denoted as Q_π, for a given policy π is used to determine the best action to take in a given state. The best action is chosen by selecting the action that yields the highest state-action value, that is, $\arg\max_a Q_\pi(s, a)$. However, this does not provide information on how much better one action is compared to others. To address this issue, the advantage function was introduced by Baird in 1993. The advantage function measures the gain or advantage that an agent would receive by taking a particular action a in a given state s. The advantage function is calculated by subtracting the state value $V_\pi(s)$ from the state-action value $Q_\pi(s, a)$, as shown in Eq. (8.10).

$$A_\pi(s, a) = Q_\pi(s, a) - V_\pi(s) \tag{8.10}$$

The advantage function helps us measure how much better or worse an action is compared to other actions. This is useful because the state value function V_π and state-action value function Q_π may be interested in different features based on the given state s. Since V_π takes the average over all actions, while Q_π is only interested in one specific action, we can rewrite the state-action value function using the advantage function as shown in Eq. (8.11).

$$Q_\pi(s, a) = V_\pi(s) + A_\pi(s, a) \tag{8.11}$$

When using neural networks to approximate Q_π or V_π, Equation (8.11) can be rewritten as Eq. (8.12), where $\boldsymbol{\beta}, \boldsymbol{\delta}$ represent the parameters of the approximation functions $\hat{V}, \hat{A}$.

$$\hat{Q}(s, a; \boldsymbol{\beta}, \boldsymbol{\delta}) = \hat{V}(s; \boldsymbol{\beta}) + \hat{A}(s, a; \boldsymbol{\delta}) \tag{8.12}$$

In practice, the neural network architecture can be designed so that $\hat{V}, \hat{A}$ share the same weights $\boldsymbol{\theta}$ to some extent, while $\hat{V}$ and $\hat{A}$ still have their own weights β, δ. The equation can be written in the following form:

$$\hat{Q}(s, a; \boldsymbol{\theta}, \beta, \delta) = \hat{V}(s; \boldsymbol{\theta}, \beta) + \hat{A}(s, a; \boldsymbol{\theta}, \delta) \tag{8.13}$$

However, when adapting Eq. (8.11) with DQN, there is no guarantee that the state value $\hat{V}$ will be accurate since it is not reflected in the objective function of Q-learning. To address this problem, the original dueling DQN paper proposed a solution to force the agent to learn the state value function. The resulting equation is shown in Eq. (8.14)

$$\hat{Q}(s, a; \boldsymbol{\theta}, \beta, \delta) = \hat{V}(s; \boldsymbol{\theta}, \beta) + \left(\hat{A}(s, a; \boldsymbol{\theta}, \delta) - \max_{a \in \mathcal{A}} \hat{A}(s, a; \boldsymbol{\theta}, \delta) \right) \tag{8.14}$$

More precisely, for the best action a_* where $a_* = \arg\max_{a \in \mathcal{A}} \hat{A}(s, a; \boldsymbol{\theta}, \delta)$, the advantage term $\hat{A}(s, a; \boldsymbol{\theta}, \delta) - \max_{a \in \mathcal{A}} \hat{A}(s, a; \boldsymbol{\theta}, \delta)$ in Eq. (8.14) becomes zero; thus, we get $\hat{Q}(s, a_*; \boldsymbol{\theta}, \beta, \delta) = \hat{V}(s; \boldsymbol{\theta}, \beta)$.

But in practice, we adapt a slightly different version where we use the average instead of maximum value in the advantage term as shown in Eq. (8.15), which achieves much better stability during training.

$$\hat{Q}(s, a; \boldsymbol{\theta}, \beta, \delta) = \hat{V}(s; \boldsymbol{\theta}, \beta) + \left(\hat{A}(s, a; \boldsymbol{\theta}, \delta) - \frac{1}{|\mathcal{A}|} \sum_{a \in \mathcal{A}} \hat{A}(s, a; \boldsymbol{\theta}, \delta) \right) \qquad (8.15)$$

The dueling network architecture proposed by Wang et al. [4] is a modification of the Deep Q-Network (DQN) that improves its ability to estimate state-action values for reinforcement learning problems. In the standard DQN, the neural network takes an image of the game screen as input and outputs the estimated state-action values for all possible actions. The dueling network architecture splits this network into two separate heads, one for estimating the state value and the other for estimating the advantage values for each action.

Figure 8.3 shows the architecture of the dueling network. The network takes a state s, which is an $84 \times 84 \times 4$ image, as input. The first three layers of the network are convolutional layers, and the parameters of these layers are shared (corresponding to $\boldsymbol{\theta}$ in Eq. (8.15)). The feature maps from the last hidden layer are then fed into two separate fully connected layers: one for estimating the state value function $\hat{V}$ and the other for estimating the advantage function $\hat{A}$.

The state value function head consists of a fully connected layer with one unit and no activation function to produce the estimated state value. The advantage function head consists of a fully connected layer with 18 units (the number of possible actions) and no activation function to produce the estimated advantage values for each action.

The final estimated state-action value is computed using a simple mathematical operation that combines the estimated state value and advantage values as shown in Eq. (8.15). This computation is indicated by the red box in Fig. 8.3.

The dueling network architecture can be used with the standard DQN or double DQN algorithm to sample mini-batches of transitions and update the network parameters during training. It can also be used with prioritized experience replay, which prioritizes transitions based on their estimated temporal difference error to improve the learning efficiency.

Figure 8.4 shows the performance of dueling double DQN and DQN agents on the Atari River Raid game. The results display the average episode return (total undiscounted rewards) and a 95% confidence interval. To evaluate the agent's performance, we ran 200,000 evaluation steps on a separate testing environment with an ϵ-greedy policy and a fixed exploration epsilon ($\epsilon = 0.05$) at the end of each training iteration, which consisted of 250,000 training steps or 1 million frames, where no reward clipping or soft termination on loss of life is applied to the evaluation environment. The results were averaged over five independent runs and smoothed using a moving average with a window size of five.

Similar to the preceding experiment, we use learning rate $\alpha = 0.00025$, discount rate $\gamma = 0.99$, batch size 32, and replay capacity 100,000 in this experiment. We update the target network every 5000 updates.

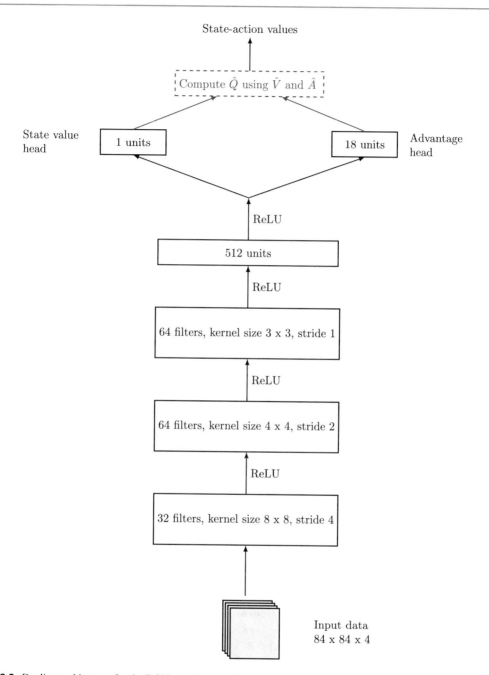

Fig. 8.3 Dueling architecture for the DQN neural network

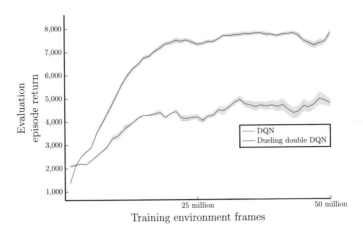

Fig. 8.4 Dueling double DQN vs. DQN on the Atari game River Raid. The results display the average episode return (total undiscounted rewards) and a 95% confidence interval. The results were averaged over five independent runs, then further smoothed using a moving average with a window size of five

8.4   Summary

In this chapter, we explored several enhancements to the standard DQN agent to address its limitations. While the DQN agent has achieved human-level performance in Atari games, it faces challenges with more complex and difficult games. By incorporating these enhancements, researchers have demonstrated significant improvements in the overall performance of the agent.

We began by introducing double Q-learning for DQN, a straightforward and simple adaptation. Double Q-learning, originally proposed by Hado Hasselt [1], and later adapted for DQN [2], helps mitigate the maximization bias associated with Q-learning. By integrating double Q-learning into the DQN agent, we observed notable performance enhancements.

Next, we delved into the concept of prioritized experience replay proposed by Schaul et al. [3], a technique that emphasizes training the agent using more important transitions rather than uniformly sampling experiences. This approach prioritizes urgent transitions and contributes to improved learning.

Furthermore, we discussed the advantage function and its incorporation into the neural network architecture of the standard DQN agent. This led us to the discussion of the dueling network architecture for the DQN agent, which exhibits remarkable performance improvements, an idea originally proposed by Wang et al. [4].

In practice, all the discussed improvements in this chapter are combined to create a significantly more robust agent. It is worth noting that, apart from the enhancements mentioned here, numerous other improvements exist. For instance, the adaptation of the N-step return (bootstrapping) method [5], as well as the exploration of using neural networks to predict the distribution of state-action values instead of single values [6,7], has also brought people's attention.

This chapter concludes Part II of the book, where we focused on value function approximation, particularly nonlinear value function approximation using neural networks. While both Parts I and II emphasized value-based methods (such as learning the state-action value function), the subsequent chapter will shift our focus to policy-based methods, an equally powerful technique for solving RL problems.

References

[1] Hado Hasselt. Double q-learning. In J. Lafferty, C. Williams, J. Shawe-Taylor, R. Zemel, and A. Culotta, editors, *Advances in Neural Information Processing Systems*, volume 23. Curran Associates, Inc., 2010.

[2] Hado van Hasselt, Arthur Guez, and David Silver. Deep reinforcement learning with double q-learning, 2015.

[3] Tom Schaul, John Quan, Ioannis Antonoglou, and David Silver. Prioritized experience replay, 2016.

[4] Ziyu Wang, Tom Schaul, Matteo Hessel, Hado van Hasselt, Marc Lanctot, and Nando de Freitas. Dueling network architectures for deep reinforcement learning, 2016.

[5] Matteo Hessel, Joseph Modayil, Hado van Hasselt, Tom Schaul, Georg Ostrovski, Will Dabney, Dan Horgan, Bilal Piot, Mohammad Azar, and David Silver. Rainbow: Combining improvements in deep reinforcement learning, 2017.

[6] Marc G. Bellemare, Will Dabney, and Rémi Munos. A distributional perspective on reinforcement learning, 2017.

[7] Will Dabney, Georg Ostrovski, David Silver, and Rémi Munos. Implicit quantile networks for distributional reinforcement learning, 2018.

Part III

Policy Approximation

Policy Gradient Methods

In this chapter, we will explore policy-based methods, which is another category of family of reinforcement learning algorithms. In previous chapters, we focused on value-based methods, which estimate the optimal state-action value function. Value-based methods can become computationally expensive for large state or action spaces, and they can struggle with environments where the dynamics of the environment are stochastic.

Policy-based methods, on the other hand, directly learn the optimal policy without estimating the state-action value function. We often use approximation techniques to parameterize the policy, such as a neural network, and use policy gradients to update the parameters. Policy gradients measure how much the expected return changes with respect to changes in the policy parameters. By optimizing the policy parameters to increase the expected return, policy-based methods learn to take actions that maximize the reward.

Policy-based methods have several advantages over value-based methods. First, they can handle continuous and high-dimensional action spaces, which value-based methods may struggle with. Second, they can learn stochastic policies, which can be useful in environments where there is uncertainty or multiple good actions.

In summary, policy-based methods provide an alternative approach to reinforcement learning that can handle large, continuous action spaces. Policy gradient algorithms like REINFORCE and Actor-Critic offer powerful ways to learn optimal policies directly from experience and can be combined with other methods to further improve performance.

9.1   Policy-Based Methods

Thus far, we have only introduced value-based reinforcement learning algorithms in this book. These algorithms aim to learn the optimal state-action value function Q_* for the optimal policy π_*, using either Monte Carlo methods or TD learning methods. We can then construct a deterministic or greedy policy based on the learned optimal state-action value function. However, the final goal is to learn the optimal policy π_*, which is what the agent need in order to make decisions when interacting with a environment. Is it possible for the agent to learn an optimal policy directly without first learning the value functions?

© The Author(s), under exclusive license to APress Media, LLC, part of Springer Nature 2023
M. Hu, *The Art of Reinforcement Learning*,
https://doi.org/10.1007/978-1-4842-9606-6_9

The answer is yes. Policy-based methods allow us to learn a parameterized policy, which is represented by a set of parameters optimized through training, rather than being defined explicitly. In this section, we will explore how to use policy gradient methods to learn such a policy.

Before we delve into policy approximation and policy gradient methods, let's take a look at some of the advantages of learning a parameterized policy directly, rather than deriving a policy from the learned value functions:

- Better convergence properties: Value-based reinforcement learning methods may oscillate during the training process for some problems, while policy-based methods tend to have a smoother learning curve, leading to better convergence properties.
- Stochastic policies: The learned policy can be stochastic, meaning that it can output a probability distribution over the available actions. This is different from value-based methods, which typically result in a deterministic policy that always selects the action with the highest value. While it is possible to add randomness to a deterministic policy such as using the ϵ-greedy policy approach, this is not the same as learning a stochastic policy with a proportional distribution over actions. A learned stochastic policy can be especially useful in situations where the optimal policy involves taking different actions with different probabilities, such as when the agent is facing stochastic or partially observable environments.
- Exploration: There is no need to use an ϵ-greedy policy to do exploration explicitly, as the nature of the stochastic policy already accomplishes this. A learned stochastic policy can guide the agent to take a range of actions, including suboptimal ones, and help the agent discover the optimal policy.
- Efficient handling of large action spaces: Policy-based methods are better suited for problems with large action spaces, even continuous action spaces. Value-based methods need to find the action with the highest value, and computing this max operation over a large number of actions can be computationally expensive. This is especially troublesome when the action space is infinite, such as problems with continuous action spaces.

Policy-based methods offer several advantages over value-based methods in reinforcement learning, such as the ability to learn a stochastic policy that assigns probabilities to actions rather than always selecting the action with the highest value. This can be particularly useful in situations where the optimal policy may involve taking different actions with different probabilities, such as when there is uncertainty in the outcome of actions or when the optimal policy changes over time. A stochastic policy can also be valuable in problems where there are multiple optimal policies or where the agent needs to take into account the behavior of other agents.

For example, in games with hidden information or bluffing, such as poker or other card games, a stochastic policy can be particularly useful, as the agent cannot see the opponent's cards, in such cases the optimal policy may involve assigning different probabilities to different actions, rather than always following a deterministic policy. Similarly, in games like Go, where the optimal policy may depend on the opponent's strategy or intentions, which may be uncertain or changing over time, a stochastic policy can be advantageous.

Furthermore, policy-based methods are better suited to continuous action spaces because they can learn to output a probability distribution over the action space, rather than trying to find the single action with the highest value. This can be easier to work with in problems with a large or infinite action space, as the agent can explore and select actions based on the probability distribution output by the policy. On the other hand, value-based methods may struggle to find the maximum value of an infinite action space or may have difficulty exploring a large action space efficiently. Additionally, value-based methods often rely on discretizing the action space, which can lead to loss of information and reduce the effectiveness of the algorithm. Policy-based methods can avoid this issue by directly parameterizing the policy, which can take into account the full range of actions available to the agent.

However, it is worth noting that policy-based methods have some disadvantages as well. For instance, they can be more computationally expensive than value-based methods, and they may struggle with high-dimensional state spaces. Moreover, policy-based methods can be more difficult to train and may require more data to converge to an optimal policy.

In summary, policy-based methods have several advantages over value-based methods, including the ability to learn a stochastic policy, which can be particularly useful in problems with multiple optimal policies, stochastic or partially observable environments, or continuous or large action spaces. Nonetheless, both policy-based and value-based methods have their strengths and weaknesses, and the choice between them will depend on the specific problem at hand.

Policy Approximation

In reinforcement learning, a policy is a function that maps a state to a probability distribution over the entire action space. Like value function approximation, we can use parameterized functions to approximate policies. By doing so, we can learn a policy that performs well in a given environment, without needing to explicitly learn the optimal value functions first.

To represent a policy using a parameterized function, we can use linear or nonlinear functions with parameters $\boldsymbol{\theta}$. However, since the output of the policy must be a probability distribution, we need to ensure that it satisfies the properties of a probability distribution, that is, non-negativitiy.

One common solution to ensure that the output of the policy is a valid probability distribution is to use the softmax function. For reinforcement learning problems with discrete action spaces, the softmax function can be used to obtain a probability distribution over the valid actions. The softmax function is defined as follows:

$$\pi(a|s; \boldsymbol{\theta}) = \frac{\exp(\phi(s, a)^{\top}\boldsymbol{\theta})}{\sum_{a' \in \mathcal{A}} \exp(\phi(s, a')^{\top}\boldsymbol{\theta})} \tag{9.1}$$

Here, $\phi(s, a)$ is a feature vector or function that maps states and actions to real-valued features. The softmax function takes the preference of each action in the current state, $\phi(s, a)^{\top}\boldsymbol{\theta}$, and exponentiates it to obtain a positive value. The resulting values are then normalized by the sum of the exponentiated preference values for all valid actions, ensuring that the output of the softmax function is a probability distribution that is non-negative.

For example, consider a robot trying to navigate through a maze. The robot's current state is its current location in the maze, and its possible actions are moving up, down, left, or right. The robot's policy is a function that maps its current location to a probability distribution over these four actions. By using the softmax function with a parameterized feature function $\phi(s, a)$, we can learn a policy that allows the robot to navigate the maze efficiently.

For reinforcement learning problems with continuous action spaces, the Gaussian function can be used to obtain a probability distribution over the action space. The Gaussian function is defined as follows:

$$\pi(a|s; \boldsymbol{\theta}) = \frac{1}{\sqrt{2\pi}\sigma(s)} \exp\left(-\frac{(a - \mu(s, \boldsymbol{\theta}))^2}{2\sigma^2(s)}\right) \tag{9.2}$$

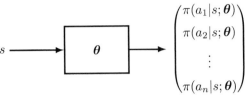

Fig. 9.1 An idea of policy approximation using neural networks, where the network takes in the environment state and outputs an action probability distribution for all valid actions

Here, $\mu(s, \boldsymbol{\theta})$ is the mean of the Gaussian distribution, $\sigma(s)$ is the standard deviation, and a is some action. The Gaussian function takes the mean and standard deviation of the distribution as input and produces a probability density function (PDF) over the action space. To obtain a probability distribution, we can integrate the PDF over the valid action space to obtain a total probability of 1.

For example, consider a drone trying to fly to a specific location in a 3D space. The drone's current state is its current position and velocity, and its possible actions are changing its velocity in three dimensions. The drone's policy is a function that maps its current state to a probability distribution over these continuous actions. By using the Gaussian function with a parameterized mean function $\mu(s, \boldsymbol{\theta})$ and standard deviation function $\sigma(s)$, we can learn a policy that allows the drone to navigate through the 3D space smoothly and reach its destination efficiently.

Similarly, we can also use neural networks to approximate a policy. As shown in Fig. 9.1, the network takes in the environment state and outputs an action probability distribution for all valid actions.

In summary, policy approximation is an important technique in reinforcement learning that allows us to learn a policy that performs well in a given environment, without needing to explicility learn the optimal value function. By using parameterized functions to represent policies, we can use the softmax or Gaussian functions to ensure that the output of the policy is a valid probability distribution. These functions are particularly useful in problems with discrete or continuous action spaces, respectively, and can be applied to a wide range of real-world problems, from maze navigation to drone flight.

9.2   Policy Gradient

In reinforcement learning, neural networks can be used to approximate both value functions and policies. However, while value-based methods use the Bellman equations to define the value update rules to the value function for a given policy, there is no such equivalent for policies parameterized by $\boldsymbol{\theta}$. This means that in order to use gradient methods to find the best parameters for a policy, we first need to define an objective function that measures its performance. This section derives the equations for the policy gradient, drawing inspiration from the remarkable work of Professor Emma Brunskill in her comprehensive course on reinforcement learning [1].

In the episodic reinforcement learning setting, we can measure the performance of a policy π_θ by its expected total reward starting from the initial state s_0. We can define the value of a policy parameterized by $\boldsymbol{\theta}$ as $V_{\pi_\theta}(s_0)$. Our goal is to find the parameters $\boldsymbol{\theta}$ that maximize this value, so we can define our objective function as

$$J(\boldsymbol{\theta}) = V_{\pi_\theta} = V_{\pi_\theta}(s_0) \tag{9.3}$$

Our task now becomes an standard optimization problem: we need to find the θ that maximizes $J(\theta)$. We can use the gradient ascent method to achieve this. Unlike gradient descent, which is used to minimize some objective function, gradient ascent involves taking steps in the direction of the steepest ascent of the function. In this case, we take a step in the direction of the gradient of $J(\theta)$, with a step size of α:

$$\theta = \theta + \alpha \nabla_\theta J(\theta) \tag{9.4}$$

The gradient $\nabla_\theta J(\theta)$ tells us how much we need to adjust each parameter θ_i in order to increase the value of the objective function. By repeatedly updating the parameters using Eq. (9.4), we can converge to the optimal policy parameters that maximize the expected total reward.

To compare the performance of different policies, we can use the state value function $V_\pi(s)$, which measures the expected total reward starting from state s when following policy π. For any two policies π and π', we say that π' is as good as or better than π, if $V_{\pi'}(s) \geq V_\pi(s)$ for all states s. This concept can be extended to the episode setting by focusing solely on the initial state s_0.

Let's expand the value function for policy π_θ. We know that the state value s_0 is just the expected return G_t (where $t = 0$) when starting from the initial state s_0 and following policy π_θ until reaching the terminal state s_T (we're only focusing on episodic cases). So the value of π_θ can be written as

$$V_{\pi_\theta} = V_{\pi_\theta}(s_0)$$

$$= \mathbb{E}_{\pi_\theta}[G_t | s_t = s_0] \tag{9.5}$$

We know the return for an episode sequence τ is the sum of discounted rewards $G(\tau) = r_0 + \gamma r_1 + \gamma^2 r_2 + \cdots + \gamma^{T-1} r_{T-1}$ (for the episodic case). The expectation sign in Eq. (9.5) is a reminder that the agent's behavior when acting in the environment and the dynamics of the environment will greatly impact the total rewards the agent can receive over the course of an episode sequence. This means the return $G(\tau)$ of an arbitrary episode sequence is influenced by two factors: the policy π_θ the agent is following and the dynamics of the environment. To keep things simple, we use τ to denote a sequence of states, actions, and rewards for a single episode, where

$$\tau = s_0, a_0, r_0, s_1, a_1, r_1, \cdots, s_{T-1}, a_{T-1}, r_{T-1}, s_T$$

We then use $G(\tau)$ to denote the sum of discounted rewards for the entire episode sequence τ, where

$$G(\tau) = \sum_{t=0}^{T-1} \gamma^t r_t$$

We use $P(\tau; \theta)$ to denote the probability of the agent obtaining an arbitrary episode sequence τ when following the policy π_θ. Specifically, $P(\tau; \theta)$ is the product of the probability of selecting actions under the policy, $\pi(a_t | s_t; \theta)$, and the probability of transitioning between states, $P(s_{t+1} | s_t, a_t)$, for each time step t in the sequence. The probability of the sequence is obtained by multiplying these probabilities across all time steps, as well as the probability of starting in a particular initial state s_0, which we represent as $\mu(s_0)$. Thus, we can rewrite the equation for the value of a policy π_θ using $P(\tau; \theta)$ and the sum of discounted rewards for the entire episode sequence, $G(\tau)$, as shown in Eq. (9.6):

$$V_{\pi_\theta} = \mathbb{E}_{\pi_\theta}[G(\tau)] = \sum_\tau P(\tau; \theta) G(\tau) \tag{9.6}$$

Here, $G(\tau)$ is the sum of discounted rewards for the entire episode sequence, and $G(\tau) = \sum_{t=0}^{T-1} \gamma^t r_t$ as defined earlier. This equation is important because it allows us to compute the value of a policy, which is a key concept in reinforcement learning.

The objective function $J(\theta)$ for the policy can then be written using the terms in Eq. (9.6). Specifically, $J(\theta)$ represents the expected return of the policy π_θ, which we want to maximize in order to find the best policy. To compute the gradients with respect to θ, we use the likelihood ratio trick, which is a well-known result in probability theory that is often used in reinforcement learning. Specifically, we have

$$
\begin{aligned}
\nabla_\theta J(\theta) &= \nabla_\theta \mathbb{E}_{\pi_\theta}[G(\tau)] \\
&= \mathbb{E}_{\pi_\theta}[\nabla_\theta G(\tau)] \\
&= \mathbb{E}_{\pi_\theta}\left[\frac{P(\tau;\theta)}{P(\tau;\theta)} \nabla_\theta G(\tau) \right] \\
&= \mathbb{E}_{\pi_\theta}\left[G(\tau) \frac{\nabla_\theta P(\tau;\theta)}{P(\tau;\theta)} \right] \qquad \text{(Likelihood ratio)} \\
&= \mathbb{E}_{\pi_\theta}\left[G(\tau) \nabla_\theta \log P(\tau;\theta) \right] \qquad\qquad (9.7)
\end{aligned}
$$

Equation (9.7) shows that to compute the gradients with respect to θ, we just need to find the derivatives for $\log P(\tau;\theta)$. However, it's important to note that the probability of starting in a particular state s_0 is often unknown and must be estimated from data. This is why we use $\mu(s_0)$ to represent the probability distribution for the initial state s_0.

We can then expand $P(\tau;\theta)$ as follows:

$$
P(\tau;\theta) = \mu(s_0) \cdot \prod_{t=0}^{T-1} \pi(a_t|s_t;\theta) \cdot \prod_{t=0}^{T-1} P(s_{t+1}|s_t, a_t) \qquad (9.8)
$$

In model-free reinforcement learning, the agent does not have access to the distribution $\mu(s_0)$ for the initial state or the dynamics of the environment $P(s_{t+1}|s_t, a_t)$. This raises the question of how we can compute the gradients with respect to θ. Fortunately, we can use the laws of logarithms to overcome this issue.

Specifically, we can apply the logarithmic identity that states that the sum of the logarithms of two numbers is equal to the logarithm of their product. By applying this identity twice, we can split Eq. (9.8) into sums of products and then obtain Eq. (9.9). This allows us to compute the gradients without having access to $\mu(s_0)$ or $P(s_{t+1}|s_t, a_t)$.

$$
\begin{aligned}
\nabla_\theta \log P(\tau;\theta) &= \nabla_\theta \log \left[\mu(s_0) \cdot \prod_{t=0}^{T-1} \pi(a_t|s_t;\theta) \cdot \prod_{t=0}^{T-1} P(s_{t+1}|s_t, a_t) \right] \\
&= \nabla_\theta \log \mu(s_0) + \nabla_\theta \log \prod_{t=0}^{T-1} \pi(a_t|s_t;\theta) + \nabla_\theta \log \prod_{t=0}^{T-1} P(s_{t+1}|s_t, a_t) \\
&= \nabla_\theta \log \mu(s_0) + \sum_{t=0}^{T-1} \nabla_\theta \log \pi(a_t|s_t;\theta) + \sum_{t=0}^{T-1} \nabla_\theta \log P(s_{t+1}|s_t, a_t) \qquad (9.9)
\end{aligned}
$$

In Eq. (9.9), we can see that the first term $\nabla_\theta \log \mu(s_0)$ does not depend on θ. This allows us to treat it as a constant. As a result, we know that the gradient with respect to a constant is zero according to basic calculus. Similarly, the last term $\sum_{t=0}^{T-1} \nabla_\theta \log P(s_{t+1}|s_t, a_t)$ is also independent of θ due to the dynamics of the environment being unaffected by changes in θ. Therefore, we can simplify the gradients of $\nabla_\theta \log P(\tau; \theta)$ with respect to θ as follows:

$$\nabla_\theta \log P(\tau; \theta) = \sum_{t=0}^{T-1} \nabla_\theta \log \pi(a_t|s_t; \theta) \tag{9.10}$$

In summary, we can simplify the gradients in Eq. (9.9) by treating the first and last terms as constants. This results in the simplified gradients shown in Eq. (9.10).

By plugging in the preceding results into Eq. (9.7), we obtain the final equation to compute the gradients (for the episode case) with respect to θ:

$$\begin{aligned}
\nabla_\theta J(\theta) &= \mathbb{E}_{\pi_\theta}\left[G(\tau)\nabla_\theta \log P(\tau; \theta)\right] \\
&= \mathbb{E}_{\pi_\theta}\left[G(\tau)\sum_{t=0}^{T-1} \nabla_\theta \log \pi(a_t|s_t; \theta)\right] \\
&= \mathbb{E}_{\pi_\theta}\left[\sum_{t=0}^{T-1} G(\tau)\nabla_\theta \log \pi(a_t|s_t; \theta)\right] \\
&= \mathbb{E}_{\pi_\theta}\left[\sum_{t=0}^{T-1} \left(\sum_{k=0}^{T-1} \gamma^k r_k\right) \nabla_\theta \log \pi(a_t|s_t; \theta)\right] \\
&= \mathbb{E}_{\pi_\theta}\left[\sum_{t=0}^{T-1} \left(\gamma^t \sum_{k=t}^{T-1} \gamma^{k-t} r_k\right) \nabla_\theta \log \pi(a_t|s_t; \theta)\right] \\
&= \mathbb{E}_{\pi_\theta}\left[\sum_{t=0}^{T-1} \gamma^t G_t \nabla_\theta \log \pi(a_t|s_t; \theta)\right] \tag{9.11}
\end{aligned}$$

The intuition behind the policy gradient as shown in Eq. (9.11) is that we want to increase the probability (or log-likelihood) of $\pi(a_t|s_t; \theta)$ in proportion to how well that action actually performed (i.e., how much reward the agent received after taking the action).

To get rid of the expectation sign in Eq. (9.11), we can use stochastic gradient ascent methods to compute sample gradients over episode sequences generated using Monte Carlo methods. Then, Eq. (9.11) is something we can compute analytically.

For example, if we're using linear methods to construct the action preference $\phi(s, a)^T \theta$ for a softmax policy distribution (as shown in Eq. (9.1)), then the policy gradients $\nabla_\theta \log \pi(a|s; \theta)$ become

$$\nabla_\theta \log \pi(a|s; \theta) = \nabla_\theta \log e^{\phi(s,a)^T \theta} - \nabla_\theta \log \left(\sum_{b \in \mathcal{A}(s)} e^{\phi(s,b)^T \theta}\right)$$

$$= \nabla_\theta \left(\phi(s, a)^T \theta\right) - \frac{\nabla_\theta \sum_b e^{\phi(s,b)^T \theta}}{\sum_b e^{\phi(s,b)^T \theta}}$$

$$= \phi(s, a) - \frac{\sum_b \phi(s, b) e^{\phi(s,b)^T \theta}}{\sum_b e^{\phi(s,b)^T \theta}}$$

$$= \phi(s, a) - \sum_b \pi(b|s; \boldsymbol{\theta}) \phi(s, b) \tag{9.12}$$

Policy Gradient Theorem

The policy gradient theorem developed by Sutton et al. [2] is a fundamental result in reinforcement learning that provides a way to compute the gradient of the performance objective with respect to the policy parameters. It is a key tool for optimizing policies in reinforcement learning problems, especially in the context of deep reinforcement learning.

The policy gradient theorem states that for any differentiable policy $\pi(a|s; \boldsymbol{\theta})$, the gradient of the performance objective $J(\boldsymbol{\theta})$ with respect to $\boldsymbol{\theta}$ is given by (without discount)

$$\nabla_{\boldsymbol{\theta}} J(\boldsymbol{\theta}) = \mathbb{E}_{\pi_{\theta}}\left[Q_{\pi_{\theta}}(s_t, a_t) \nabla_{\boldsymbol{\theta}} \log \pi(a_t|s_t; \boldsymbol{\theta})\right] \tag{9.13}$$

where $Q_{\pi_{\theta}}(s, a)$ is the state-action value function for the policy π_{θ}, and $\nabla_{\boldsymbol{\theta}} \log \pi(a|s; \boldsymbol{\theta})$ is the score function for the policy. The score function is simply the derivative of the logarithm of the policy with respect to the policy parameters:

$$\nabla_{\boldsymbol{\theta}} \log \pi(a|s; \boldsymbol{\theta}) = \frac{\nabla_{\boldsymbol{\theta}} \pi(a|s; \boldsymbol{\theta})}{\pi(a|s; \boldsymbol{\theta})} \tag{9.14}$$

The official proof of Eq. (9.13) is more involved than the derivation of Eq. (9.11) (refer to Chapter 13.2 of *Reinforcement Learning: An Introduction* by Sutton and Barto [3]). However, we can think of the policy gradient theorem as a generalized version of Eq. (9.11), where we replace the sample return G_t with the state-action value function $Q_{\pi_{\theta}}(s_t, a_t)$. This is possible since, in essence, G_t and $Q_{\pi_{\theta}}(s_t, a_t)$ measure the same thing: the expected returns.

We can define the performance objective for both episodic and continuous reinforcement learning problems as

$$\nabla_{\boldsymbol{\theta}} J(\boldsymbol{\theta}) = \mathbb{E}_{\pi_{\theta}}\left[\sum_{t=0}^{\infty} \gamma^t Q_{\pi_{\theta}}(s_t, a_t) \nabla_{\boldsymbol{\theta}} \log \pi(a_t|s_t; \boldsymbol{\theta})\right] \tag{9.15}$$

Note that Eqs. (9.13) and (9.15) are powerful results, as they allow us to optimize a policy directly using gradient-based optimization methods. In practice, we estimate the policy gradient using samples collected by interacting with the environment, using techniques such as Monte Carlo method.

9.3   REINFORCE

In this section, we'll introduce the REINFORCE algorithm, one of the earliest policy gradient algorithms developed to solving reinforcement learning problems. The algorithm was originally proposed by Williams in 1992 [4].

REINFORCE is an on-policy, model-free, and offline learning algorithm designed to solve episodic problems. The algorithm employs the Monte Carlo method to generate sample episode sequences. At the end of each episode, the algorithm updates the parameters of the policy using the rule given by Eq. (9.16):

$$\boldsymbol{\theta} \leftarrow \boldsymbol{\theta} + \alpha \gamma^t G_t \nabla_{\boldsymbol{\theta}} \log \pi(a_t|s_t; \boldsymbol{\theta}) \tag{9.16}$$

It is important to note that, unlike value-based methods, we do not need to use an ϵ-greedy policy to encourage exploration in REINFORCE. During training, the agent samples an action $a \sim \pi_{\boldsymbol{\theta}}$ according to the probabilities of each action in the action space, rather than following an ϵ-greedy policy to make decisions. This random sampling method serves as a form of exploration, allowing the agent to learn which actions are most effective. As training progresses, the agent samples the best actions more frequently and the worst actions less frequently. After training is complete, it is possible to let the trained agent act deterministically by always choosing the action with the highest probability. However, this should depend on the specific problem, since the true optimal policy may be stochastic.

Algorithm 1: REINFORCE

Input: Discount rate γ, learning rate α, number of episodes K
Initialize: $i = 0$, initialize the parameters $\boldsymbol{\theta}$ for the policy
Output: The approximated optimal policy $\pi_{\boldsymbol{\theta}}$

1 **while** $i < K$ **do**
2   Generate a sample episode τ by following policy $\pi_{\boldsymbol{\theta}}$, where $\tau = s_0, a_0, r_0, s_1, a_1, r_1, \cdots$
3   **for** $t = 0, 1, 2, \cdots, T-1$ **do**
4           $G_t = \sum_{k=t}^{T-1} \gamma^{k-t} r_k$
5   Update policy parameters $\boldsymbol{\theta} \leftarrow \boldsymbol{\theta} + \alpha \gamma^t G_t \nabla_{\boldsymbol{\theta}} \log \pi(a_t|s_t; \boldsymbol{\theta})$
6       $i = i + 1$

In practice, if we're using neural networks to approximate the policy and using the deep learning software to perform the stochastic gradient ascent step, then all we need to do is to compute policy gradient loss, which is the product between the sample return G_t and the log-likelihood of the action probability $\log \pi(a_t|s_t; \boldsymbol{\theta})$, then plug in the loss into some neural network optimizer (like SGD or Adam) to update the parameters. One thing that's worth to mention is that by default, most of these optimizer are designed to minimize some objective function (like minimize MSE or squared error), in practice we often compute the *negative* log-likelihood $-\log \pi(a_t|s_t; \boldsymbol{\theta})$ for policy gradient, so that when we use these tools to update the parameters of the neural network, it's actually doing gradient ascent (because $y = x - (-b) = x + b$), not gradient descent.

A simple neural network architecture for the policy network is shown in Fig. 9.2. The architecture consists of multiple fully connected layers, with each layer followed by a nonlinear activation function. The final output of the neural network is the predicted action probabilities for all the actions in the action space.

Note, this example is designed for very simple reinforcement learning problems. In practice, the architecture can be more complex, with multiple layers, skip connections, and other advanced techniques to improve the performance of the RL agent. Therefore, the choice of an appropriate architecture for the policy network should be carefully considered based on the specific requirements of the problem and the available resources.

Fig. 9.2 A simple
example of a neural
network architecture for
the policy network

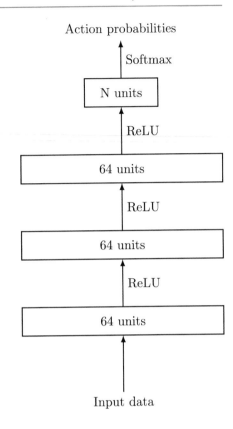

Action probabilities

Softmax

N units

ReLU

64 units

ReLU

64 units

ReLU

64 units

Input data

9.4   REINFORCE with Baseline

The REINFORCE algorithm we introduced in the last section can suffer from high variance during training, which can make it slow and inefficient. Variance arises because the algorithm uses Monte Carlo methods to estimate returns, which can vary significantly between episodes. To address this issue, we can introduce a baseline value $b(s)$ that's subtracted from the estimated returns to reduce variance.

The baseline value $b(s)$ can be any function that doesn't depend on the actions. For example, we can use a constant value or a function of the state s. By subtracting the baseline value from the estimated returns, we can reduce the impact of random fluctuations in the reward signal.

The parameter update rule for the REINFORCE algorithm with a baseline function $b(s)$ is

$$\boldsymbol{\theta} = \boldsymbol{\theta} + \alpha\gamma^t\big(G_t - b(s_t)\big)\nabla_{\boldsymbol{\theta}}\log\pi(a_t|s_t;\boldsymbol{\theta}) \tag{9.17}$$

This equation is also known as the parameter update rule for the *vanilla policy gradient algorithm*. We can replace $b(s_t)$ with any function that's suitable for reducing variance.

One common choice for the baseline function is the state value function $V_{\pi_\theta}(s_t)$, which measures the expected return starting from the state s_t when following policy π_θ. In practice, we often use the value function approximation method introduced in Part II of this book to learn an estimate of $V_{\pi_\theta}(s_t)$. This involves minimizing the squared error loss between the sample return and the estimated value, using a function approximator such as a neural network. Using the state value function as a baseline can help reduce variance because it captures the expected return from the current state.

It's important to note that using a baseline value is not a panacea and can sometimes lead to biased estimates. This is because we're subtracting a fixed value from the estimated returns, which can introduce a systematic error if the baseline value is not well chosen. Therefore, it's important to choose a baseline function that's appropriate for the problem at hand and to evaluate its effectiveness carefully.

In summary, introducing a baseline value in the REINFORCE algorithm can help reduce noise and variance during training, which makes the algorithm more efficient and effective. By choosing the right baseline function, we can further improve the performance of the algorithm. However, using a baseline value requires careful consideration and experimentation to ensure that it doesn't introduce bias or other issues.

Algorithm 2 presents the pseudocode for the REINFORCE algorithm with a state value function as the baseline, where we use linear value function approximation for the baseline function. During training, the agent learns a parameterized policy function and a parameterized value function simultaneously. It is important to note that using different learning rates for these two tasks is beneficial, as their natures are fundamentally different. Therefore, we use separate learning rates α for the policy and β for the baseline function.

Algorithm 2: REINFORCE with baseline [linear-VFA]

Input: Discount rate γ, learning rate α for policy, learning rate β for baseline function, number of episodes K
Initialize: $i = 0$, initialize the parameters θ for the policy, initialize weight vector w for the baseline $\hat{V}$
Output: The approximated optimal policy π_θ

1   **while** $i < K$ **do**
2   Generate a sample episode τ by following policy π_θ, where $\tau = s_0, a_0, r_0, s_1, a_1, r_1, \cdots$
3   **for** $t = 0, 1, 2, \cdots, T - 1$ **do**
4         $G_t = \sum_{k=t}^{T-1} \gamma^{k-t} r_k$
5         $\delta_t = G_t - \hat{V}(s_t; w)$
6   Update baseline parameters: $w \leftarrow w + \beta \delta_t \nabla_t w \hat{V}(s_t; w)$
7   Update policy parameters: $\theta \leftarrow \theta + \alpha \gamma^t \delta_t \nabla_t \theta \log \pi(a_t | s_t; \theta)$
8      $i = i + 1$

In practical applications, we often use neural networks to approximate the policy and baseline functions. Here, we use θ to denote the parameters of the policy and use ψ to denote the parameters for the baseline function. In this case, we compute the squared error loss $(G_t - \hat{V}(s_t; \psi))^2$ between the sample return G_t and the estimated value for the baseline network, and the negative log-likelihood loss $-\log \pi(a_t | s_t; \theta)$ for the policy network. It is important to accumulate gradients over a sequence of time steps when using neural networks, rather than updating the parameters on a step-by-step basis. We can then use a neural network optimizer, such as stochastic gradient descent (SGD) or Adam, to update the parameters based on the accumulated gradients.

Overall, using a parameterized policy and value function and leveraging neural networks to approximate them, can lead to more efficient and effective reinforcement learning algorithms.

Figure 9.3 shows the performance of the REINFORCE algorithm with and without baseline on the cart pole classic control task. Performance is measured in terms of the average reward obtained. To evaluate the agent's performance, we ran 20,000 evaluation steps on a separate testing environment at the end of each training iteration, which consisted of 20,000 training steps. The results were averaged over five independent runs and smoothed using a moving average with a window size of five.

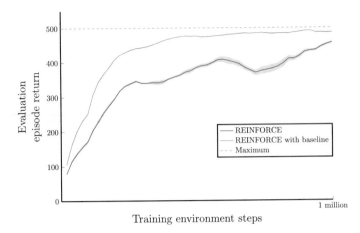

Fig. 9.3 REINFORCE with and without baseline on the cart pole control task. The results were averaged over five independent runs. We use the same discount rate $\gamma = 0.99$, learning rate $\alpha = 0.0002$ for the policy neural network for these two agents, and $\beta = 0.0005$ for the baseline network

---

Algorithm 3: REINFORCE with baseline [using neural networks]

Input: Discount rate γ, learning rate α for policy, learning rate β for baseline function, number of episodes K
Initialize: $i = 0$, initialize the parameters θ for the policy, initialize parameters ψ for the baseline $\hat{V}$
Output: The approximated optimal policy π_θ

1 while $i < K$ do
2   Generate a sample episode τ by following policy π_θ, where $\tau = s_0, a_0, r_0, s_1, a_1, r_1, \cdots$
3 $g_v = 0, g_\pi = 0$
4   for $t = 0, 1, \cdots, T-1$ do
5 $G_t = \sum_{k=t}^{T-1} \gamma^{k-t} r_k$
6   Accumulate gradients w.r.t. ψ for baseline: $g_v \leftarrow g_v + \beta \nabla_\psi \big(G_t - \hat{V}(s_t; \psi)\big)^2$
7   Accumulate gradients w.r.t. θ for policy: $g_\pi \leftarrow g_\pi + \alpha \gamma^t \big(G_t - \hat{V}(s_t; \psi)\big) \nabla_\theta \log \pi(s_t|s_t; \theta)$
8   Do stochastic gradient descent step for ψ based on g_v
9   Do stochastic gradient ascent step for θ based on g_π
10 $i = i + 1$

In this experiment, we used a very simple neural network architecture consisting of fully connected neural networks with two hidden layers. Each hidden layer has 64 units and is followed by a ReLU activation function. The output of each hidden layer is then fed to the final output layer, where we use a softmax activation function for policy and no activation function for baseline. This simple architecture was chosen to keep the focus on the reinforcement learning algorithm and to reduce computational requirements.

For both agents, we use the same discount rate of $\gamma = 0.99$ and a learning rate of $\alpha = 0.0002$ for the policy neural network. We also use a learning rate of $\beta = 0.0005$ for the baseline network. We trained the neural networks using the Adam optimizer.

The values for these hyper-parameters were chosen based on prior experiments and may not represent the best ones for the specific tasks.

9.5   Actor-Critic

One drawback of the REINFORCE algorithm and its variant using a baseline function is that it's an offline learning algorithm. This is because the agent has to wait until the end of an episode to update the policy parameters. As a result, this algorithm is only capable of solving episodic reinforcement learning problems. In this section, we introduce a new family of algorithms called *Actor-Critic*, originally proposed by Sutton and Barto [5]. It is a model-free, online, on-policy learning algorithm. With Actor-Critic, the agent doesn't have to wait until the end of the episode to update the policy parameters, which means we can use it to solve both episodic and continuing reinforcement learning problems. In Actor-Critic, we call the learned policy the *Actor* and the baseline function the *Critic*. The agent only acts according to the learned policy, but during training, the baseline function is also used to help reduce variance. In this chapter, we'll focus on using Actor-Critic algorithms to solve episodic reinforcement learning problems with discrete action spaces.

Let's first review the Bellman equation for the state value function V_π for an MDP. We can write it recursively using the immediate reward r_t and the discounted value of the successor state $\gamma V_\pi(s_{t+1})$ as shown in Eq. (9.18):

$$
\begin{aligned}
V_\pi(s) &= \mathbb{E}_\pi\left[G_t \mid s_t = s\right] \\
&= \mathbb{E}_\pi\left[r_t + \gamma V_\pi(s_{t+1}) \mid s_t = s\right]
\end{aligned}
\tag{9.18}
$$

This means we can use $r_t + \gamma V_\pi(s_{t+1})$ to replace the return G_t in the equation for policy gradients with respect to the policy parameters $\boldsymbol{\theta}$. Note that the following equation covers both episodic and continuing problems, so we use time steps $t = 0, 1, \cdots, \infty$, which is why the upper limit in the summation is ∞ in Eq. (9.19).

$$
\nabla_\theta J(\boldsymbol{\theta}) = \mathbb{E}_{\pi_\theta}\left[\sum_{t=0}^{\infty} \gamma^t G_t \nabla_\theta \log \pi(a_t \mid s_t; \boldsymbol{\theta})\right]
\tag{9.19}
$$

$$
= \mathbb{E}_{\pi_\theta}\left[\sum_{t=0}^{\infty} \gamma^t \left(r_t + \gamma V_\pi(s_{t+1})\right) \nabla_\theta \log \pi(a_t \mid s_t; \boldsymbol{\theta})\right]
\tag{9.20}
$$

With Eq. (9.20), the algorithm now becomes fully online, as we only need a single transition (s, a, r, s') to update the policy parameters. To help reduce variance, we can also include the baseline term, just like in REINFORCE with baseline. In this case, the baseline function is just the same state value function V_π we use to compute the return G_t, so Eq. (9.20) can be written as follows:

$$
\nabla_\theta J(\boldsymbol{\theta}) = \mathbb{E}_{\pi_\theta}\left[\sum_{t=0}^{\infty} \gamma^t \left(G_t - V_\pi(s_t)\right) \nabla_\theta \log \pi(a_t \mid s_t; \boldsymbol{\theta})\right]
\tag{9.21}
$$

$$
= \mathbb{E}_{\pi_\theta}\left[\sum_{t=0}^{\infty} \gamma^t \left(r_t + \gamma V_\pi(s_{t+1}) - V_\pi(s_t)\right) \nabla_\theta \log \pi(a_t \mid s_t; \boldsymbol{\theta})\right]
\tag{9.22}
$$

We can also write the policy gradient using the advantage function, in the case of using the state value function V_π as the baseline function, and using state-action value function Q_π to estimate the return G_t, as shown:

$$\nabla_\theta J(\theta) \approx \mathbb{E}_{\pi_\theta} \left[\sum_{t=0}^{\infty} \gamma^t \left(Q_\pi(s_t, a_t) - V_\pi(s_t) \right) \nabla_\theta \log \pi(a_t | s_t) \right]$$

$$\approx \mathbb{E}_{\pi_\theta} \left[\sum_{t=0}^{\infty} \gamma^t \hat{A}_\pi(s_t, a_t) \nabla_\theta \log \pi(a_t | s_t) \right] \tag{9.23}$$

where $\hat{A}_\pi$ is the advantage function, and $\hat{A}_\pi(s, a) = Q_\pi(s, a) - V_\pi(s)$.

This is possible because

$$Q_\pi(s, a) = r_t + \gamma \sum_{s' \in \mathcal{S}} P(s' \mid s, a) V_\pi(s')$$

$$\approx r_t + \gamma V_\pi(s') \tag{9.24}$$

Overall, the Actor-Critic algorithm is a powerful tool for solving reinforcement learning problems. It provides a way to learn a policy online, without waiting for the end of an episode, while also reducing variance using a baseline function. The estimated advantage function helps measure the advantage of taking actions in a given state, which can improve the performance of the algorithm.

When using neural networks θ and ψ to approximate the policy π and state value function V_π, respectively, we can express Eq. (9.20) as follows:

$$\nabla_\theta J(\theta) = \mathbb{E}_{\pi_\theta} \left[\sum_{t=0}^{\infty} \gamma^t \left(r_t + \gamma \hat{V}(s_{t+1}; \psi) - \hat{V}(s_t; \psi) \right) \nabla_\theta \log \pi(a_t | s_t; \theta) \right] \tag{9.25}$$

In practice, we often simplify this equation by removing the summation, which gives us a much simpler equation:

$$\nabla_\theta J(\theta) = \mathbb{E}_{\pi_\theta} \left[\left(r_t + \gamma \hat{V}(s_{t+1}; \psi) - \hat{V}(s_t; \psi) \right) \nabla_\theta \log \pi(a_t | s_t; \theta) \right] \tag{9.26}$$

When using neural networks to approximate the policy in reinforcement learning, updating the policy parameters on a step-by-step basis may not be ideal. One solution is to wait until the agent has collected a sequence of transitions before updating the policy parameters. In practice, we often collect a fixed length of N transitions over some time steps $t, t + 1, t + 2, \cdots, t + N$ while following the policy π_θ, where t is the starting time step of the current sequence.

The sequence of transitions is given by

$$s_t, a_t, r_t, s_{t+1}, a_{t+1}, r_{t+1}, \cdots, s_{t+N-1}, a_{t+N-1}, r_{t+N-1}, s_{t+N}$$

After the agent has collected a sequence of N transitions, we use the rewards $r_t, r_{t+1}, \cdots, r_{t+N-1}$ in the sequence to compute the return G_t for each time step $t = 0, 1, \cdots, N$ in the sequence, for example, using TD(0) methods or N-step methods, where we use the discounted value of the successor state to bootstrap.

In our case of a sequence of N transitions, we can compute the return G_t for each of the transition in the sequence, as shown in Eq. (9.27):

$$G_t = r_t + \gamma r_{t+1} + \cdots + \gamma^{N-1} r_{t+N-1} + \gamma^N \hat{V}(s_{t+N}; \boldsymbol{\psi}) \tag{9.27}$$

where G_t is the finite horizon return at time step t, r_t is the reward at time step t, γ is the discount factor, N is the number of transitions, and $\hat{V}(s_{t+N}; \boldsymbol{\psi})$ is the estimated value function for the state s_{t+N} using parameters $\boldsymbol{\psi}$.

To update the policy parameters, we follow a procedure similar to the REINFORCE with baseline algorithm. We accumulate gradients over these N transitions in the sequence, and the policy gradient can be computed using Eq. (9.28), which is adapted from Eq. (9.21).

$$\nabla_{\boldsymbol{\theta}} J(\boldsymbol{\theta}) = \sum_{t=0}^{N} \gamma^t \left(G_t - V_\pi(s_t) \right) \nabla_{\boldsymbol{\theta}} \log \pi(a_t | s_t; \boldsymbol{\theta}) \tag{9.28}$$

The objective for the value network is also similar to the REINFORCE with baseline algorithm, which is to minimize the squared error loss between the estimated return and the predicted state value, $\left(G_t - \hat{V}(s_t; \boldsymbol{\psi}) \right)^2$.

We're now ready to introduce the Actor-Critic algorithm which uses neural networks to approximate the policy and state value function. The training of the policy network (Actor) and the value network (Critic) is similar to that of the REINFORCE with baseline. Specifically, we aim to minimize the squared error loss for the estimated state values coming from the value network and maximize the log-likelihood for the policy gradient of the specific action taken by the agent, weighted by the advantage of taking the action. However, in Actor-Critic, we update the parameters based on a sequence of N transitions rather than a complete episode sequence. This allows the algorithm to learn more efficiently and update its parameters in a more timely manner.

Actor-Critic is a popular family of reinforcement learning algorithms that combine the advantages of value-based and policy-based methods. The Critic learns the value function, which estimates the expected cumulative reward given the current state and action. The Actor then leverages this information to select actions that are more likely to lead to higher rewards. The combination of these two components allows the agent to learn both the value function and the policy simultaneously.

The choice of neural network architecture for Actor-Critic depends on the complexity of the problem. In practice, we can use separate networks for the policy and value functions, or we can use a shared weights architecture.

Using two separate neural networks allows for greater flexibility in terms of applying different optimization techniques or step sizes for learning the policy and value functions respectively. It also allows for more specialized architectures tailored to each function and may perform better in scenarios where the state space is not high-dimensional. However, using two separate networks can be computationally demanding, as both networks must take in the same input data. It can also be more difficult to train and maintain due to the need for two separate networks.

Using a single neural network with two separate heads can be more computationally efficient since the network only needs to take in the input data once. It can also be easier to train and maintain since there is only one network to work with and may perform better in scenarios where the state space is visually complex. However, it may not allow for as much flexibility in terms of specialized architectures for each function and may not perform as well as using separate networks in scenarios where the policy and value functions have very different requirements.

Algorithm 4: Actor-Critic

Input: Discount rate γ, learning rate α for policy, learning rate β for value function, sequence length N, number of environment steps K

Initialize: Initialize the parameters θ for the policy, initialize parameters ψ for the state value function $\hat{V}$

Output: The approximated optimal policy π_θ

1 **while** train environment steps $< K$ **do**
2   Collect τ a sequence of N transitions (s_t, a_t, r_t);
3     $g_v = 0, g_\pi = 0$
4   **for** $i = 0, 1, \cdots, N - 1$ **do**
5   Compute return $G_i = \sum_{k=i}^{N-2} \gamma^{k-i} r_k$ from sequences in τ
6        $G_i = \begin{cases} G_i & \text{if } s_{N-1} \text{ is terminal state} \\ G_i + \gamma^{N-1} \hat{V}(s_{N-1}; \psi) & \text{otherwise bootstrapping} \end{cases}$
7   Compute advantage estimate $\delta_i = G_i - \hat{V}(s_i; \psi)$
8   Accumulate gradients w.r.t. ψ for baseline: $g_v \leftarrow g_v + \beta \nabla_\psi \delta_i^2$
9   Accumulate gradients w.r.t. θ for policy: $g_\pi \leftarrow g_\pi + \alpha \delta_i \nabla_\theta \log \pi(a_i|s_i; \theta)$
10   Do stochastic gradient descent step for ψ based on g_v
11   Do stochastic gradient ascent step for θ based on g_π

A simple example of a shared weights architecture is shown in Fig. 9.4. Here, a few shared layers are used to compute hidden features or embeddings, which are then fed into the two heads for computing the value and policy, respectively. Note that these two heads do not share any weights.

Figure 9.5 shows the performance of Actor-Critic vs. REINFORCE with baseline on the cart pole classic control task. Performance is measured in terms of the average reward obtained. To evaluate the agent's performance, we ran 20,000 evaluation steps on a separate testing environment at the end of each training iteration, which consisted of 20,000 training steps. The results were averaged over five independent runs and smoothed using a moving average with a window size of five.

We use the same neural network architecture and learning rate for these agents, and we use two separate neural networks for the Actor-Critic agent. For Actor-Critic, we let the agent collect 64 transitions before updating the parameters of the policy and baseline networks. We trained the neural networks using the Adam optimizer.

We can see that the Actor-Critic agent almost matches the performance of the REINFORCE with baseline agent.

9.6   Using Entropy to Encourage Exploration

Entropy is a concept from information theory and statistics that measures the randomness or uncertainty of a probability distribution. In general, entropy is defined as the average amount of information needed to describe the outcomes of a random variable.

For example, consider a coin flip with two possible outcomes: heads or tails. If the coin is fair, then the probability of heads is 0.5 and the probability of tails is also 0.5. The entropy of this distribution is 1 bit, which means that on average, 1 bit of information is needed to describe the outcome of a single coin flip.

In policy-based algorithms for reinforcement learning, the policy is typically represented by a neural network that outputs a probability distribution over actions given a state. The entropy of this distribution can be used as a measure of the policy's uncertainty or randomness.

The ideal of using entropy to encourage exploration was first proposed by Williams, R. J. [4]. In such case, the policy is encouraged to explore more and take more diverse actions, which can be

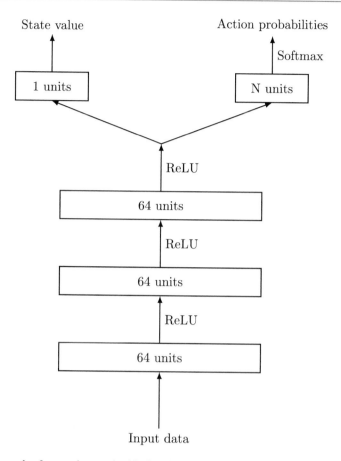

Fig. 9.4 A simple example of a neural network with shared weights architecture for the Actor-Critic algorithm

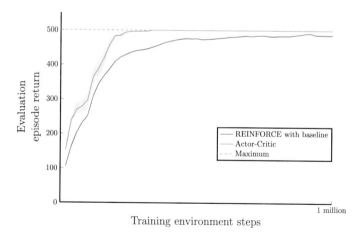

Fig. 9.5 Actor-Critic vs. REINFORCE with baseline on the cart pole control task. The results were averaged over five independent runs. We use the same discount rate $\gamma = 0.99$, learning rate $\alpha = 0.0002$ for the policy neural network for these two agents, and $\beta = 0.0005$ for the baseline network. For Actor-Critic, we use sequence length $N = 64$

useful for learning a robust policy that performs well in a variety of situations. On the other hand, low entropy means that the policy is more deterministic and less exploratory, which can lead to suboptimal policies that get stuck in local optimal.

To encourage exploration in policy gradient algorithms, a negative entropy term can be added to the policy gradient loss. This encourages the policy to take more diverse actions while still favoring actions with higher expected returns.

For reinforcement learning problems with discrete action space, we can use Eq. (9.29) to compute the entropy H for some policy π, where N is the number of actions in the action space, and $\pi(a_t|s_t)$ is the probability of taking action a_t in state s_t when following policy π.

$$H = -\sum_{k=1}^{N} \pi(a_k|s_t) \log \pi(a_k|s_t) \tag{9.29}$$

To understand why using entropy can help encourage the agent to do exploration, let's look at a simple experiment. Suppose for a particular state s, there are four actions the agent can choose from. And we have three completely different policies: π_1, π_2, and π_3. Policy π_1 is a optimal deterministic policy, which means when in state s it will always choose the best action (assume it's the first action), so the probability distribution for the different actions under policy π_1 will be something like $[1, 0, 0, 0]$. The second policy π_2 is a stochastic policy with different probability distributions under policy π_2, and the probability distribution is something like $[0.6, 0.4, 0, 0]$. The third policy π_3 is a uniform random policy, which means the probability is evenly distributed over all available actions $[0.25, 0.25, 0.25, 0.25]$. We can then use Eq. (9.29) to compute the entropy for these different policies for state s.

The entropy for these different policies is shown as follows:

- The entropy for π_1 is $H_{\pi_1} = -(1.0 * \log(1.0)) = 0$
- The entropy for π_2 is $H_{\pi_2} = -(0.6 * \log(0.6) + 0.4 * \log(0.4)) = 0.67$
- The entropy for π_3 is $H_{\pi_3} = -(0.25 * \log(0.25)) * 4) = 1.39$

We can see that for the deterministic policy π_1, the entropy is 0, since it will always choose the best action, so there's no randomness in the decision process. For stochastic policy π_2, the entropy is 0.67. And for the uniform random policy π_3, the entropy is 1.39. From this example, we can clearly see that the more randomness we have in the process, the higher the entropy.

To incorporate entropy into the policy-based method algorithm, it is added as an additional term to the objective function that is optimized during training. The objective function typically includes the policy gradient term that maximizes the expected cumulative reward, and the entropy term penalizes policies with low entropy. The relative weighting of these terms can be controlled through a coefficient w to balance exploration and exploitation, as shown in Eq. (9.30), where $H(\pi(s_t; \theta))$ is the entropy computed using Eq. (9.29) for a particular state s_t with respect to the policy π_θ.

$$\nabla_\theta J(\theta) = \sum_{t=0}^{N} \gamma^t \Big(G_t - V_\pi(s_t)\Big) \nabla_\theta \log \pi(a_t|s_t; \theta) + w H\big(\pi(s_t; \theta)\big) \tag{9.30}$$

The value of the entropy coefficient w in the entropy term depends on the specific problem at hand and requires experimentation to find the optimal value. The entropy coefficient determines the balance between exploration and exploitation in the policy, and a larger entropy coefficient encourages more exploration.

A good starting point for the entropy coefficient could be around 0.01–0.1, and we could gradually increase or decrease the value based on the performance of the algorithm. If the entropy coefficient is too small, the agent may not explore enough, and if it is too large, the agent may explore too much and not converge to an optimal policy.

In practice, sometimes it's also possible to use a decay schedule for the entropy coefficient, where the entropy coefficient starts high and gradually decreases over time as the agent learns a better policy. This approach can help the agent explore more in the early stages of learning and converge to a more optimal policy later on.

The training of the Actor-Critic algorithm with entropy loss to encourage exploration is almost the same as the standard Actor-Critic algorithm shown in Algorithm 4. The only difference is that when we compute the policy gradient we'll have to add the entropy loss by using Eq. (9.30). More precisely, we would use the following update rule to accumulate gradients with respect to θ for policy:

$$g_\pi \leftarrow g_\pi + \alpha \delta_j \nabla_\theta \log \pi(a_j|s_j; \theta) + w H\big(\pi(s_j; \theta)\big)$$

Figure 9.6 shows the performance of the Actor-Critic agent on the Atari visual complex game Pong. The results display the average episode return (total undiscounted rewards) and a 95% confidence interval. To evaluate the agent's performance, we ran 200,000 evaluation steps on a separate testing environment with a greedy policy at the end of each training iteration, which consisted of 250,000 training steps or 1 million frames, where no reward clipping or soft termination on loss of life is applied to the evaluation environment. The results were averaged over three independent runs and smoothed using a moving average with a window size of five.

We use a neural network architecture similar to DQN; however, we use the shared neural network architecture. The network has two output heads: one for the action probability and one for state value. We use discount rate $\gamma = 0.99$, learning rate $\alpha = 0.00025$, entropy weights 0.025, and sequence length 128.

We use the same environment processing as DQN, which involves resizing the frame to 84×84 and converting it to grayscale. We also apply the skip action technique, where we only process every fourth frame, and we stack the last four frames to create a final state image of size $84 \times 84 \times 4$.

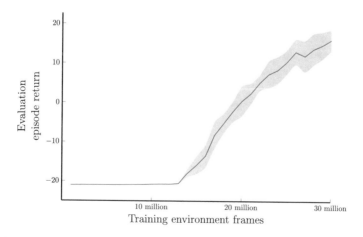

Fig. 9.6 Actor-Critic on the Atari video game Pong. The results display the average episode return (total undiscounted rewards) and a 95% confidence interval. The results were averaged over three independent runs and then smoothed using a moving average with a window size of five

Additionally, we clip the reward values to the range of -1 to 1, and we treat the loss of a life as a soft-terminal state.

Entropy loss is particularly useful in improving the performance of a reinforcement learning agent when the environment is complex. In such scenarios, the agent may struggle to explore and discover the optimal policy because the reward signal is not informative enough to guide the agent toward the goal.

In such situations, using entropy loss as a regularization term can encourage the agent to explore more widely, take more diverse actions, and learn a more robust policy. This exploration can be particularly useful for discovering regions of the state space that are critical for achieving the ultimate goal.

Moreover, entropy loss can be particularly useful in situations where the optimal policy is stochastic or requires a certain level of randomness to be effective. In these cases, enforcing high entropy can help the agent learn a more optimal policy that takes into account the inherent randomness of the environment.

9.7   Summary

In this chapter, we delved into policy-based methods for solving RL problems. Contrasting the value-based approaches covered earlier in the book, policy-based methods offer distinct advantages. They allow us to learn stochastic policies and effectively handle problems with large action spaces.

We began with a brief introduction to policy-based methods, highlighting their advantages over value-based methods. Next, we explored the concept of approximation methods for representing policies. Subsequently, we delved into the theory behind policy gradient methods, which forms the foundation for using approximation methods in policy representation. Specifically, we examined the REINFORCE algorithm, a prominent policy gradient approach that utilizes Monte Carlo sampling to update the learned policy. Moreover, we investigated the integration of a baseline function to alleviate bias and variance issues associated with the standard REINFORCE algorithm.

Furthermore, we introduced the Actor-Critic algorithm, a powerful technique that combines policy-based methods with value-based methods. This hybrid approach has demonstrated its significance in solving real-world RL problems.

To conclude the chapter, we discussed the incorporation of entropy as a mechanism to encourage exploration in policy-based methods like Actor-Critic. In the subsequent chapter, we will delve into the application of policy-based methods to solve RL problems with continuous action spaces, with a specific focus on tasks involving robotic control.

References

[1] Professor Emma Brunskill and Stanford University. Cs234: Reinforcement learning. http://web.stanford.edu/class/cs234/, 2021.

[2] Richard S Sutton, David McAllester, Satinder Singh, and Yishay Mansour. Policy gradient methods for reinforcement learning with function approximation. In S. Solla, T. Leen, and K. Müller, editors, *Advances in Neural Information Processing Systems*, volume 12. MIT Press, 1999.

[3] Richard S. Sutton and Andrew G. Barto. *Reinforcement Learning: An Introduction*. The MIT Press, second edition, 2018.

[4] Ronald J. Williams. Simple statistical gradient-following algorithms for connectionist reinforcement learning. *Machine Learning*, 8(3):229–256, May 1992.

[5] R.S. Sutton and A.G. Barto. Reinforcement learning: An introduction. *IEEE Transactions on Neural Networks*, 9(5):1054–1054, 1998.

Problems with Continuous Action Space

10

In the previous chapters, we covered reinforcement learning problems with discrete action spaces, where actions are typically represented using discrete numbers, such as integers. However, in the real world, many problems require continuous actions that cannot be represented by integers. Continuous action spaces are common in real-world applications such as robotics, where precise and continuous control is essential.

This chapter focuses on using policy-based methods to solve reinforcement learning problems with continuous action spaces. We will explore classic robotic control problems that can be solved using reinforcement learning, such as controlling the position and orientation of a robotic ant and balancing a simple human-like robot. We will see how policy-based methods can be used to achieve precise and continuous control in these systems.

By the end of this chapter, you will have a good understanding of how to use policy-based methods to solve reinforcement learning problems with continuous action spaces. Additionally, you will have learned how these methods can be applied to solve classic robotic control problems.

10.1 The Challenges of Problems with Continuous Action Space

In many real-world reinforcement learning problems, the action space is continuous, which means that the actions are represented by continuous values instead of discrete numbers. For example, in the case of controlling the angle movement of a robotic arm, we cannot use integers to represent the actions since the angle movement is often a continuous value. This presents unique challenges that require different approaches from those used for problems with discrete action space.

One of the main challenges of problems with continuous action space is that the action space is typically infinite, which makes it difficult to apply value-based methods such as Q-learning, since these methods require discretizing the action space. Discretization of the action space can result in a large loss of information and can make it difficult to learn the optimal policy. In addition, discretization of continuous action space often leads to enormous number of possible actions, which can make the learning process computationally infeasible.

Policy gradient methods are a class of reinforcement learning algorithms that are well suited for problems with continuous action space. Instead of learning an explicit action-value function, policy gradient methods learns a parameterized policy that maps from states to actions. This policy can be a neural network, for example, with the weights serving as the parameters to be learned.

Policy gradient methods are able to learn a policy that can output continuous actions directly, without the need for discretization of the action space. This allows for more fine-grained control and better performance. However, policy gradient methods can be more difficult to train than value-based methods, and convergence to the optimal policy can be slower. In addition, the variance in the gradient estimate can be high, which can make the learning process unstable. To address these challenges, various techniques such as baselines and variance reduction have been developed.

To better understand the challenges of problems with continuous action space, let's consider a simplified locomotion task. In this task, the goal is for the robotic arm to pick up objects laying on the table. The robot has multiple moving parts or joints, each with its own range of motion. All joints can and often need to be controlled simultaneously, requiring a high degree of coordination and precise control of the robot's movements. This is analogous to the movement of a human arm. To successfully complete the task of picking up an object, the robot has to coordinate the movement of its joints precisely at every time step.

Designing a suitable reward function for problems with continuous action space can be challenging as well. In the locomotion task, moving different joints will often have different results, which makes it more challenging to define a suitable reward function. The reward function needs to be carefully designed to encourage the robot to perform the desired actions while avoiding undesired actions. For example, the reward function could reward the robot for picking up the object while penalizing it for knocking over other objects on the table.

In summary, problems with continuous action space pose unique challenges that require different approaches from those used for problems with discrete action space. Policy gradient methods are suitable for continuous action spaces since they can learn a policy that can output continuous actions directly. However, designing a suitable reward function and coordinating the movement of joints precisely remains a challenge. By understanding these challenges and using appropriate techniques, we can develop effective reinforcement learning solutions for problems with continuous action space.

10.2   MuJoCo Environments

Building a robotic control environment from scratch can be a challenging and time-consuming task, requiring extensive knowledge of physics, mechanics, and robotics. Fortunately, prebuild simulation environments based on open-source software available that can help researchers and developers test their reinforcement learning algorithms without the need for extensive domain knowledge or expensive hardware.

One such tool is the Multi-Joint Dynamics with Contact (MuJoCo) physics engine, developed by Todorov et al. [1]. MuJoCo provides a fast and accurate simulation environment for a wide range of robotic systems, including manipulators and humanoid robots. MuJoCo is particularly useful for problems with continuous action spaces, which are difficult to simulate accurately using traditional physics engines.

By using prebuild simulation environments such as those available in MuJoCo and OpenAI Gym [2], researchers and developers can focus on developing and testing their algorithms without the need for extensive domain knowledge or expensive hardware. This can help accelerate the pace of research in reinforcement learning and bring us closer to developing intelligent systems that can operate in complex real-world environments.

Humanoid

The Humanoid environment[1] from the OpenAI Gym and MuJoCo is a standard benchmark for evaluating reinforcement learning algorithms that can control a 3D bipedal robot designed to simulate a human. The task is to coordinate the motion of the robot's 17 hinge joints so that it can walk forward as fast as possible without falling over.

Each action taken by the agent represents the torques applied to the joints of the humanoid, and the observations consist of positional values and velocities for different body parts of the humanoid. The torso has a pair of legs and arms, with each leg and arm consisting of two links.

The reward function for the Humanoid environment is composed of four components. The agent receives a fixed value reward every time step that the humanoid is alive, known as the "healthy reward." It also receives a reward for walking forward, measured using predefined rules, known as the "forward reward." However, the agent is penalized for using too much torque to control the joints, known as the "control cost," and for experiencing too much external contact force, known as the "contact cost."

The total reward for the agent is calculated as the sum of the healthy reward, forward reward, minus the control cost and the contact cost, given by reward = healthy reward + forward reward − control cost − contact cost. By maximizing this total reward, the agent learns to coordinate the motion of the humanoid's joints to achieve the goal of walking forward without falling over.

Ant

The Ant[2] is another 3D robot designed for simple locomotion tasks in OpenAI Gym and MuJoCo. It consists of one torso, which is a free rotational body, with four legs attached to it. Each leg has two links, and there are eight hinges connecting the two links of each leg to the torso. The goal of the Ant task is to coordinate the legs to move the ant forward by applying torques on these eight hinges.

The action space of the Ant task consists of the torques applied to the hinge joints. The observation space consists of positional values and velocities for different body parts of the ant.

The final reward for the Ant task is computed as the sum of four components: healthy reward, forward reward, minus the control cost and contact cost. The healthy reward encourages the ant to maintain its stability, the forward reward encourages the ant to move forward, the control cost penalizes the magnitude of the control inputs, and the contact cost penalizes collisions with the environment.

MuJoCo offers a wide variety of robotic control environments that we can use to study and test our algorithms for solving continuing reinforcement learning problems. In addition to the two environments we will focus on in this book, some other examples of MuJoCo environments include

- Walker2d: A bipedal robot designed to walk and run on two legs
- Hopper: A one-legged robot designed to jump and hop
- HalfCheetah: A quadrupedal robot designed to run and sprint quickly
- Swimmer: A robot designed to swim and move through water

These environments provide a diverse set of challenges for our reinforcement learning algorithms, allowing us to test their robustness and effectiveness in a variety of scenarios.

---

[1] More details of the Humanoid environment from the OpenAI Gym: https://www.gymlibrary.dev/environments/mujoco/humanoid/

[2] More details of the Ant environment from the OpenAI Gym: https://www.gymlibrary.dev/environments/mujoco/ant/

10.3   Policy Gradient for Problems with Continuous Action Space

We're now going to focus on the algorithms that can solve the classic robotic control problem we introduced earlier. So far in this book, we've covered two families of algorithms: value-based reinforcement learning and policy-based reinforcement learning. While we can use these two families of algorithms to solve reinforcement learning problems with discrete action spaces, for problems with continuous action spaces, we often use policy-based reinforcement learning methods because they are more effective as explained earlier in this chapter.

In essence, for problems with continuous action spaces, we need to learn a normal (or Gaussian) distribution, as the actions are real numbers, not just integers. The probability density function (PDF) for the normal distribution is shown in Eq. (10.1), where μ and σ are the mean and standard deviation of the normal distribution, and $\pi \approx 3.1415926$ is a constant.

$$p(x|\mu, \sigma) = \frac{1}{\sigma\sqrt{2\pi}} \exp -\frac{(x-\mu)^2}{2\sigma^2} \tag{10.1}$$

The probability density function for a normal distribution has a bell-shaped curve, like the ones shown in Fig. 10.1. The normal distribution is a good approximation for many real-world distributions because it describes real-valued random variables that cluster around a single mean value. In theory, for a normal distribution, 68% of the population or samples will be plus or minus one standard deviation σ from the population or sample mean μ, and 95% of the population or samples will stretch out between plus and minus two standard deviations (precisely 1.96) on either side of the mean. For a continuous random variable, the probability of obtaining exactly any particular value is always zero because there are an infinite number of possible values.

In the context of reinforcement learning, we know that a policy $\pi(a|s)$ is just a mapping from a state s to the probability of taking action a. To construct a policy for continuous actions based on Eq. (10.1), we need to take into consideration the state s and action a. We can replace the x in Eq. (10.1) with the action a. Next, we use parametric functions $\sigma(s, \boldsymbol{\theta}_\sigma)$ and $\mu(s, \boldsymbol{\theta}_\mu)$ to approximate the mean μ and standard deviation σ. The parameters $\boldsymbol{\theta}_\sigma$ and $\boldsymbol{\theta}_\mu$ are learned during the training process, and they represent the policy's ability to generate actions based on the observed states. The complete equation for such a policy is then shown as follows:

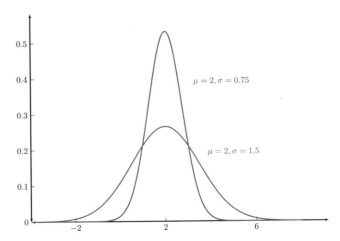

Fig. 10.1 Example of a Gaussian probability density function with different mean μ and standard deviation σ

$$\pi(a|s, \boldsymbol{\theta}_\sigma, \boldsymbol{\theta}_\mu) = \frac{1}{\sigma(s, \boldsymbol{\theta}_\sigma)\sqrt{2\pi}} \exp -\frac{(a - \mu(s, \boldsymbol{\theta}_\mu))^2}{2\sigma(s, \boldsymbol{\theta}_\sigma)^2} \tag{10.2}$$

Equation (10.2) represents a Gaussian distribution centered around the mean $\mu(s, \boldsymbol{\theta}_\mu)$ and with a standard deviation of $\sigma(s, \boldsymbol{\theta}_\sigma)$. If we know the mean and standard deviation for the policy, we can then sample actions from this distribution.

We can use any of the approximation methods introduced in Part II of the book (such as linear or nonlinear methods) to construct the mean $\mu(s, \boldsymbol{\theta}_\mu)$ and standard deviation $\sigma(s, \boldsymbol{\theta}_\sigma)$ of a normal distribution. It is important to keep in mind that the standard deviation must always be positive, so in practice we often take the exponential after the approximation. In this book, we focus specifically on using neural networks to approximate μ and σ.

In practice, it is also possible to use a single neural network with partially shared weights and two separate heads to approximate μ and σ, as shown on the left side of Fig. 10.2. By sharing some of the network parameters, we can reduce the overall number of parameters in the network, thus save computation, this could also improve the convergence speed for some tasks as we have lesser parameters to optimize. Here, N represents the dimension of the action space in the task, such as the number of joints in a robotic arm.

In some cases, the standard deviation is not directly connected to any of the hidden layers of the policy network, as shown on the right side of Fig. 10.2. Instead, it is usually treated as a separate output that is computed independent of the output of the hidden layers of the network.

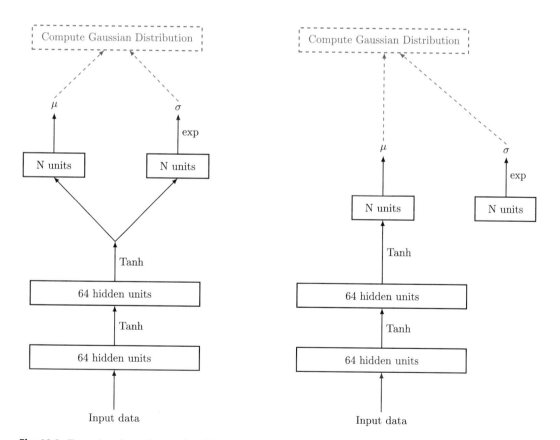

Fig. 10.2 Examples of neural network architecture to approximate the mean μ and standard deviation σ for a Gaussian distribution

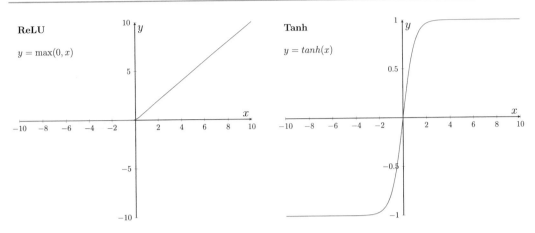

Fig. 10.3 Example of ReLU vs. Tanh activation function

One reason for this design choice is that the standard deviation is typically modeled as a non-negative scalar value, which can be more easily handled as a separate output rather than being incorporated into the activations of the hidden layers. Additionally, by separating the standard deviation from the hidden layers, the policy network can learn to adjust the level of stochasticity in the output actions without affecting the underlying structure of the network's hidden layers.

Another advantage of this approach is that it can allow for more efficient computation during training and inference. By decoupling the standard deviation from the hidden layers, the network can generate different levels of stochasticity without having to recompute the hidden layers for each level of variance.

For many locomotion tasks, we often use the Hyperbolic Tangent (Tanh) as the activation function instead of the rectified linear unit (ReLU), because Tanh has a more symmetric range that covers both positive and negative values, as shown in Fig. 10.3.

As mentioned in the previous chapter, using entropy as a regularization term can encourage the agent to explore more, which is especially beneficial for complex problems, such as those involving continuous action spaces.

Entropy is a measure of the uncertainty or randomness in a probability distribution. By adding entropy to the objective function of a reinforcement learning algorithm, we can encourage the agent to take actions that are less predictable and explore more of the state space.

For a continuous action space, the agent samples actions from a learned Gaussian distribution, where the learned network outputs the mean and standard deviation of the Gaussian distribution for each continuous action. The entropy of the Gaussian distribution is calculated using the following formula:

$$H = \frac{1}{2} \sum_{i=1}^{n} \left(\log(2\pi e \sigma_i^2) \right) \tag{10.3}$$

where n is the dimension of the continuous action space, and σ_i is the standard deviation of the action at i dimension. This formula reflects the fact that the entropy of a Gaussian distribution increases as its variance increases, which encourages exploration in the policy.

In theory, these are the only changes we need to make in order to use policy-based methods to solve reinforcement learning problems with continuous action space when compared to using the policy-based methods to solve reinforcement learning problems with discrete action space. The overall training procedure is the same when compared to the cases for discrete action space. However, the

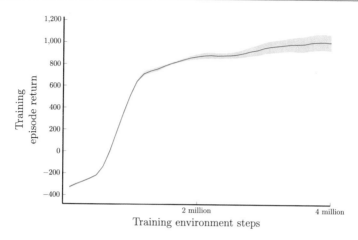

Fig. 10.4 Actor-Critic on the Ant locomotion task. The results display the average episode return (total undiscounted rewards) and a 95% confidence interval. The results were averaged over five independent runs and then smoothed using a moving average with a window size of five

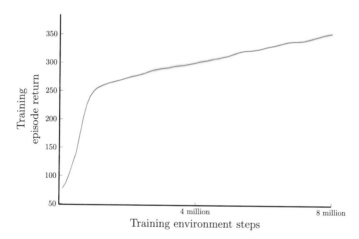

Fig. 10.5 Actor-Critic on the Humanoid locomotion task. The results display the average episode return (total undiscounted rewards) and a 95% confidence interval. The results were averaged over five independent runs and then smoothed using a moving average with a window size of five

training hyperparameters like learning rate might need to be fine-tuned for the continuous case. In practice, we may also want to do some additional work, such as normalize the environment state before feeding it into the neural network and normalize the reward signal.

Figures 10.4 and 10.5 show the performance of the Actor-Critic agent on the Ant and Humanoid locomotion tasks, respectively. The results display the average episode return (total undiscounted rewards) and a 95% confidence interval. To evaluate the agent's performance, we aggregate the average episode return at the end of each training iteration, which consisted of 100,000 training steps. The results were averaged over five independent runs and smoothed using a moving average with a window size of five.

We used separate neural networks for the policy (Actor) and the value function (Critic), with no shared weights. The policy network architecture is shown in Fig. 10.2 (left), and both the policy and value function networks had two hidden layers with Tanh activation functions.

We used a learning rate of 0.0002 for the Actor and 0.0003 for the Critic, a discount factor of 0.99, and a sequence length of 2048 before updating the parameters of the neural network. To encourage exploration, we used entropy with a weight of 0.1 in the loss function. We trained the neural networks using the Adam optimizer.

Note that the figures show the sum of episode rewards collected in the training environment, which is the same environment that the agent collects training transitions in. It is crucial to evaluate the agent's performance on the same environment used for training because even small differences between the two environments can have a significant impact on the agent's behavior due to the complexity of locomotion tasks.

10.4   Summary

In this chapter, we delved into the application of policy-based methods to tackle RL problems with continuous action spaces. While previous examples primarily focused on RL problems with discrete action spaces, certain tasks such as robotic control necessitate specialized considerations.

We started with a brief exploration of the challenges posed by continuous action spaces, highlighting why traditional value-based methods like Q-learning are ill-suited for addressing such problems. Subsequently, we introduced MuJoCo, a test bed equipped with prebuilt, straightforward robotic control tasks that serve as suitable environments for testing a variety of RL algorithms.

Next, we shifted our focus to discussing the necessary modifications required for standard policy-based methods like Actor-Critic in order to effectively train RL agents for robotic control tasks. The key alteration involves constructing neural networks capable of accurately predicting the policy's distribution and enabling the sampling of actions from this policy. Our examination of policy gradient methods will continue in the subsequent chapter, where we will introduce a highly effective algorithm called Proximal Policy Optimization (PPO), known for its state-of-the-art performance in real-world scenarios.

References

[1] Emanuel Todorov, Tom Erez, and Yuval Tassa. Mujoco: A physics engine for model-based control. In *2012 IEEE/RSJ International Conference on Intelligent Robots and Systems*, pages 5026–5033, 2012.
[2] Greg Brockman, Vicki Cheung, Ludwig Pettersson, Jonas Schneider, John Schulman, Jie Tang, and Wojciech Zaremba. Openai gym, 2016.

Advanced Policy Gradient Methods

11

One of the primary challenges associated with policy gradient methods is their instability and sensitivity to hyperparameters, such as the learning rate. This can lead to oscillations in the agent's performance, resulting in slow convergence or even divergence. Furthermore, these methods often suffer from high variance in gradient estimates, which hampers convergence speed. Moreover, standard policy gradient methods exhibit poor sample efficiency, as they only use the generated sample transitions once to train the policy.

To address these issues, this chapter explores advanced policy gradient methods, specifically focusing on the concept of surrogate objective functions. We will delve into the underlying principles of the Proximal Policy Optimization (PPO) algorithm, an exceptional state-of-the-art method that demonstrates remarkable efficiency in solving a wide range of problems. PPO incorporates the idea of a trust region to ensure stable updates and employs a surrogate objective function to reduce variance in gradient estimation, resulting in faster convergence and improved sample efficiency.

11.1 Problems with the Standard Policy Gradient Methods

Unstable Policy Updates

Policy gradient methods are a class of reinforcement learning algorithms that optimize a parameterized policy directly to maximize the expected return, which is the sum of discounted rewards received from the environment. One common issue with policy gradient methods is that they can be unstable and converge slowly, especially in high-dimensional or complex environments. In this context, "converge slowly" means that it can take a lot of iterations or training steps before the policy gradient method finds a good policy, which can lead to long training times or even failure to converge.

The main reason behind this issue is because the policy gradient method uses stochastic gradient ascent methods to update the parameters of the policy, where it uses a step size α to adjust the parameters:

$$\theta = \theta + \alpha \nabla_\theta J(\theta)$$

And choosing a good step size for the update is not very easy. If the step is too large, performance collapse is possible, where the policy becomes worse than the initial policy, and the optimization fails to converge to a good solution. This happens because the policy can move too far away from the

M. Hu, *The Art of Reinforcement Learning*,
https://doi.org/10.1007/978-1-4842-9606-6_11

optimal policy, resulting in a lower expected return. One good example of a too large step size is when the policy can oscillate around the optimal policy, never converging to a stable solution.

This is because in reinforcement learning, bad policy means bad training samples. This is especially fatal for on-policy learning methods such as standard policy gradient methods. Because once the agent starts to following some really bad policy to act in the environment, it may not be able to recover from a bad choice and collapse in performance.

Using a smaller step size in the policy optimization process is a common solution to address the stability issue. However, this approach also has its drawbacks. When using a smaller step size, the update to the policy is slower and requires more iterations to converge to the optimal solution. This results in a longer training time and increased computational resources. In addition, using a small step size can also slow down the convergence rate, making it more difficult to find the optimal policy.

Furthermore, using a small step size may also prevent the policy from exploring the state and action spaces effectively. If the step size is too small, the policy may get stuck in a local optima and fail to find the global optima. This can result in suboptimal policies that do not perform well in the given environment.

In short, using a smaller step size can reduce the risk of performance divergence, but it also slows down the learning process and can lead to suboptimal policies. A trade-off must be made between stability and efficiency in the policy optimization process.

However, the issue is not just about the right step size. The reason is the distance between two policies in policy space, defined by the space of possible policies, may not correspond to the distance between the corresponding policy parameters in parameter space. Or simply put, small changes to the policy parameters could result into very large change in the action probability.

To illustrate this problem, consider a parameterized policy $\pi(a|\theta)$ where θ is just a single scalar. The policy is defined as follows:

$$\pi(a = 1|\theta) = \frac{1}{1 + e^{-\theta}}$$

$$\pi(a = 0|\theta) = 1 - \pi(a = 1|\theta)$$

Now, let's evaluate the policy at two different parameter values, $\theta_1 = 0$ and $\theta_2 = 1$. At θ_1, the policy selects action $a = 0$ with probability $\pi(a = 0|\theta_1) = \frac{1}{2}$ and action $a = 1$ with probability $\pi(a = 1|\theta_1) = \frac{1}{2}$.

At θ_2, the policy selects action $a = 1$ with probability $\pi(a = 1|\theta_2) \approx 0.73$ and action $a = 0$ with probability $\pi(a = 0|\theta_2) \approx 0.27$.

We can see that a small change in the policy parameter from θ_1 to θ_2 results in a large change in the action probability. In fact, the probability of selecting action $a = 1$ has increased by more than 45% from θ_1 to θ_2. This highlights the sensitivity of the policy to small changes in the policy parameters and underscores the importance of careful selection of optimization methods and step sizes in policy gradient algorithms.

Poor Data Efficiency

Another problem with the standard policy gradient methods is that it has poor data efficiency. This is because policy gradient methods like Actor-Critic are on-policy learning methods. On-policy methods update the policy parameters using trajectories generated by the current policy, which can lead to slow convergence and poor sample efficiency.

Furthermore, policy gradient algorithms require a large number of trajectories to estimate the gradient accurately. This can be problematic in environments where data is expensive or time-consuming to obtain, as it can lead to poor sample efficiency. For example, in real-world applications such as robotics, it may be expensive or time-consuming to obtain a large number of trajectories due to the physical constraints of the robot.

One possible solution is to use the importance sampling ratio to address this issue. The importance sampling weight is the ratio of the probability of taking an action under the current policy to the probability of taking the same action under the distribution from which the sample was drawn (old policy).

The importance sampling ratio can be computed using the following equations:

$$\rho_t = \frac{\pi'(a_t|s_t)}{\pi(a_t|s_t)} \tag{11.1}$$

where π' is the new policy and π is the old policy used to generate the samples.

Although importance sampling seems to be useful, they might introduce other issues like exploding or vanishing importance sampling weights. This is because when the distribution from which the sample was drawn assigns a low probability to the action taken by the current policy, the importance sampling weight can become very large, leading to instability and slow convergence of the policy gradient. Similarly, when the distribution assigns a high probability to the action taken by the current policy, the importance sampling weight can become very small, leading to a vanishing gradient.

For example, consider a simple reinforcement learning problem where the agent can take two actions, a_1 and a_2, with equal probability. The reward for taking action a_1 is 1, and the reward for taking action a_2 is -1. The agent's policy is to take action a_1 with probability p and action a_2 with probability $1 - p$. Let's assume that we want to estimate the gradient of the expected reward with respect to p using importance sampling. We can sample from a uniform distribution, which assigns a probability of 0.5 to both actions.

The importance sampling weight for action a_1 is

$$w_1 = \frac{p}{0.5} = 2p$$

The importance sampling weight for action a_2 is

$$w_2 = \frac{1-p}{0.5} = 2(1-p)$$

However, if the agent's policy assigns a low probability to taking action a_1, say $p = 0.1$, then the importance sampling weight for action a_1 becomes very small:

$$w_1 = 2(0.1) = 0.2$$

Similarly, if the agent's policy assigns a high probability to taking action a_1, say $p = 0.9$, then the importance sampling weight for action a_1 becomes very large:

$$w_1 = 2(0.9) = 1.8$$

In both cases, the importance sampling weight is not stable and can lead to slow convergence of the policy gradient. This instability is known as the high variance problem. The problem can get worse if we use a large step size to update the policy parameters.

11.2   Policy Performance Bounds

We want to find a way to ensure that the updated policy is not too far from the current policy, thereby avoiding large updates that can cause instability, but not compromising the training speed, or avoiding getting stuck in local optima like what we could face when using smaller step size.

We first introduce the relative policy performance equation, as shown in Eq. (11.2). Essentially, the equation states that for any two policies π' and π, for example, π' could be the new policy, and π being the current policy, then the performance difference between these policies can be expressed as

$$J(\pi') - J(\pi) = \mathbb{E}_{\tau \sim \pi'}\left[\sum_{t=0}^{\infty} \gamma^t \hat{A}_{\pi}(s_t, a_t)\right] \tag{11.2}$$

Equation (11.2) is known as the policy improvement theorem in reinforcement learning, first introduced by Russell and Norvig [1]. The left-hand side of the equation, $J(\pi') - J(\pi)$, represents the difference in expected cumulative reward between the new policy π' and the old policy π. This is the quantity we want to maximize as we update our policy.

The right-hand side of the equation, $\mathbb{E}_{\tau \sim \pi'}\left[\sum_{t=0}^{\infty} \gamma^t \hat{A}_{\pi}(s_t, a_t)\right]$, is the expected value of the sum of the advantages under the new policy π', where τ is a trajectory of states, actions, and rewards collected following the new policy π', and $\hat{A}\pi(s_t, a_t)$ is the advantage function at time step t. The advantage function measures how much better an action is compared to the average action under the current policy π, in terms of expected cumulative reward.

The advantage function is defined as $\hat{A}_{\pi}(s, a) = Q_{\pi}(s, a) - V_{\pi}(s)$, where $Q_{\pi}(s, a)$ is the state-action value function under policy π, and $V_{\pi}(s)$ is the state value function under policy π. The state-action value function represents the expected cumulative reward of taking action a in state s and following policy π thereafter, while the state value function represents the expected cumulative reward starting from state s and following policy π thereafter.

The proof for Eq. (11.2) is shown here:

$$J(\pi') - J(\pi) = \mathbb{E}_{\tau \sim \pi'}\left[\sum_{t=0}^{\infty} \gamma^t \hat{A}_{\pi}(s_t, a_t)\right]$$

$$= \mathbb{E}_{\tau \sim \pi'}\left[\sum_{t=0}^{\infty} \gamma^t \left(R(s_t, a_t) + \gamma V_{\pi}(s_{t+1}) - V_{\pi}(s_t)\right)\right]$$

$$= J(\pi') + \mathbb{E}_{\tau \sim \pi'}\left[\sum_{t=0}^{\infty} \gamma^{t+1} V_{\pi}(s_{t+1}) - \sum_{t=0}^{\infty} \gamma^t V_{\pi}(s)\right]$$

$$= J(\pi') + \mathbb{E}_{\tau \sim \pi'}\left[\sum_{t=1}^{\infty} \gamma^t V_{\pi}(s) - \sum_{t=0}^{\infty} \gamma^t V_{\pi}(s)\right]$$

$$= J(\pi') - \mathbb{E}_{\tau \sim \pi'}[V_{\pi}(s_0)]$$

$$= J(\pi') - J(\pi)$$

Equation (11.2) suggests we can rewrite the objective for optimizing the new policy π', with respect to the current or old policy π, such that

$$\max_{\pi'} J(\pi') = \max_{\pi'} J(\pi') - J(\pi)$$

$$= \max_{\pi'} J(\pi') \mathbb{E}_{\tau \sim \pi'} \left[\sum_{t=0}^{\infty} \gamma^t \hat{A}_\pi(s_t, a_t) \right] \tag{11.3}$$

However, in the equation, we still need the trajectories sampled when following the new policy π'; this requirement makes it impossible to compute.

But let's rewrite Eq. (11.2) in terms of the discounted future state distribution d_π and use the property that the sum of an infinite geometric series with common ratio γ is $\frac{1}{1-\gamma}$, such as

$$J(\pi') - J(\pi) = \mathbb{E}_{\tau \sim \pi'} \left[\sum_{t=0}^{\infty} \gamma^t \hat{A}_\pi(s_t, a_t) \right]$$

$$= \frac{1}{1-\gamma} \mathbb{E}_{\substack{s \sim d_{\pi'} \\ a \sim \pi'}} \left[\hat{A}_\pi(s, a) \right]$$

$$= \frac{1}{1-\gamma} \mathbb{E}_{\substack{s \sim d_{\pi'} \\ a \sim \pi}} \left[\frac{\pi'(a|s)}{\pi(a|s)} \hat{A}_\pi(s, a) \right] \tag{11.4}$$

where d_π is the expected discounted frequency of visiting a state s under the policy π, and d_π is defined as

$$d_\pi = (1-\gamma) \sum_{t=0}^{\infty} \gamma^t P(s_t = s | \pi)$$

If the discounted future state distributions $d_{\pi'}$ and d_π are close, that is, $d_{\pi'} \approx d_\pi$, then we can approximate Eq. (11.4) using d_π, such as

$$J(\pi') - J(\pi) \approx \frac{1}{1-\gamma} \mathbb{E}_{\substack{s \sim d_\pi \\ a \sim \pi}} \left[\frac{\pi'(a|s)}{\pi(a|s)} \hat{A}_\pi(s, a) \right] \tag{11.5}$$

$$\doteq \mathcal{L}_\pi(\pi')$$

It turns out this approximation of $\mathcal{L}_\pi(\pi') \approx J(\pi') - J(\pi)$ produces really good results, as proved by Achiam et al. [2], it has a performance bound, if the new policy π' and old policy π are close in the policy space.

$$|J(\pi') - (J(\pi) + \mathcal{L}_\pi(\pi'))| \leq C \sqrt{\mathbb{E}_{s \sim d_\pi}[D_{KL}(\pi'||\pi)[s]]} \tag{11.6}$$

where D_{KL} is the Kullback-Leibler (KL) divergence, which measures the distance between two probability distributions. For example, given two probability distributions P and Q over some random variable, the KL divergence is computed using the following equation:

$$D_{KL}(P||Q) = \sum_x P(x) \log \frac{P(x)}{Q(x)} \tag{11.7}$$

Note the KL divergence distance is zero if the two probability distributions are the same, that is, $D_{KL}(P||Q) = 0$, if $P = Q$.

Specifically, in the context of reinforcement learning, given two policies π' and π, the KL divergence between them for a given state s can be computed using the equation:

$$D_{KL}(\pi'||\pi)[s] = \sum_{a \in \mathcal{A}} \pi'(a|s) \log \frac{\pi'(a|s)}{\pi(a|s)} \tag{11.8}$$

Monotonic Improvement Theory

We can rearrange Eq. (11.6) as

$$J(\pi') - J(\pi) \geq \mathcal{L}_\pi(\pi') - C\sqrt{\mathbb{E}_{s \sim d_\pi}[D_{KL}(\pi'||\pi)[s]]} \tag{11.9}$$

This equation is known as the performance improvement bound in reinforcement learning. On the left-hand side of the equation, $J(\pi') - J(\pi)$ represents the difference in expected cumulative reward between the new policy π' and the old policy π. This is the quantity we want to improve as we update our policy.

On the right-hand side of the equation, $\mathcal{L}_\pi(\pi') - C\sqrt{\mathbb{E}_{s \sim d_\pi}[D_{KL}(\pi'||\pi)[s]]}$ is the surrogate objective function, where $\mathcal{L}_\pi(\pi')$ is the advantage function of the policy π', and C is a hyperparameter that balances the trade-off between the expected improvement and the KL divergence penalty.

The first term, $\mathcal{L}_\pi(\pi')$, measures how much better the new policy π' is than the old policy π in terms of expected reward, based on the advantage function. The advantage function represents how much better each action is compared to the average action under the current policy and is often used to estimate the policy gradient.

The second term, $C\sqrt{\mathbb{E}_{s \sim d_\pi}[D_{KL}(\pi'||\pi)[s]]}$, penalizes the KL divergence between the new policy π' and the old policy π, weighted by the constant C. This term ensures that the new policy is not too different from the old policy, to avoid making large and potentially harmful changes.

The performance improvement bound provides a theoretical guarantee that by optimizing the surrogate objective function, we can improve the expected cumulative reward of the policy. It also gives us a nice property such that if we can maximize the terms on the right side, then we're guaranteed that the new policy π' is an improvement over the old policy π.

Surrogate Objective Function

By using this approximation of $\mathcal{L}_\pi(\pi') \approx J(\pi') - J(\pi)$, we can rearrange our initial relative policy performance equation as

$$\mathcal{L}_\pi(\pi') = \frac{1}{1-\gamma} \mathbb{E}_{\substack{s \sim d_\pi \\ a \sim \pi}} \left[\frac{\pi'(a|s)}{\pi(a|s)} \hat{A}_\pi(s, a) \right]$$

$$= \mathbb{E}_{\tau \sim \pi} \left[\sum_{t=0}^{\infty} \gamma^t \frac{\pi'(a|s)}{\pi(a|s)} \hat{A}_\pi(s, a) \right] \tag{11.10}$$

Equation (11.10) is often referred to as the surrogate objective function for policy gradient methods. This is something we can compute and optimize using trajectories sampled from the old policy π. The probability ratio term in the equation is the importance sampling, but because it only depends on the current time step and not preceding history, it does not suffer the vanish or explode weights problems as we've discussed at the beginning of this chapter.

With Eq. (11.10), our objective is then to find the new policy π', which maximizes the relative performance difference between the new policy π' and old policy π, such as

$$\underset{\pi'}{\arg\max}\, J(\pi') = \underset{\pi'}{\arg\max}\, J(\pi') - J(\pi)$$

$$= \underset{\pi'}{\arg\max}\, \mathbb{E}_{\tau\sim\pi}\left[\sum_{t=0}^{\infty} \gamma^t \frac{\pi'(a|s)}{\pi(a|s)} \hat{A}_\pi(s, a)\right] \tag{11.11}$$

However, since in Eq. (11.10) we're using an approximation solution $\mathcal{L}_\pi(\pi')$, where we use the trajectories generated from the old policy π instead of the new policy π', this only works if the new policy and old policy are close, as shown in the monotonic improvement theory.

This makes the surrogate objective function a constrained optimization problem:

$$\underset{\pi'}{\arg\max}\, \mathcal{L}_\pi(\pi') = \underset{\pi'}{\arg\max}\, \mathbb{E}_{\tau\sim\pi}\left[\sum_{t=0}^{\infty} \gamma^t \frac{\pi'(a|s)}{\pi(a|s)} \hat{A}_\pi(s, a)\right] \tag{11.12}$$

$$\text{subject to} \quad \max_{s\in\mathcal{S}} D_{KL}(\pi'||\pi) \leq \delta$$

where the constraint is defined as the maximum distance D_{KL} divergence between the new policy π' and old policy π, and $D_{KL}(\pi'||\pi) \leq \delta$.

One solution to this constrained optimization problem as defined in Eq. (11.12) is to adapt the trust region optimization methods. Figure 11.1 illustrates the concept of trust region methods, where the current (old) policy π is at the central point. A trust region is then defined around the current policy, which can be defined as the D_{KL} divergence. The trust region method would then find a step size and direction to update the policy within the constraint of this trust region.

Several algorithms have been developed to solve the surrogate objective function in the context of policy optimization in reinforcement learning, such as Natural Policy Gradient by Kakade, Sham M [3], and Trust Region Policy Optimization (TRPO) proposed by Schulman et al. [4]. However,

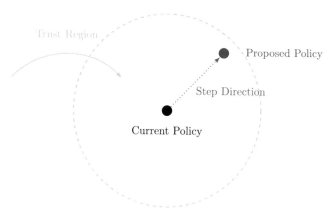

Fig. 11.1 The concept of trust region methods

these algorithms require additional approximations for $\mathcal{L}_\pi(\pi')$ and $D_{KL}(\pi'||\pi)$, respectively. This introduces additional complexity, such as the inability to use stochastic gradient ascent to optimize the policy parameters. Instead, other methods must be used to compute the gradients, which are required to update the parameters. This also reduces computation efficiency, as more resources are required to update the policy at each iteration.

Fortunately, there is a simpler and more efficient algorithm available: Proximal Policy Optimization (PPO). PPO avoids the additional complexity of other algorithms and provides a straightforward approach to optimizing the surrogate objective function.

11.3   Proximal Policy Optimization

Proximal Policy Optimization (PPO), proposed by Schulman et al. [5], is a state-of-the-art policy gradient algorithm that improves upon the stability and data efficiency of previous policy gradient methods, such as Actor-Critic [6] algorithms, while maintaining a simple implementation. PPO has been successfully applied to various reinforcement learning problems and can be scaled to handle large, complex environments that require more complex neural network architectures.

One of the key benefits of PPO is its use of stochastic gradient ascent to optimize the policy, eliminating the need for complex gradient computation process and search for suitable optimization directions, which was required in its predecessor TRPO. This approach results in faster convergence and better performance in practice.

PPO is a powerful and versatile algorithm that has proven to be a reliable and effective choice for many reinforcement learning tasks, offering a simpler implementation compared to previous methods while still achieving state-of-the-art performance.[1]

Clipped Surrogate Objective Function

PPO builds on the idea of the surrogate objective function as shown in Eq. (11.12). However, it does not require the computation of the KL divergence D_{KL} directly. To understand why this is possible, let's examine the relationship between the probability ratio and the KL divergence.

Recall that given two policies π' and π, the KL divergence between them for a given state s can be computed using the equation:

$$D_{KL}(\pi'||\pi)[s] = \sum_{a \in \mathcal{A}} \pi'(a|s) \log \frac{\pi'(a|s)}{\pi(a|s)}$$

The KL divergence measures the log difference between the probability distributions of the old policy π and the new policy π'. If the new policy is the same as the old policy, that is, $\pi' = \pi$, then the probability ratio is 1.0. As the new policy moves away from the old policy, this ratio can change dramatically. However, what is the relationship between this ratio and the KL divergence?

---

[1] The PPO algorithm has been widely utilized in various research studies. During the development of this book, OpenAI has published a research paper called "Training language models to follow instructions with human feedback" [8], in which they applied a modified version of the PPO algorithm to train a large language model (LLM) known as InstructGPT. InstructGPT is an extension of GPT-3 [9]. It is also speculated that the highly popular ChatGPT was trained using a similar approach, although OpenAI has not officially published any research paper detailing the construction of ChatGPT.

As an example, consider a simple Markov decision process (MDP) with a single state s and two actions a_1 and a_2. Suppose we have a new policy π' and an old policy π, where the probabilities of choosing actions a_1 and a_2 in state s under the old policy are both 0.5, that is, $\pi(a_1|s) = 0.5$ and $\pi(a_2|s) = 0.5$.

We can calculate the probability ratio of π' and π as $\frac{\pi'(a|s)}{\pi(a|s)}$ for each action a_1, a_2 in state s. Table 11.1 shows the probability ratios and the corresponding KL divergences for different new policies.

The results show some very interesting pattern, that is, as two policies π' and π move further apart, meaning their probability ratio deviates more from 1.0, the KL divergence also increases.

This gives us an idea that instead of computing the KL divergence, maybe we can directly apply the constraint on the probability ratio to keep the ratio close to 1.0, like shown in Eq. (11.13), where ϵ is a small variable that controls the amount of constraint we want to apply to the probability ratio.

$$clip \left(\frac{\pi'(a|s)}{\pi(a|s)}, 1 - \epsilon, 1 + \epsilon \right) \hat{A}_t \qquad (11.13)$$

However, in some cases, using the clipped probability ratio can lead to an overestimation of the advantage function, resulting in a policy that is too conservative and fails to explore new actions that could be better than those chosen by the old policy. To understand why, let's consider a simple example.

Suppose the old policy assigns a probability of 0.8 to taking action a at state s, and the advantage estimate is 0.5. The new policy assigns a probability of 0.3 to taking action a at state s. And we use $\epsilon = 0.2$, so the clipping range should be [0.8, 1.2].

Using the clipped ratio would result in a ratio of $0.3/0.8 = 0.375$, which is less than the lower bound of the clipping range. Therefore, the clipped ratio would be 0.8, and the objective function would evaluate to $0.8 * 0.5 = 0.4$.

However, if we use the original unclipped probability ratio, the objective function evaluates to $0.375 * 0.5 = 0.1875$. In this case, the clipped probability ratio results in a much higher estimate of the advantage function than using the unclipped ratio. This can lead to a policy that over estimates the values of some action, while in reality ignore other actions that yields better expected return chose by the old policy.

To address this issue, PPO also takes the element-wise minimum of the clipped objective and nonclipped objective to mitigate the problem of overestimation. This helps to mitigate the problem of overestimation and allows for better exploration of new actions. By using the minimum of the two objectives, the update will be more conservative in regions where the clipped ratio would have resulted

Table 11.1 Example of the probability ratio and KL divergence between different new policies π' and the old policy π

$\pi'(\cdot|s)$	$\pi(\cdot|s)$	$\frac{\pi'(a_1|s)}{\pi(a_1|s)}$	$\frac{\pi'(a_2|s)}{\pi(a_2|s)}$	$D_{KL}(\pi'||\pi)[s]$
[0.1, 0.9]	[0.5, 0.5]	0.2	1.8	0.37
[0.2, 0.8]	[0.5, 0.5]	0.4	1.6	0.19
[0.3, 0.7]	[0.5, 0.5]	0.6	1.4	0.08
[0.4, 0.6]	[0.5, 0.5]	0.8	1.2	0.02
[0.5, 0.5]	[0.5, 0.5]	1.0	1.0	0.0
[0.6, 0.4]	[0.5, 0.5]	1.2	0.8	0.02
[0.7, 0.3]	[0.5, 0.5]	1.4	0.6	0.08
[0.8, 0.2]	[0.5, 0.5]	1.6	0.4	0.19
[0.9, 0.1]	[0.5, 0.5]	1.8	0.2	0.37

in an overestimate, but will still allow for exploration in regions where the unclipped ratio would have resulted in a significant change in the policy.

The final clipped surrogate objective function $\mathcal{L}_\pi^{CLIP}(\pi')$ for PPO is shown in Eq. (11.14).

$$\arg\max_{\pi'} \mathcal{L}_\pi^{CLIP}(\pi') = \arg\max_{\pi'} \mathbb{E}_{\tau \sim \pi}\left[\min\left(\frac{\pi'(a|s)}{\pi(a|s)}\hat{A}_t,\ clip\left(\frac{\pi'(a|s)}{\pi(a|s)}, 1-\epsilon, 1+\epsilon\right)\hat{A}_t\right)\right]$$

(11.14)

With this clipped surrogate objective function, we can use stochastic gradient ascent methods to train the policy network as we'd do for standard policy gradient methods like Actor-Critic. There's no need to manually compute gradients and then searching for the optimization direction, as required by other algorithms like TRPO.

Generalized Advantage Estimation

PPO uses the generalized advantage estimation (GAE) method proposed by Schulman et al. [10] to compute the advantages, which is used to estimate the expected future rewards and the advantage of taking a particular action in a given state. GAE provides a more stable and low-variance estimate of the advantage function by combining the N-step returns with a bootstrapped estimate of the value function. This approach allows PPO to handle tasks with varying length, as it can estimate the advantage function over a fixed horizon, which can be adapted to the length of the task.

The advantage function measures how much better to take certain action instead of other actions, and is a key quantity in policy gradient methods like PPO. The advantage function is usually estimated using the difference between the observed rewards and the expected rewards under the current policy.

The advantage function is calculated using a truncated version of the finite horizon return method, as shown in Eq. (11.15), where λ is a parameter that controls the trade-off between bias and variance in the estimation of the advantage function. When $\lambda = 1$, GAE reduces to the standard N-step TD methods.

$$\hat{A}_t = \delta_t + (\gamma\lambda)\delta_{t+1} + (\gamma\lambda)^2\delta_{t+2} + \cdots + (\gamma\lambda)^{T-t-1}\delta_{T-1}$$

(11.15)

Here, δ_t is the temporal difference error, which is the difference between the sum of the rewards and the estimated value of the current state and the estimated value of the successor state, as shown in Eq. (11.16).

$$\delta_t = r_t + \gamma V_\pi(s_{t+1}) - V_\pi(s_t)$$

(11.16)

To recover the returns for a specific state in the sequence, we can use the GAE advantages plus the estimated state value, as shown in Eq. (11.17).

$$G_t = \hat{A}_t + V_\pi(s_t)$$

(11.17)

The GAE method has been shown to improve the stability and convergence of PPO while maintaining a good balance between bias and variance in the estimation of the advantage function. It is a powerful tool that can be used to estimate returns and advantages in a variety of reinforcement learning settings.

Log Probability Ratio

In practice, we often use the log probability ratio instead of the probability ratio $\frac{\pi'(a|s)}{\pi(a|s)}$ in the clipped surrogate objective function, which can lead to more stable and efficient optimization, especially when dealing with small or large probability ratios. One reason for this is that taking the logarithm of the probability ratio can help to reduce the range of possible values. In practice, the probability ratio can sometimes take on very small or very large values, which can cause numerical instability or underflow/overflow errors. By taking the logarithm, we can transform these values into a more manageable range.

The log probability ratio can be rewritten as

$$\frac{\pi'(a|s)}{\pi(a|s)} = \exp(\log \pi'(a|s) - \log \pi(a|s)) \tag{11.18}$$

PPO then performs M epoch parameter updates, reusing the same sequence of transitions for each update; thus, it achieves much better data efficiency. During each update, the policy parameters are optimized to maximize the clipped surrogate objective function, while the parameters of the state value function is optimized to minimize the squared error. The clip epsilon ϵ is a hyperparameter that controls the extent to which the probability ratio is clipped to prevent large policy updates. A common value for ϵ is 0.1 or 0.2. By clipping the probability ratio, PPO can prevent large policy updates that may destabilize the training process.

Algorithm 1: Proximal Policy Optimization algorithms with clipped surrogate objective

---

Input: Discount rate γ, learning rate α for policy, learning rate β for value function, sequence length N, clip epsilon ϵ, number of update epochs M, number of environment steps K

Initialize: Initialize the parameters θ, θ_{old} for the two policies, where $\theta_{old} = \theta$, initialize parameters ψ for the state value function $\hat{V}$

Output: The approximated optimized policy π_θ

1 **while** train environment steps $< K$ **do**
2   Collect τ a sequence of N transitions $(s_t, a_t, r_t, \pi(a_t|s_t; \theta_{old}))$
3   Compute returns G_t and generalized advantage estimate $\hat{A}_t$ for every transition in τ, where $t = 0, 1, \ldots, N - 1$;
4   **for** epoch $0, 1, \cdots, M$ **do**
5   Update the policy parameters by maximizing the clipped surrogate objective:
6          $\theta \leftarrow \theta + \alpha \nabla_\theta \sum_{t=0}^{N-1} \min\left(\frac{\pi(a_t|s_t; \theta)}{\pi(a_t|s_t; \theta_{old})} \hat{A}_t, \; clip\left(\frac{\pi(a_t|s_t; \theta)}{\pi(a_i|s_t; \theta_{old})}, 1 - \epsilon, 1 + \epsilon \right) \hat{A}_t \right)$
7   Update value parameters by minimizing the squared error:
8          $\psi \leftarrow \psi + \beta \nabla_\psi \sum_{t=0}^{N-1} \left(G_t - \hat{V}(s_t; \psi) \right)^2$
9      $\theta_{old} \leftarrow \theta$

In the Algorithm 1, the updates are performed in one go for all the transitions in the sequence. However, in practice, it is more efficient to perform updates using a small mini-batch of transitions at a time. This is because due to computation resource limitations, sometimes it's hard to processing all the transitions in one go, for example due to limited GPU memory.

In addition, PPO can also include an entropy term in the clipped surrogate policy gradient objective function. This can help promote exploration by encouraging the policy to choose actions with more uncertainty. Moreover, PPO can decay the clip epsilon ϵ over the training session. For example, a common strategy is to use 0.2 as the initial value for ϵ, then decay it to some fixed small value, for example annealing it linearly to 0.02 over the course of the training process. This is because when the agent starts training, it might perform randomly, so using a larger trust region for the parameter updates is necessary. However, as the agent progresses and becomes more confident, smaller trust regions are preferable to prevent overly large policy updates that may disrupt the learned policy.

Adaptive KL Penalty

The adaptive KL penalty is a variation of the Proximal Policy Optimization (PPO) algorithm that aims to prevent excessively large policy updates that could result in instability. In the standard clipped version of PPO, the probability ratio between the new and old policies is clipped within a certain range. This approach constrains the surrogate objective function and has the same effect as applying the KL divergence as a constraint.

In contrast, the adaptive KL divergence penalty approach proposes adding the KL divergence as a penalty to the surrogate objective function instead of using it as a constraint. Furthermore, the adaptive KL divergence penalty adjusts the penalty based on the magnitude of the policy update.

The surrogate objective function with the KL divergence as a penalty can be written as

$$\arg\max_{\pi'} \mathcal{L}_\pi(\pi') - \beta \bar{D}_{KL}(\pi'||\pi) \tag{11.19}$$

where β controls the level of penalty, and $\bar{D}_{KL}(\pi'||\pi)$ is the estimated mean KL divergence between the new and old policies.

Unlike the clipped version, the KL divergence is not used as a constraint on the surrogate objective function. Instead, it's used as a penalty controlled by the parameter β. If the KL divergence is high, the penalty term will also be high. Here, β is adjusted on every iteration during the learning session, making it an adaptive KL penalty. Additionally, a target KL divergence δ can be used to adjust β.

To estimate the mean KL divergence between the new policy and the old policy, an approximation is used. This approximation is necessary because computing the exact KL divergence is often infeasible due to the large state space of some environments. The target KL divergence δ can be used to adjust the β parameter, making it an adaptive KL penalty.

Algorithm 2 shows the pseudocode for the PPO algorithm with the adaptive KL divergence penalty: While both versions of PPO can work well, the clipped surrogate objective function is generally preferred in practice due to its simplicity and stability.

Simplicity refers to the fact that the clipped surrogate objective function is easier to implement and tune compared to the KL penalty version. The KL penalty version requires tuning an additional hyperparameter for the penalty term, which can be time-consuming and difficult to optimize.

Improved stability refers to the fact that the clipped surrogate objective function is less prone to instability or divergence during training compared to the KL penalty version. The KL penalty version is more sensitive to the choice of the penalty coefficient and can lead to unstable training when the coefficient is set incorrectly. The clipped surrogate objective function directly limits the size of the policy update, preventing large changes that could destabilize the training process.

In summary, while both versions of PPO's surrogate objective function can be effective, the clipped surrogate objective function is generally preferred in practice due to its simplicity and stability.

Figure 11.2 shows the performance between the PPO agent with clipped surrogate objective function and the Actor-Critic agent on the Atari video game Pong. Performance is measured in terms of the average reward obtained. To evaluate the agent's performance, we ran 200,000 evaluation

---

Algorithm 2: Proximal Policy Optimization algorithms with adaptive KL penalty

Input: Discount rate γ, learning rate α for policy, learning rate β for value function, sequence length N, initial value β, target KL divergence δ, number of update epochs M, number of environment steps K

Initialize: Initialize the parameters θ, θ_{old} for the two policies, where $\theta_{old} = \theta$, initialize parameters ψ for the state value function $\hat{V}$

Output: The approximated optimized policy π_θ

1   **while** train environment steps $< K$ **do**
2   Collect τ a sequence of N transitions $\left(s_t, a_t, r_t, \pi(a_t|s_t; \theta_{old})\right)$
3   Compute returns G_t and generalized advantage estimate $\hat{A}_t$ for every transition in τ, where $t = 0, 1, \ldots, N-1$;
4   **for** epoch $0, 1, \cdots, M$ **do**
5   Update the policy parameters by maximizing the surrogate objective:
6       $\theta \leftarrow \theta + \alpha \nabla_\theta \sum_{t=0}^{N-1} \dfrac{\pi(a_t|s_t; \theta)}{\pi(a_t|s_t; \theta_{old})} \hat{A}_t - \beta \bar{D}_{KL}(\pi_\theta \| \pi_{\theta_{old}})$
7   Update value parameters by minimizing the squared error:
8       $\psi \leftarrow \psi + \beta \nabla_\psi \sum_{t=0}^{N-1} \left(G_t - \hat{V}(s_t; \psi)\right)^2$
9
10     $\theta_{old} \leftarrow \theta$
11   **if** $\bar{D}_{KL}(\pi_\theta \| \pi_{\theta_{old}}) \geq 1.5\delta$ **then**
12       $\beta = 2\beta$
13   **else if** $\bar{D}_{KL}(\pi_\theta \| \pi_{\theta_{old}}) \leq \frac{\delta}{1.5}$ **then**
14       $\beta = \frac{\beta}{2}$

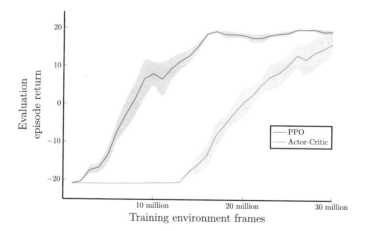

Fig. 11.2 PPO vs. Actor-Critic on Atari Pong. The results display the average episode return (total undiscounted rewards) and a 95% confidence interval. The results were averaged over three independent runs and then smoothed using a moving average with a window size of five

steps on a separate testing environment with a greedy policy at the end of each training iteration, which consisted of 250,000 training steps or 1 million frames, where no reward clipping or soft termination on loss of life is applied to the evaluation environment. The results were averaged over three independent runs and smoothed using a moving average with a window size of five.

We use the same neural network architecture and hyperparameters for both PPO and Actor-Critic agents. Specifically, we use discount rate $\gamma = 0.99$, learning rate 0.00025, and sequence length 128, and for PPO we reuse the same transition sequences to update the policy parameters over 4 epochs.

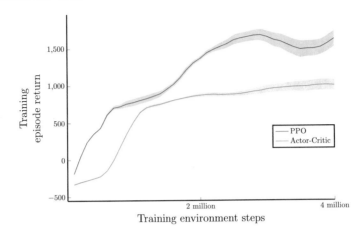

Fig. 11.3 PPO vs. Actor-Critic on the Ant classic robotic control task. The results display the average episode return (total undiscounted rewards) and a 95% confidence interval. The results were averaged over five independent runs and then smoothed using a moving average with a window size of five

We also use entropy to encourage exploration, with an entropy weight of 0.025. To keep the settings similar, we use generalized advantage estimation as introduced in Eq. (11.15) for both agents, and we use GAE lambda 0.95. The neural networks were trained using the Adam optimizer.

We use the same environment processing as DQN, which involves resizing the frame to 84×84 and converting it to grayscale. We also apply the "skip action" technique, where we only process every fourth frame, and we stack the last four frames to create a final state image of size $84 \times 84 \times 4$.

Additionally, we clip the reward values to the range of -1 to 1, and we treat the loss of a life as a "soft-terminal" state.

We can see that the PPO agent clearly has its advantage.

To show the true strength of the PPO algorithm, we set the agent to run the robotic control tasks. Figure 11.3 shows the performance between the PPO agent with the clipped surrogate objective function and the Actor-Critic agent on the Ant classic robotic control task. The results display the average episode return (total undiscounted rewards) and a 95% confidence interval. To evaluate the agent's performance, we aggregate the average episode return at the end of each training iteration, which consisted of 100,000 training steps. The results were averaged over five independent runs and smoothed using a moving average with a window size of five.

We use the same neural network architecture and hyperparameters for both PPO and Actor-Critic agents. Specifically, we use discount rate $\gamma = 0.99$, learning rate 0.0002 for actor and 0.0003 for critic, and sequence length 2048, and for PPO we use 4 update epochs. We also use entropy to encourage exploration, with an entropy weight of 0.1. To keep the settings similar, we use generalized advantage estimation as introduced in Eq. (11.15) for both agents, and we use GAE lambda 0.95. The neural networks were trained using the Adam optimizer.

Similar results can be observed in the more challenging Humanoid robotic control task, as shown in Fig. 11.4. The training and evaluation methods are identical to the previous Ant task.

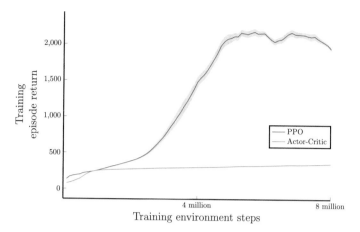

Fig. 11.4 PPO vs. Actor-Critic on the Humanoid classic robotic control task. The results display the average episode return (total undiscounted rewards) and a 95% confidence interval. The results were averaged over five independent runs and then smoothed using a moving average with a window size of five

11.4 Summary

In this chapter, we explored advanced policy gradient methods as an enhancement to standard techniques, addressing their limitations and the need for a more robust approach in policy-based methods.

We began by examining the challenges encountered when using standard policy gradient methods, such as instability during policy updates and suboptimal data efficiency.

Next, we introduced the concept of policy performance bounds, which forms the core theory underlying this chapter. We then laid the groundwork by presenting the surrogate objective function for policy gradient methods, which serves as the basis for a set of algorithms aimed at improving standard policy gradient techniques.

Our primary focus in this chapter is the Proximal Policy Optimization (PPO) algorithm. PPO represents a state-of-the-art advanced policy gradient method that demonstrates remarkable efficiency in solving a wide range of problems, encompassing both discrete and continuous action spaces. We delved into the intuition behind PPO and explained how the algorithm utilizes a clipped surrogate objective function to avoid explicit computation of the KL divergence.

The chapter concluded with a brief discussion on the adaptive KL version of PPO, a variant that offers an alternative to the clipped version. Additionally, we compared the performance of the PPO algorithm with that of standard policy gradient methods like Actor-Critic.

This chapter marks the end of our exploration into policy-based methods. We have discussed the theory behind policy gradients and various algorithms built upon this foundation. In the following chapters, we will delve into more advanced topics in RL.

References

[1] Stuart J. Russell and Peter Norvig. *Artificial Intelligence: a modern approach*. Pearson, 3rd edition, 2009.
[2] Joshua Achiam, David Held, Aviv Tamar, and Pieter Abbeel. Constrained policy optimization, 2017.
[3] Sham M Kakade. A natural policy gradient. In T. Dietterich, S. Becker, and Z. Ghahramani, editors, *Advances in Neural Information Processing Systems*, volume 14. MIT Press, 2001.

[4] John Schulman, Sergey Levine, Pieter Abbeel, Michael Jordan, and Philipp Moritz. Trust region policy optimization. In Francis Bach and David Blei, editors, *Proceedings of the 32nd International Conference on Machine Learning*, volume 37 of *Proceedings of Machine Learning Research*, pages 1889–1897, Lille, France, 07–09 Jul 2015. PMLR.

[5] John Schulman, Filip Wolski, Prafulla Dhariwal, Alec Radford, and Oleg Klimov. Proximal policy optimization algorithms, 2017.

[6] R.S. Sutton and A.G. Barto. Reinforcement learning: An introduction. *IEEE Transactions on Neural Networks*, 9(5):1054–1054, 1998.

[7] Ronald J. Williams. Simple statistical gradient-following algorithms for connectionist reinforcement learning. *Machine Learning*, 8(3):229–256, May 1992.

[8] Long Ouyang, Jeff Wu, Xu Jiang, Diogo Almeida, Carroll L. Wainwright, Pamela Mishkin, Chong Zhang, Sandhini Agarwal, Katarina Slama, Alex Ray, John Schulman, Jacob Hilton, Fraser Kelton, Luke Miller, Maddie Simens, Amanda Askell, Peter Welinder, Paul Christiano, Jan Leike, and Ryan Lowe. Training language models to follow instructions with human feedback, 2022.

[9] Tom B. Brown, Benjamin Mann, Nick Ryder, Melanie Subbiah, Jared Kaplan, Prafulla Dhariwal, Arvind Neelakantan, Pranav Shyam, Girish Sastry, Amanda Askell, Sandhini Agarwal, Ariel Herbert-Voss, Gretchen Krueger, Tom Henighan, Rewon Child, Aditya Ramesh, Daniel M. Ziegler, Jeffrey Wu, Clemens Winter, Christopher Hesse, Mark Chen, Eric Sigler, Mateusz Litwin, Scott Gray, Benjamin Chess, Jack Clark, Christopher Berner, Sam McCandlish, Alec Radford, Ilya Sutskever, and Dario Amodei. Language models are few-shot learners, 2020.

[10] John Schulman, Philipp Moritz, Sergey Levine, Michael Jordan, and Pieter Abbeel. High-dimensional continuous control using generalized advantage estimation, 2018.

Distributed Reinforcement Learning

12

This chapter explores the use of distributed reinforcement learning, which involves multiple agents running in parallel to interact with the environment to generate sample trajectories or transitions, and use samples to train the agent (e.g., to learn the optimal policy or value function). This approach offers several benefits over single-agent architectures, including faster convergence, better exploration, improved robustness, and increased scalability.

By running multiple agents in parallel, these agents can explore different parts of the environment simultaneously, leading to faster convergence and improved exploration. This can help overcome limitations of single-agent architectures, such as slow convergence and overfitting to specific parts of the environment.

In the following sections of this chapter, we will explore the specific techniques and algorithms used in distributed reinforcement learning. These techniques enable agents to share information and learn from each other's experiences, leading to more efficient and effective training of reinforcement learning agents. By the end of this chapter, you will have a better understanding of how distributed reinforcement learning can be used to train reinforcement learning agents more efficiently and effectively.

12.1   Why Use Distributed Reinforcement Learning

Reinforcement learning is a type of machine learning where an agent learns to make decisions by interacting with an environment. However, in a single-agent architecture, training an agent can be slow and may result in the agent getting stuck in local optima or overfitting to a specific part of the environment. This is because, in a single-agent architecture, the agent's ability to learn is often constrained on the experience it has gained in the specific environment, if the agent keeps making bad decisions, then it may never recover from these bad situation (state).

To overcome these challenges, distributed reinforcement learning can be used. In this approach, multiple agents work in parallel to interact with the environment. These agents can explore different parts of the environment simultaneously and share their experiences with each other, leading to faster convergence and improved overall performance.

Distributed reinforcement learning is especially useful when training deep neural networks for reinforcement learning tasks that require a large amount of data. This approach can help to efficiently utilize resources by generating sample transitions and updating neural network parameters in parallel.

For instance, if we have multiple robots learning to navigate a complex environment while avoiding obstacles, they can explore different areas of the environment using distributed reinforcement learning. This can lead to a much faster and more efficient learning process than training a single robot at a time. Moreover, this exploration can help to find novel solutions to the problem.

Overall, distributed reinforcement learning can lead to significant improvements in performance and speed when training deep neural networks for reinforcement learning tasks. It also has additional benefits, such as avoiding local optima and improving the stability of the learning process.

12.2   General Distributed Reinforcement Learning Architecture

Distributed reinforcement learning is a powerful training architecture that enables efficient use of resources in machine learning. In this architecture, two types of agents are involved: environment agents (or actor agents), which generate sample transitions by interacting with the environment, and learning agents, which are responsible for learning, for example learn the optimal policy or value function, typically through approximation using neural networks.

By separating these agents, the architecture allows for more efficient use of computation resources. Learning agents can focus solely on updating the neural network without needing to interact with the environment, which reduces computational load and speeds up the learning process. This is because, in general, accelerated hardware cannot significantly speed up interactions with the environment, whereas training a deep neural network can be beneficial if we use the accelerated hardware solutions. Additionally, the architecture is scalable to more complex and challenging problems. We can increase the number of agents on demand, allowing us to tackle problems that would otherwise be too computationally expensive to solve.

The general distributed reinforcement learning architecture is designed to speed up the process of generating training samples and training. Multiple actors (sometimes also called observers or workers) interact with the environment and send their transitions to a central server. The learner then samples a mini-batch of transitions from the central server, computes the loss based on these transitions, and updates the parameters of the neural network using stochastic gradient descent. The new parameters are then sent back to the central server for distribution to the actors.

This architecture is suitable for both on-policy and off-policy methods, but is particularly useful for on-policy methods. On-policy methods are reinforcement learning algorithms that learn the optimal policy while following the current policy. An example of an on-policy method is Proximal Policy Optimization (PPO). In contrast, off-policy methods learn the optimal policy while following a different policy. An example of an off-policy method is the Deep Q-Network (DQN), which uses experience replay to store and reuse recent sample transitions.

It is important to note that distributed reinforcement learning is not the same as multi-agent reinforcement learning. In multi-agent reinforcement learning, the agents all acts in the same environment simultaneously in a synchronized manner, while each agent might have it's own policy and objective function. In contrast, distributed reinforcement learning the agents would acts in it's own separated copy of the environment, which are often not synchronized.

Here's an example to better understand how the architecture works. Suppose we have a task of training an agent to play a game. Multiple actors interact with the game environment and generate transitions (state, action, reward). These transitions are sent to a central server. The learner samples a mini-batch of transitions from the central server, computes the loss based on these transitions, and updates the parameters of the neural network. The new parameters are then sent back to the central server for distribution to the actors. As training progresses, the learner's neural network improves and

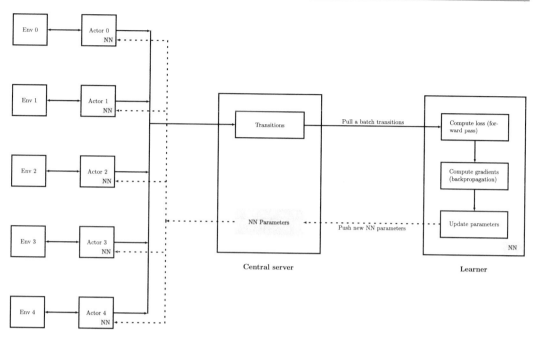

Fig. 12.1 Simple diagram to illustrate the idea of using multiple actors and a single learner agent to do distributed reinforcement learning

the actors generate better transitions. This iterative process continues until the agent learns to play the game effectively.

Here's a simplified explanation of the general distributed reinforcement learning training architecture as shown in Fig. 12.1.

In distributed reinforcement learning, we have multiple agents (called actors) that interact with their own environments. Each agent has its own copy of neural network to make decisions based on the state of the environment. We also have a learner agent that has a copy of the neural network but does not have an environment attached to it. The central server acts as a coordinator between the actors and learner agent, collecting newly generated experiences and distributing the most up-to-date parameters of the neural network to the actors.

The architecture is designed to make training more efficient by distributing the workload across multiple processors or machines. The actors can run on moderate hardware like CPUs, while the learner runs on more powerful hardware like GPUs.

During training, the actors may act differently, especially in the early stages when the policy is more random. This exploration helps to discover new and better policies for the reinforcement learning agent. The central server can also store training logs and statistics for monitoring and analysis.

For example, let's say we have a reinforcement learning agent that needs to learn how to play a video game. Instead of having one agent play the game and learn from its experiences, we can have multiple agents, each play the game with slightly different strategies. The central server collects the experiences and distributes the most up-to-date strategy to each agent. This way, the agents can learn from each other and explore different strategies in parallel, leading to faster and better learning.

As an example, the pseudocode for the general distributed reinforcement learning is shown in Algorithm 1.

Algorithm 1: General distributed reinforcement learning

Initialize: Neural network parameters θ_l for learner and actors $k = 0, 1, \ldots$, where $\theta_k = \theta_l$

1 for actor $k = 0, 1, \ldots$ **do**
2 | Collect transitions D_k, by interacting with the environment env_k while following the policy θ_k
3 | Send D_k to the central server
4 | Pulling parameters from the central server from time to time (for example, every N actor's steps), so
   | $\theta_k \leftarrow \theta_{latest}$
5 |
6 while not converge **do**
7 | Sample a mini-batch of transitions M from the central server
8 | Compute the loss based on transitions M
9 | Compute the gradients with respect to neural network θ_l
10 | Update the parameters of the neural network θ_l by using stochastic gradient descent
11 | Send the new parameters θ_{latest} to the central server

---

When using multiple agents in reinforcement learning, there are some challenges that need to be addressed. One of these challenges is ensuring that the agents are generating diverse and useful training samples. To tackle this issue, one technique is to randomly initialize the environment the actor is interacting with. This approach helps encourage exploration of different areas of the state-action space, which in turn can lead to the discovery of new and better policies for the agent.

Finally, there are other challenges such as how to handle the communication overhead across different machines or agents. Despite these challenges, the distributed reinforcement learning training architecture is a powerful method for training reinforcement learning agents. By using multiple actors running in parallel, the architecture speeds up the process of generating training samples and encourages exploration of different areas of the state-action space, leading to the discovery of new and better policies for the agent.

Applications of Distributed Reinforcement Learning

Two popular distributed reinforcement learning architectures are Ape-X and IMPALA. Ape-X was proposed by Horgan et al. [1], which is based on value-based methods similar to the DQN. It uses multiple actors to generate environment transitions, while a single central agent, called the learner, is responsible for learning. The central agent updates the parameters of the neural network based on a prioritized replay buffer, without any environment interactions because the samples are generated by the actors, allowing it to focus on learning. IMPALA was proposed by Espeholt et al. [2]; it's a policy-based method. It also uses multiple actors to generate environment transitions, while a single learner updates the neural network parameters. Both Ape-X and IMPALA can scale to large numbers of agents and can be used for on-policy and off-policy reinforcement learning.

Another example of distributed reinforcement learning is AlphaGo developed by Silver et al. [3], which uses multiple self-play actors across hundreds of servers to generate self-play games, while multiple agents run on multiple servers to update the neural network parameters. This approach allows for efficient exploration of the game tree and effective learning.

Distributed Reinforcement Learning for On-Policy vs. Off-Policy Learning

When designing distributed reinforcement learning algorithms, it's important to distinguish whether they use on-policy or off-policy learning. On-policy learning agents learn the optimal policy or value function based on the experiences generated by following it's own policy, while off-policy learning agents use data collected from other policies.

For off-policy learning agents, such as DQN, large amounts of historical data can be used for learning. This allows actors and learner agents to work in parallel, with actors generating samples without needing to use the latest policy. For example, DQN uses experience replay to learn from a buffer of transitions that are stored and sampled independently of the current policy. Actors can use a policy that's a few iterations behind the latest policy, with parameters synchronized every few hundred environment steps to reduce communication overhead.

On the other hand, for on-policy learning agents, such as Actor-Critic or PPO, it's crucial for all actors to use the latest policy to make decisions. This can be challenging in a distributed computing architecture, requiring proper synchronization mechanisms between agents to ensure that all actors use the latest policy. One common approach is to use parameter servers to store and update policy parameters that actors and learner agents can access. Alternatively, a message passing scheme can be used to synchronize policies in real time. However, these approaches may introduce additional communication overhead and computational costs.

In summary, understanding the differences between on-policy and off-policy learning agents is essential for designing effective distributed reinforcement learning algorithms. By carefully considering these factors and designing appropriate algorithms and architectures, we can enable efficient and scalable distributed reinforcement learning in a wide range of applications.

Exploration with Distributed Reinforcement Learning

Reinforcement learning agents need to explore the environment in order to learn an optimal policy. For policy-based agents, like Actor-Critic or PPO, the nature of the policy already encourages exploration. But we can also use the concept of entropy to encourage the agent to try new actions. This applies to both single-agent and distributed settings.

For value-based agents, like DQN, a common strategy to encourage exploration is to use an ϵ-greedy policy. With this approach, the agent chooses the best action with probability $1 - \epsilon$ and acts randomly with probability ϵ. By gradually reducing ϵ over time, we can shift the agent's focus from exploration to exploitation.

In distributed settings, each actor is given its own exploration rate, ϵ_i, using an equation that depends on the number of actors and a tuning parameter. Such solution is also implemented in Ape-X. The equation is $\epsilon_i = \epsilon^{1+\frac{i}{N-1}a}$, where ϵ is the initial exploration rate, N is the total number of actors, i is the index of the current actor, and a controls how quickly the exploration rate decays as i increases. By adjusting the value of a, we can control the distribution of exploration rates across actors and find a balance between exploration and exploitation that works well for the problem at hand.

For example, if we have four actors and set $\epsilon = 0.4$ and $a = 1$, the exploration rates for each actor would be $[0.4, 0.04716, 0.00556, 0.00066]$. This means that the first actor will explore much more than the others, while the last actor will hardly explore at all.

Distributed PPO

As an example, the pseudocode in Algorithm 2 presents a high-level distributed PPO algorithm where we have multiple actors $k = 0, 1, \ldots$ running in parallel to generate transitions and one learner agent to update the parameters of the neural network. For each actor, once the sequence reaches a predefined length N, we compute the finite horizon returns G_t and estimated advantages $\hat{A}_t$ using the same mechanism described in Chap. 10 and store these transition sequences in the central server. In the meantime, the learner will constantly run some epochs to update the parameters of the neural network. After that, the latest neural network parameters are sent to the central server, and the actors will periodically update it's own copy of the neural network, and use this updated network to make decisions.

Algorithm 2: Distributed Proximal Policy Optimization

Initialize: Neural network parameters θ_l for learner and actors $k = 0, 1, \ldots$, where $\theta_k = \theta_l$

1 **while** not converge **do**
2   **for** actor $k = 0, 1, \ldots$ **do**
3   Pull the newest parameter from the central server so $\theta_k \leftarrow \theta_{latest}$
4   Collect sequence of N transitions D_k, by interacting with the environment env_k and following the policy θ_k
5   Compute finite horizon returns G_t for each time step in D_k
6   Compute estimated advantages $\hat{A}_t$ for each time step in D_k
7   Add trajectories D_k along with G_t, $\hat{A}_t$ to central server
8
9   **for** epoch $0, 1, \cdots, M$ **do**
10   Update the policy parameters by maximizing (stochastic gradient ascent) the clipped surrogate objective
11   Update baseline parameters by minimizing (stochastic gradient descent) the squared error
12   Send the new parameters θ_{latest} to the central server

---

Figure 12.2 shows the performance comparison of the distributed PPO algorithm with different numbers of actors on the Ant robotic control task. The results display the average episode return (total undiscounted rewards) and a 95% confidence interval. To evaluate the agent's performance, we aggregate the average episode return at the end of each training iteration, which consisted of 100,000 training steps. The results were averaged over three independent runs and smoothed using a moving average with a window size of five.

We use the same neural network architecture and hyperparameters for different runs. Specifically, we use discount rate $\gamma = 0.99$, learning rate 0.0002 for actor and 0.0003 for critic, sequence length 2048, GAE lambda 0.95, and 4 update epochs. We also use entropy to encourage exploration, with an entropy weight of 0.1. The neural networks were trained using the Adam optimizer.

As we can see from the chart, the performance gain with more actors is significant when compared to lesser actors. And the agent tends to converge to the optimal policy much faster. However we should not confuse faster converge with lesser training samples, as the figure only shows the training environment steps per actor agent, not the sum of steps across the actors.

Figure 12.3 shows the experiments result in the more challenging Humanoid robotic control task.

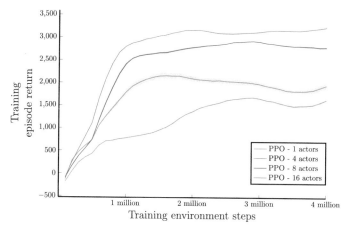

Fig. 12.2 The performance comparison of a distributed PPO algorithm with different numbers of actors on the Ant robotic control task. The results display the average episode return (total undiscounted rewards) and a 95% confidence interval. The results were averaged over three independent runs and then smoothed using a moving average with a window size of five

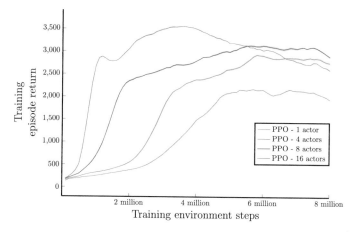

Fig. 12.3 The performance comparison of a distributed PPO algorithm with different numbers of actors on the Humanoid robotic control task. The results display the average episode return (total undiscounted rewards) and a 95% confidence interval. The results were averaged over three independent runs and then smoothed using a moving average with a window size of five

12.3 Data Parallelism for Distributed Reinforcement Learning

The distributed reinforcement learning architecture shown in Fig. 12.1 is a powerful tool that allows us to tackle large-scale or more complex reinforcement learning problems; it can improve the training speed of the agent by using multiple actors running in parallel. However, if we have a problem that requires hundreds or even thousands of actors all running in parallel, then relying on just on learner agent to update the neural network can become a bottleneck. To overcome this issue, we can use a technique called data parallelism, which is commonly used in deep learning.

Data parallelism involves dividing a large dataset into smaller batches and distributing these batches across multiple processors or machines. Each processor or machine trains a model on its

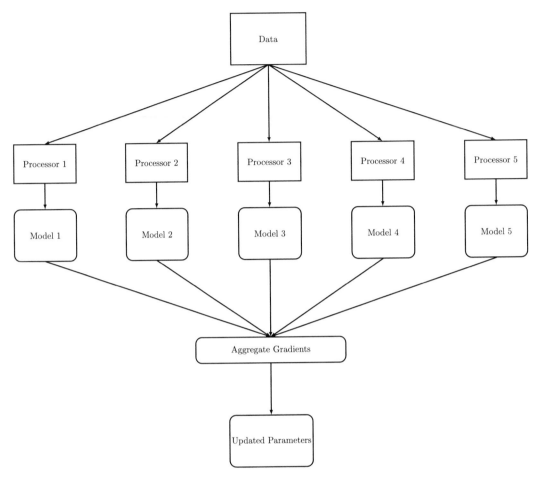

Fig. 12.4 Simple diagram to illustrate the concept of data parallelism in deep learning, where we have multiple processors compute the gradients for its own copy of the model and a final processor to perform the gradient aggregation and parameter update

batch of data, and the resulting gradients are combined to produce a final set of gradients. These gradients are then used to update the model parameters (Fig. 12.4).

For example, to train Google's BERT language model developed by Devlin et al. [4] or OpenAI's GPT-3 developed by Brown et al. [5], hundreds of processors were used in parallel to divide the data and train the models.

In distributed reinforcement learning, we can use multiple parallel learner agents, each computing the loss and gradients locally, but not updating the parameters of the neural network. Instead, they send their local gradients to a central parameter server, which collects and aggregates the gradients (e.g., by taking the mean) before performing the parameter update. This ensures that the parameter update only happens in a central place.

Once the parameter server updates the parameters of the neural network, it broadcasts the newest parameters to all the actor agents, not just the learner agents. This kind of architecture is particularly useful for large-scale and complex reinforcement learning problems, such as training world-class Go players [3, 6, 7], where multiple parallel learners can significantly reduce training time.

A simple diagram to illustrate the idea is shown in Fig. 12.5.

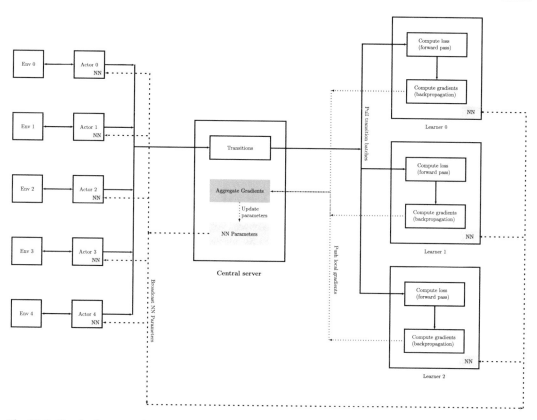

Fig. 12.5 Simple diagram to illustrate the idea of distributed reinforcement learning with multiple learner agents to compute gradients and a central parameter server to do gradient aggregation and parameter update

However, there are challenges with this approach. Communication overhead between the learner agents can become a bottleneck, and proper gradient aggregation techniques are required to make it work effectively.

In summary, data parallelism is a powerful technique for distributed reinforcement learning, enabling multiple processors or machines to work on different subsets of the data in parallel and aggregating the results to update the model parameters. By using multiple parallel learners, we can significantly improve the efficiency and effectiveness of reinforcement learning, but careful consideration and proper techniques are required to overcome the challenges that come with this approach.

12.4 Summary

In this chapter, we delved into the advantages of utilizing distributed reinforcement learning (RL) to tackle complex problems more efficiently. We introduced the general architecture for distributed RL training, which involves running multiple actors simultaneously. Each actor interacts with its own environment, generating training samples. Additionally, a learner continuously updates the neural network, such as a value network or a policy. This distributed training architecture has proven to be highly efficient, as the increased number of actors facilitates extensive exploration of the environment, leading to improved performance.

Furthermore, we provided a brief overview of the data parallel approach within the distributed RL training architecture. This approach draws inspiration from the broader field of deep learning. Multiple learners work in parallel to compute gradients for the neural network, while a central parameter server aggregates these gradients and updates the parameters accordingly.

In the subsequent chapter, we will explore how the general distributed RL training architecture, in conjunction with other techniques, can be applied to overcome more demanding RL problems more effectively.

References

[1] Dan Horgan, John Quan, David Budden, Gabriel Barth-Maron, Matteo Hessel, Hado van Hasselt, and David Silver. Distributed prioritized experience replay, 2018.

[2] Lasse Espeholt, Hubert Soyer, Remi Munos, Karen Simonyan, Volodymir Mnih, Tom Ward, Yotam Doron, Vlad Firoiu, Tim Harley, Iain Dunning, Shane Legg, and Koray Kavukcuoglu. Impala: Scalable distributed deep-RL with importance weighted actor-learner architectures, 2018.

[3] David Silver, Aja Huang, Chris J. Maddison, Arthur Guez, Laurent Sifre, George van den Driessche, Julian Schrittwieser, Ioannis Antonoglou, Veda Panneershelvam, Marc Lanctot, Sander Dieleman, Dominik Grewe, John Nham, Nal Kalchbrenner, Ilya Sutskever, Timothy Lillicrap, Madeleine Leach, Koray Kavukcuoglu, Thore Graepel, and Demis Hassabis. Mastering the game of go with deep neural networks and tree search. *Nature*, 529(7587):484–489, Jan 2016.

[4] Jacob Devlin, Ming-Wei Chang, Kenton Lee, and Kristina Toutanova. Bert: Pre-training of deep bidirectional transformers for language understanding, 2019.

[5] Tom B. Brown, Benjamin Mann, Nick Ryder, Melanie Subbiah, Jared Kaplan, Prafulla Dhariwal, Arvind Neelakan-tan, Pranav Shyam, Girish Sastry, Amanda Askell, Sandhini Agarwal, Ariel Herbert-Voss, Gretchen Krueger, Tom Henighan, Rewon Child, Aditya Ramesh, Daniel M. Ziegler, Jeffrey Wu, Clemens Winter, Christopher Hesse, Mark Chen, Eric Sigler, Mateusz Litwin, Scott Gray, Benjamin Chess, Jack Clark, Christopher Berner, Sam McCandlish, Alec Radford, Ilya Sutskever, and Dario Amodei. Language models are few-shot learners, 2020.

[6] David Silver, Julian Schrittwieser, Karen Simonyan, Ioannis Antonoglou, Aja Huang, Arthur Guez, Thomas Hubert, Lucas Baker, Matthew Lai, Adrian Bolton, Yutian Chen, Timothy Lillicrap, Fan Hui, Laurent Sifre, George van den Driessche, Thore Graepel, and Demis Hassabis. Mastering the game of go without human knowledge. *Nature*, 550(7676):354–359, Oct 2017.

[7] David Silver, Thomas Hubert, Julian Schrittwieser, Ioannis Antonoglou, Matthew Lai, Arthur Guez, Marc Lanctot, Laurent Sifre, Dharshan Kumaran, Thore Graepel, Timothy Lillicrap, Karen Simonyan, and Demis Hassabis. Mastering chess and shogi by self-play with a general reinforcement learning algorithm, 2017.

Curiosity-Driven Exploration

13

Reinforcement learning is a type of machine learning in which an agent learns by taking actions in an environment and receiving feedback in the form of reward signals. The objective of the agent is to maximize the cumulative reward over time. However, there are certain challenges that can hinder the learning process. These challenges include difficult-to-explore environments, sparse or uncertain rewards.

When the environment is hard to explore and the rewards are scarce or uncertain, we face obstacles in training the agent effectively. This can occur due to a large state space or a task that requires a sequence of actions to reach a rewarding state. Moreover, the agent may not receive clear or frequent reward signals from the environment, making it challenging for the agent to learn and make appropriate decisions.

To tackle these challenges, this chapter introduces the concept of curiosity-driven exploration as a potential solution, focusing particularly on Random Network Distillation (RND). Curiosity-driven exploration involves motivating the agent to explore its environment and gain a deeper understanding of it, even in the absence of clear reward signals. By promoting exploration and enabling the agent to acquire knowledge about its surroundings, we can enhance its ability to make informed decisions and achieve better outcomes.

For example, imagine an agent tasked with navigating a maze to reach a goal. If the agent only receives a reward when it reaches the goal, it may struggle to learn the optimal path if the reward signal is sparse or uncertain. However, if we encourage the agent to explore the maze and learn more about its structure and layout, it may be able to navigate more effectively and ultimately reach the goal more efficiently.

In the following sections, we'll explore how to use curiosity-driven exploration with policy-based methods to improve reinforcement learning in challenging environments.

13.1   Hard-to-Explore Problems vs. Sparse Reward Problems

Hard-to-Explore Problem

When it comes to reinforcement learning, a hard-to-explore problem is one that presents a challenge for an agent to find an optimal policy due to limited information about the environment. This can happen when the environment has a large state or action space or when certain actions are rarely taken. The agent may have trouble exploring the state-action space enough to learn an effective policy.

© The Author(s), under exclusive license to APress Media, LLC, part of Springer Nature 2023
M. Hu, *The Art of Reinforcement Learning*,
https://doi.org/10.1007/978-1-4842-9606-6_13

One concrete example of a hard-to-explore problem is navigating a maze. The agent has to learn to navigate the maze and reach the goal, but the maze can have many paths, some of which lead to dead ends or suboptimal outcomes. Another example can be found in video games, where the agent needs to learn how to navigate the game world, defeat enemies, and complete tasks. The game world can be vast and complex, and the agent may not have enough information about the game to make informed decisions.

To overcome these challenges, the agent may need to explore the environment extensively to determine the optimal path. However, random exploration, such as an ϵ-greedy policy, may not be sufficient to discover hidden states or hard-to-reach parts of the environment, and the agent may not receive any reward until it reaches them. This can make it challenging for the agent to learn a successful policy.

Sparse Reward Problem

The sparse reward problem is another common challenge in reinforcement learning where the reward signal is only provided in specific states, making it difficult for the agent to understand which actions led to those rewards. This can make learning a good policy a slow and challenging process, especially in complex environments or tasks where the reward signal is infrequent or delayed.

A concrete example of a sparse reward problem is the game of chess, where the agent only receives a meaningful signal (+1, 0, -1) at the end of the game, and zero reward otherwise. This means the agent has no direct feedback on which actions are good or bad during the game and needs to explore a large state space to find a good policy. Similarly, in robotic navigation tasks, the reward signal may only be given when the robot reaches the goal, making it difficult for the agent to learn which actions contributed to the successful navigation.

It's important to note that sparse reward problems are distinct from hard-to-explore problems, which occur when the agent has to search a large state space to find the optimal policy. However, it's often the case that a sparse reward problem is also a hard-to-explore problem.

In summary, the sparse reward problem is a significant challenge in reinforcement learning that can make learning a good policy a slow and difficult process, especially in complex environments or tasks where the reward signal is infrequent or delayed.

Montezuma's Revenge

The Atari video game Montezuma's Revenge is considered to be both a hard-to-explore problem and a sparse reward problem.

In this game, the player must navigate a labyrinthine temple to collect treasures and defeat enemies[1]. The ultimate goal of the game is to reach the treasure chamber and collect the coveted "Treasure of Montezuma."

The game's high level of difficulty is due to numerous obstacles, enemies, and traps, such as spike pits and disappearing floors, that must be avoided or defeated. Additionally, the game's large state space (e.g., the game has multiple rooms and each room has different layout) requires the player to execute specific actions in the correct order to access various areas or rooms, making it a hard-to-explore problem.

Furthermore, the game only rewards the player for discovering secret rooms and hidden treasures, which makes it a sparse reward problem. For instance, in the first room, the player must first collect the key to unlock the next room and receive a reward.

---

[1] A nice visualization of the Montezuma's Revenge game from OpenAI: https://openai.com/research/reinforcement-learning-with-prediction-based-rewards

Overall, Montezuma's Revenge poses a challenging problem for reinforcement learning agents due to its complex environment and sparse rewards. It requires the agent to be curious, explore the environment thoroughly, and plan long-term strategies to successfully complete the game.

13.2 Curiosity-Driven Exploration

Curiosity-driven exploration is a powerful technique used in reinforcement learning to encourage exploration and learning in complex or sparse reward problems. The idea was first proposed by Pathak et al. [1]. In traditional RL methods, agents are motivated to take actions that result in a maximum reward provided by the environment. However, this can lead to short-sighted behavior and a lack of exploration, which can cause the agent to miss out on valuable information.

Curiosity-driven exploration takes a different approach by encouraging agents to explore the environment based on intrinsic motivation rather than extrinsic rewards. The idea is to promote intrinsic motivation by adding a curiosity bonus to the reward signal. This bonus is calculated based on how much the agent's prediction of the outcome of an action differs from the actual outcome. As a result, the agent is incentivized to explore actions that lead to new and surprising experiences, even if they do not result in immediate rewards from the environment.

This approach differs from traditional extrinsic rewards, which are provided by the environment and are directly tied to achieving specific goals or desirable actions. Intrinsic rewards, on the other hand, are generated internally by the agent and incentivize exploration and discovery of new things in the environment.

For example, consider a robot trained to navigate a room to find a target object in the shortest time. Traditional RL algorithms would encourage the robot to take the shortest path to the target, even if there are other paths that may have useful information. However, with curiosity-driven exploration, the robot is also motivated to explore different paths and learn more about the environment to maximize its curiosity bonus.

Intrinsic reward is given to the agent for exploring and discovering new things in the environment, rather than for achieving a specific goal or performing a desirable action. This reward is generated internally by the agent itself and incentivizes the agent to try out new behaviors or visit new states that it has not seen before.

Previous research has shown the effectiveness of curiosity-driven exploration in various RL tasks, such as playing Atari games, navigating mazes, and solving the Rubik's Cube. OpenAI even trained a humanoid robot to learn a diverse set of skills through curiosity-driven exploration, solely by exploring its environment and seeking novelty.

In this chapter, we'll focus on one innovative and widely adapted solution called "Exploration by Random Network Distillation" by Burda et al. [2], which uses a random network to generate the curiosity bonus. This approach can lead to more efficient computation and effective learning in complex tasks, making curiosity-driven exploration a promising technique in RL.

13.3 Random Network Distillation

Random Network Distillation (RND) is a reinforcement learning algorithm designed to encourage exploration in learning agents. This algorithm was introduced by Burda et al. [2] and is based on the idea of computing an intrinsic reward based on the novelty of the agent's experience. In other words, the algorithm rewards the agent for encountering new and unexpected experiences.

RND uses two neural networks: a fixed and randomly initialized target network, denoted by f, and a predictor network, denoted by $\hat{f}$, which is trained on data collected by the agent. The weights of the fixed network f remain constant during the training session, while the weights of the predictor network $\hat{f}$ are updated using supervised learning based on the prediction error.

The intrinsic reward signal r_t^I for state s_t at time step t is computed as

$$r_t^I = |f(s_t) - \hat{f}(s)|^2 \tag{13.1}$$

More precisely, the RND algorithm can be described in the following steps:

- The fixed network f takes a state s_t as input and outputs an embedding feature vector τ_t.
- The predictor network takes the same state s_t as input and outputs its own embedding feature vector $\hat{\tau}_t$.
- The difference between the predictions of the predictor network $\hat{f}$ and the fixed network f is then used to compute the intrinsic reward, which is $|\hat{\tau}_t - \tau_t|^2$.
- The predictor network is then trained to minimize the prediction error $|\hat{\tau}_t - \tau_t|^2$, which is done by using supervised learning methods.

And Fig. 13.1 shows how the RND works.

The intrinsic reward is combined with the reward signal from the environment, resulting in a total reward $r_t = r_t^E + r_t^I$, where r_t^E is the reward provided by the environment at time step t, and r_t^I is the exploration bonus (intrinsic reward) associated with the transition at time step t. This total reward can be plugged into any of the existing reinforcement learning algorithms, like DQN, Actor-Critic, or even PPO.

To understand the intuition behind RND, let's look at a simple example. Suppose we're tying to train an agent to navigate a maze, the environment only provides positive reward +1 to the agent when it successfully reaches a goal position, and zero reward otherwise. In such case, the agent might have trouble making adequate decisions and explore the environment, since no meaningful reward is provided unless it reaches the goal.

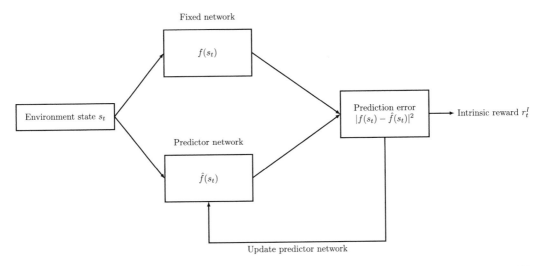

Fig. 13.1 How RND computes the intrinsic reward and updates the predictor network. Adapted from Burda et al. [2]

To encourage the agent to explore more, we can use RND. The fixed network f is randomly initialized and doesn't change throughout the training session. It takes in the agent's current state in the maze and outputs a unique embedding feature vector τ_t that represents the state. Meanwhile, the predictor network $\hat{f}$ takes in the same state and tries to predict the embedding feature vector τ_t from the fixed network.

As the agent explores the maze, the predictor network $\hat{f}$ predicts the embedding feature vector $\hat{\tau}_t$ for the current state s_t, and the fixed network f also generates its own embedding feature vector τ_t for s_t. The intrinsic reward is then calculated based on the squared Euclidean distance between these two predicted vectors: $|\hat{\tau}_t - \tau_t|^2$. If the agent encounters a new state that is significantly different from any previously seen states, the distance between the predicted vectors will be large, resulting in a high intrinsic reward. On the other hand, if the agent repeatedly visits the same states, the distance between the predicted vectors will be small, resulting in a low intrinsic reward.

This intrinsic reward is then combined with the environment reward, and the total reward is used to update the agent's policy. By encouraging the agent to explore new states and experiences, RND can help the agent learn more effectively and navigate the maze more efficiently.

The neural network architecture of the RND module for Atari games is depicted in Fig. 13.2. The module comprises a target network on the left, initialized randomly, with its parameters kept fixed during the training process. On the right is the predictor network, which is similar to the target network but consists of two additional fully connected layers. Both networks receive a normalized $84 \times 84 \times 1$ single frame as input and output an embedding vector. The objective of the predictor network is to accurately predict the embedding vector generated by the fixed target network.

In the original RND paper, Burda et al. [2] found that prediction errors in reinforcement learning can be caused by a number of factors, including the amount of training data, stochasticity, model misspecification, and learning dynamics. To leverage this information for exploration, the RND approach uses prediction error as an exploration bonus.

However, the fixed target network used in RND makes training the predictor network much easier than learning the optimal policy. As a result, the exploration bonus can become less effective as the predictor network becomes better at predicting outcomes. To address this, Burda et al. [2] proposed deliberately limiting the amount of training data used to train the predictor network, when the number of actors is very large. Specifically, they suggest using portion of the available samples according to this formula $\min(1, \frac{32}{N})$, where N is the number of actors running in parallel to generate training samples.

A major challenge of using prediction error as an exploration bonus is that the scale of the reward can vary greatly between different environments and time steps. To keep the rewards on a consistent scale, the intrinsic reward is often normalized, for example by dividing it using the running estimate of the standard deviation of the intrinsic returns.

Another issue with the RND approach is how to deal with input data for the two RND neural networks. Observation normalization is crucial when using a fixed target network since the parameters cannot adjust to the scale of different datasets. Without normalization, the variance of the inputs can become very high. The RND approach addresses this issue by using an observation normalization scheme, similar to that used in continuous control problems, whereby each dimension is whitened by subtracting the running mean and then dividing by the running standard deviation. The normalized observations are then clipped to be between -5 and 5. The running mean and standard deviation are initialized by stepping a random agent in the environment for a small number of steps before beginning optimization. The same observation normalization is used for both predictor and target networks, but not for the policy network.

Typically, the RND module is only used during training and not during the agent's interaction with the environment. However, in certain cases, it can be integrated into the agent's decision-making

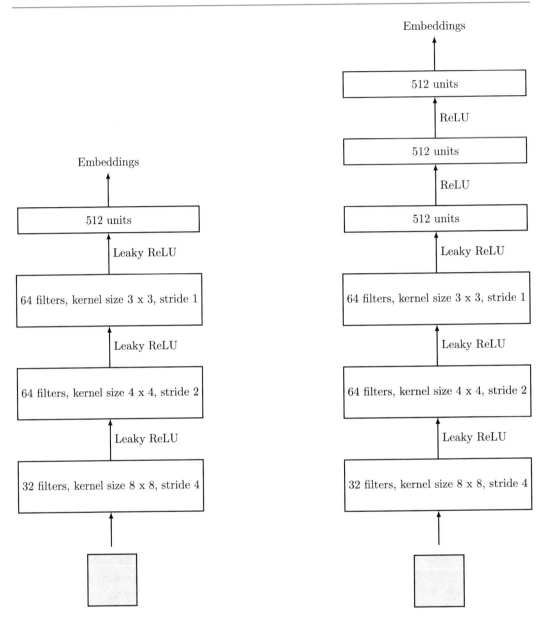

Fig. 13.2 RND neural network architecture proposed by Burda et al. [2] for the Atari video game, where the left side shows the randomly initialized target network and the right side shows the predictor network

process. For instance, the Agent57 algorithm developed by Badia et al. [3] is an advanced version of the DQN method, which uses intrinsic rewards from the RND module as input data for the neural network, which approximates the optimal state-action value function, allowing the agent to make more informed decisions.

This chapter will focus on using the RND module in conjunction with the Proximal Policy Optimization (PPO) algorithm, which is a popular policy-based method. By incorporating intrinsic rewards, we hope to improve the agent's learning efficiency and achieve better performance in complex environments.

RND with PPO

Proximal Policy Optimization (PPO) is a popular reinforcement learning algorithm that uses a surrogate objective function to update the policy. The surrogate objective function maximizes the probability of taking actions that improve the policy while also ensuring that the policy doesn't deviate too much from the previous policy.

Recall that PPO uses this clipped surrogate objective function:

$$\arg\max_{\pi'} \mathcal{L}_\pi^{CLIP}(\pi') = \arg\max_{\pi'} \mathbb{E}_{\tau \sim \pi}\left[\min\left(\frac{\pi'(a|s)}{\pi(a|s)}\hat{A}_t,\ clip\left(\frac{\pi'(a|s)}{\pi(a|s)}, 1-\epsilon, 1+\epsilon\right)\hat{A}_t\right)\right]$$

$$(13.2)$$

The advantage function is calculated using the generalized advantage estimation (GAE) method:

$$\hat{A}_t = \delta_t + (\gamma\lambda)\delta_{t+1} + (\gamma\lambda)^2\delta_{t+2} + \cdots + (\gamma\lambda)^{T-t-1}\delta_{T-1} \tag{13.3}$$

Here, δ_t is the temporal difference error, which is the difference between the sum of the rewards and the estimated value of the current state and the estimated value of the successor state:

$$\text{where } \delta_t = r_t + \gamma V_\pi(s_{t+1}) - V_\pi(s_t) \tag{13.4}$$

In order to use the intrinsic rewards produced by RND in the PPO algorithm, we need to make some modifications to the existing PPO algorithm. More precisely, we want to combine the extrinsic and intrinsic estimated advantages as a single advantage estimation. This means that we also have to estimate the intrinsic state values and advantages.

$$\hat{A}_t = \hat{A}_t^E + \hat{A}_t^I$$

To estimate the intrinsic advantages $\hat{A}_t^I$, we need the estimated intrinsic state values; that's why we also need to use the neural network (Critic) to predict the intrinsic state values. This neural network can be trained simultaneously with the network that predicts the policy and extrinsic state values, and the objective is similar to the extrinsic state value case, where we want to minimize the prediction error.

In practice, we can share the weights of the neural network used to predict the intrinsic state values with the network that predicts the policy and extrinsic state values. This can help to reduce the computational cost of training and make the algorithm more efficient. We also use a slightly more complex neural network architecture, which involves skip connection, as a concept borrowed from ResNet, as shown in Fig. 13.3.

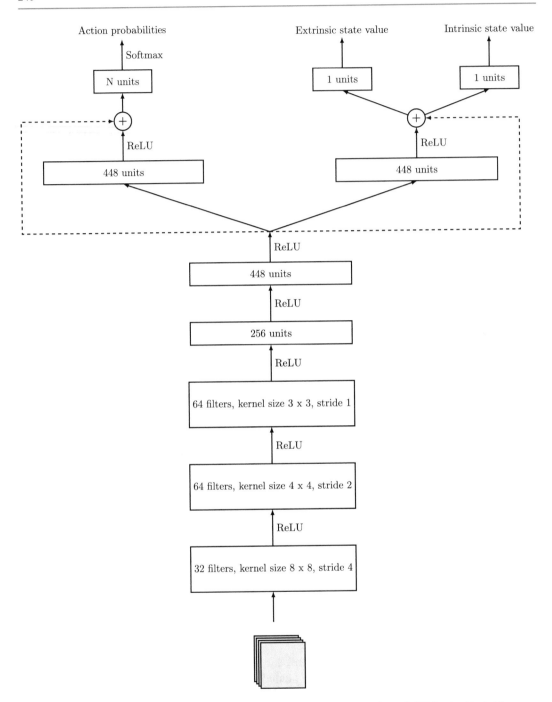

Fig. 13.3 The neural network architecture for PPO with RND proposed by Burda et al. [2] for the Atari video game, where the network has three output heads: the policy head to predict the action probabilities, the extrinsic state value head to predict the extrinsic state value, and intrinsic state value head to predict the intrinsic state value

As an example, the pseudocode for the RND combined with the PPO algorithm is shown in Algorithm 1.

Algorithm 1: RND and Proximal Policy Optimization

Input: Discount rate γ, learning rate α for policy, learning rate β for value function, sequence length N, clip epsilon ϵ, number of update epochs M, number of initial steps for initializing observation normalization L, number of environment steps K

Initialize: Initialize the parameters θ_f for the fixed network f, and $\theta_{\hat{f}}$ for the predictor network $\hat{f}$ of RND, initialize the parameters θ, θ_{old} for the two policies, where $\theta_{old} = \theta$, initialize parameters ψ for the state value function $\hat{V}$

Output: The approximated optimized policy π_θ

1   **for** 0, 1, ..., L **do**
2   Sample a_t randomly for s_t
3   Take action a_t in the environment and observe s_{t+1}
4   Update observation normalization statistics (mean and standard deviations) using s_{t+1}
5   **while** train environment steps $< K$ **do**
6   Collect τ, a sequence of N transitions $\left(s_t, a_t, r_t^E, \pi(a_t|s_t; \theta_{old})\right)$
7   Compute intrinsic rewards $r_0^I, r_1^I, \ldots, i_{N-1}^I$ for every transition in τ, where $r_t^I = |\hat{f}(s_t) - f(s_t)|^2$
8   Normalize intrinsic rewards $r_0^I, r_1^I, \ldots, i_{N-1}^I$
9   Update intrinsic reward normalization statistics using $r_0^I, r_1^I, \ldots, i_{N-1}^I$
10   Compute generalized advantage estimate $\hat{A}_t^I$ using intrinsic reward r_t^I intrinsic state values for every transition in τ
11   Compute finite-horizon returns G_t and generalized advantage estimate $\hat{A}_t^E$ using extrinsic reward r_t^E and extrinsic state values for every transition in τ
12   Compute combined advantages $\hat{A}_t = \hat{A}_t^E + \hat{A}_t^I$
13   Update observation normalization statistics using samples from τ
14   **for** epoch 0, 1, ..., M **do**
15   Update the PPO policy parameters by maximizing the PPO clipped surrogate objective
16   Update extrinsic and intrinsic state value parameters by minimizing the squared error
17   Update the RND predictor parameters by minimizing the prediction error
18      $\theta_{old} \leftarrow \theta$

Figure 13.4 shows the performance of the PPO (clipped version) agent with the RND module on the Atari video game Montezuma's Revenge.

We use a distributed reinforcement learning architecture, where we run 32 actors in parallel to collect sample sequences, and a single learner agent to perform parameter updates. We use the same neural network architecture as explained earlier in this chapter for both PPO and RND. Specifically, we use different discount rates for extrinsic (environment) and intrinsic (curiosity bonus) rewards, with 0.999 for the extrinsic reward and 0.99 for the intrinsic reward. We use a learning rate of 0.0001 for both PPO policy and RND predictor networks, GAE lambda of 0.95 for the advantages, sequence length of 128, and four update epochs. We also use entropy to encourage exploration, with an entropy weight of 0.001. The neural networks were trained using the Adam optimizer.

We use a similar environment processing as DQN, which involves resizing the frame to 84×84 and converting it to grayscale. We also apply the skip action technique, where we only process every fourth frame, and we stack the last four frames to create a final state image of size $84 \times 84 \times 4$. For RND networks, we only take the last frame as input, which is an $84 \times 84 \times 1$ image.

Additionally, we clip the reward values to the range of -1 to 1, and we set the maximum episode length to 18,000, which is 4500 steps after applying a skip action and frame skip.

Performance is measured in terms of the average reward obtained. To evaluate the agent's performance, we ran 100,000 evaluation steps on a separate testing environment with a greedy policy

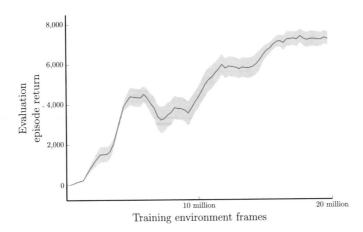

Fig. 13.4 PPO with the RND module on the Atari video game Montezuma's Revenge. The results display the average episode return (total undiscounted rewards) and a 95% confidence interval. The results were averaged over three independent runs and then smoothed using a moving average with a window size of five

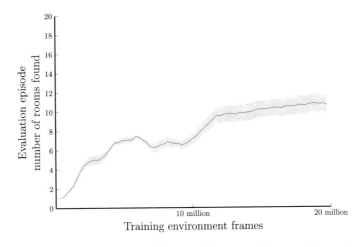

Fig. 13.5 PPO with the RND module on the Atari video game Montezuma's Revenge. The results display the average number of rooms found by the agent and a 95% confidence interval. The results were averaged over three independent runs and then smoothed using a moving average with a window size of five

at the end of each training iteration, which consisted of $128 \times 500 \times 4$ frames from each actor, where no reward clipping or soft termination on loss of life is applied to the evaluation environment. The results were averaged over three independent runs and smoothed using a moving average with a window size of five.

Figure 13.5 also shows the average number of rooms found by the agent per evaluation game. Visiting more rooms indicates that the agent has explored more states of the game, which often results in finding hidden treasures and earning large rewards.

It is important to note that the x-axis in the preceding plots represents the number of training environment frames per actor only. To obtain the total number of training environment frames, we need to multiply this value by the number of actors involved in the training process. In our case, the total number of training environment frames is 0.64 billion, which was obtained by multiplying 20 million by 32, the number of actors involved in the training process.

The results of the experiment appear promising at first glance, but upon closer inspection of the recorded gameplay, it is evident that the agent did not truly master the game. For instance, in the first room, after collecting the key, the agent failed to use the ladders to ascend to the top of the room, which would have been the reverse of the sequence it used to reach the key. Instead, it opted to jump off a high wall, resulting in a loss of life.

In another example, in some of the rooms, the agent missed the opportunity to descend the ladder to move to the next room, and, instead, it chose to fight with a devil character, which was impossible to defeat and led to the loss of another life.

These and other similar suboptimal actions throughout the gameplay reveal that while curiosity-driven exploration can encourage an agent to tackle complex and challenging problems, it does not guarantee that the learned policy will always be optimal. This is because the objective of curiosity-driven exploration is fundamentally different from the goal of the policy. Consequently, this issue has spurred active research into safe exploration for reinforcement learning [4, 5].

13.4   Summary

This chapter delved into the challenges posed by hard-to-explore problems and sparse reward problems in reinforcement learning. We explored the fundamental difficulties faced by RL agents when navigating these types of problems, where exploration and the scarcity of reward signals hinder learning progress.

To address these challenges, we introduced the concept of curiosity-driven exploration. This technique empowers agents to construct their intrinsic rewarding system, which aids in overcoming complex RL problems. We presented a particularly intriguing and practical approach known as Random Network Distillation (RND). We delved into the inner workings of RND and its underlying intuition. By combining RND with the Proximal Policy Optimization (PPO) algorithm, we demonstrated the remarkable performance achieved by agents in Montezuma's Revenge, one of the most demanding Atari video games.

Importantly, curiosity-driven exploration encourages agents to seek out novel and interesting experiences to enhance their knowledge and skills, while the policy remains focused on optimizing the agent's performance in accomplishing specific tasks or goals. It is worth noting that while curiosity-driven exploration can greatly enhance an agent's overall learning capabilities, it may not necessarily guarantee optimal performance in all scenarios.

Looking ahead, the next chapter will shift our focus to model-based learning. Specifically, we will explore the renowned AlphaZero agent, which has surpassed human mastery in games such as Go, Chess, and Shogi.

References

[1] Deepak Pathak, Pulkit Agrawal, Alexei A. Efros, and Trevor Darrell. Curiosity-driven exploration by self-supervised prediction, 2017.
[2] Yuri Burda, Harrison Edwards, Amos Storkey, and Oleg Klimov. Exploration by random network distillation, 2018.
[3] Adrià Puigdomènech Badia, Bilal Piot, Steven Kapturowski, Pablo Sprechmann, Alex Vitvitskyi, Daniel Guo, and Charles Blundell. Agent57: Outperforming the Atari human benchmark, 2020.
[4] Zhaohan Daniel Guo and Emma Brunskill. Directed exploration for reinforcement learning, 2019.
[5] Enrico Marchesini, Davide Corsi, and Alessandro Farinelli. Benchmarking safe deep reinforcement learning in aquatic navigation, 2021.

Planning with a Model: AlphaZero

14

Throughout this book, we have discussed various model-free reinforcement learning algorithms that have proven effective in solving simple reinforcement learning problems such as classic control tasks and Atari video games. However, for more complex problems, these algorithms may not perform well, even with the application of advanced techniques like distributed training and curiosity-driven exploration. Therefore, in this final chapter, we will explore how to use model-based reinforcement learning to tackle more challenging tasks.

We'll focus on employing an accurate model of the game environment and advanced search algorithms such as Monte Carlo Tree Search (MCTS) for planning. Specifically, we'll examine how the AlphaZero algorithm works for complex two-player zero-sum games like Chess and Go.

This chapter will showcase the power of model-based reinforcement learning as a tool for solving complex games, along with the use of advanced search algorithms like MCTS to enhance the planning process.

14.1 Why We Need to Plan in Reinforcement Learning

Reinforcement learning has demonstrated impressive performance in solving simple problems such as classic control tasks and Atari video games. However, its potential for solving practical and challenging real-world tasks is of greater interest. One of these challenging tasks is the game of Go, which is considered one of the most complex board games.

Go is a strategic board game played on a 19×19 grid. Two players participate, each using black or white stones as their representatives. The main goal of the game is to either capture the opponent's stones or establish territories on the board. The game concludes when all possible moves have been made, typically when both players mutually agree that no further advantageous moves remain. At the end, the player with the highest score, determined by applying standard scoring rules, is declared the winner. Additionally, if a player believes they have no chance of winning, the player may choose to resign, resulting in a victory for the opponent.

Go is a zero-sum, perfect information, deterministic, and strategy game, similar to other well-known games like chess and checkers. Despite its simple rules, Go's practical strategy is extremely complex, emphasizing the importance of balance on multiple levels.

© The Author(s), under exclusive license to APress Media, LLC, part of Springer Nature 2023
M. Hu, *The Art of Reinforcement Learning*,
https://doi.org/10.1007/978-1-4842-9606-6_14

Go is an incredibly complex game, with an astonishing 2.1×10^{170} possible board positions [2], making it the most challenging board game in the world. In comparison, the number of legal positions in chess is estimated to be between 10^{43} and 10^{50}.

In addition to its astonishing number of possible board positions, Go and other two-player zero-sum games also face the credit assignment issue. Rewards are zero during intermediate steps until reaching a terminal state at time step T, at which point a reward of $+1$ or -1 is given depending on the game's outcome. In some cases, such as Chess, the outcome can also be a draw, where neither player wins or loses, and the final reward is zero.

Unlike most capture-based games like chess or checkers, where the game's complexity decreases as pieces are removed, the game of Go becomes increasingly intricate with every move as a new piece is placed on the board, as shown in Fig. 14.1.

Playing Go requires a high degree of strategic thinking and planning, which cannot be achieved by simply exploring the game randomly. Beginners often find themselves losing to experienced players who possess the ability to create effective formations. Professional players use complex strategy to simulate possible moves and anticipate their opponent's next move, evaluating potential outcomes to select the best move. This simulation and planning process can involve thinking many moves ahead and relies on the player's experience, expertise, and intuition.

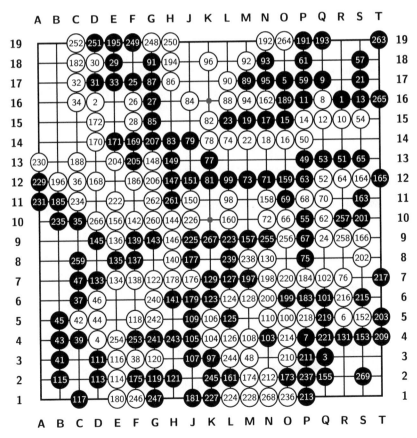

Fig. 14.1 The game of Go becomes increasingly intricate with every move as a new piece is placed on the board. Shida Tatsuya (black) vs. Suzuki Shinji (white), Go game from the 66th NHK Cup, 2018. White won by 2.5 points. Game record from CWI [1]

Although reinforcement learning has demonstrated impressive performance in simpler tasks, current algorithms face significant challenges in tackling the complexity and strategic nature of Go. Winning at this game requires advanced planning and strategic thinking, which random exploration cannot provide. Consequently, a new algorithm that can play Go at a high level is necessary.

To equip our reinforcement learning agent with the ability to plan and think strategically like a professional human player, specialized algorithms tailored specifically for Go have been developed. These algorithms rely on a model that can assist the agent in its planning and strategic decision-making.

A model of the environment consists of two components: the dynamics function (also known as the state transition function) and the reward function. The dynamics function describes how the state of the game changes in response to actions taken by the agent and its opponent. The reward function assigns a numerical value to the player, based on the game state and the action taken in that state. By using these functions, the agent can simulate possible future scenarios, evaluate potential outcomes, and select the best possible move based on its strategy and goals.

In the case of two-player zero-sum games like Go or Chess, we can create an accurate model because the games are deterministic. We know exactly what the succeeding state will be if a player makes a move, so we can develop an almost perfect dynamics function (although it's actually quite complex for games like Go). Since the rewards are always zero, except for the final state, and the game's scoring rules are well known, we can also create an accurate reward function for the game. Furthermore, information about possible legal moves for every game state is readily available since the board configuration is fully observable.

Assuming we have access to a perfect model of the game, how can we utilize the model to enable the reinforcement learning agent to plan like professional human players? One traditional method for two-player zero-sum games is to use some search algorithm to do the planning, like the Minimax algorithm proposed by JOhn von Neumann and Osakar Morgenstern [3]. This algorithm creates a tree data structure comprising nodes and edges, with nodes representing the different states of the game and edges linking the various nodes representing the legal moves or actions that lead to subsequent nodes or states. The search process's objective is to discover the move that minimizes the opponent's maximum reward in every state or minimizes the potential loss to force a draw.

IBM's Deep Blue [4] is an example of utilizing the Minimax search algorithm for playing Chess at the master level. Deep Blue employs a heuristic evaluation function and an alpha-beta search algorithm (an improvement to the standard Minimax search algorithm), in addition to parallel computing, to conduct multiple steps of planning to determine the best moves in every board configuration. The heuristic evaluation function is crucial to Deep Blue's success, and it was created based on rules and patterns defined by chess masters and computer scientists.

However, creating or learning an accurate heuristic evaluation function for Go is considerably more difficult due to the game's dynamic nature and larger branching factor. People have tried using neural networks and supervised learning to evaluate the position, for example, M. Enzenberger [5] proposed to use neural networks and temporal difference learning [6] to predict the territory during the gameplay. However, these work only achieved a weak or medium level of play. As a result, we need to identify a more robust search algorithm suitable for Go.

14.2   Monte Carlo Tree Search

Monte Carlo Tree Search (MCTS) is a widely used search algorithm for finding optimal moves in games and other decision-making domains. It combines tree search with Monte Carlo random sampling to estimate the value of states and actions and uses these estimates to make better decisions.

MCTS works by iteratively building a search tree from the current state, selecting promising nodes to explore, expanding the tree with new nodes, simulating game outcomes from these new nodes using Monte Carlo sampling, and updating the tree structure with the results. One major advantage of MCTS is that it requires little or no domain knowledge to conduct the search, making it a popular choice for planning in complex domains, such as game of Go.

One of the key advantages of MCTS is its ability to handle complex decision-making problems in a variety of domains. Unlike other search algorithms, MCTS does not require domain-specific evaluation functions or heuristics to guide the search. Instead, it relies on Monte Carlo sampling to explore the search space and evaluate potential moves, making it a flexible and general-purpose algorithm.

The success of MCTS has been most notable in the game of Go, where MCTS-based AI programs have achieved human master level play on small board sizes. For example, in 2006, the first computer programs using MCTS were implemented by Sylvain Gelly and Yizao Wang [7]. Since then, MCTS has been rapidly adapted in the computer Go. In 2008, Sylvain Gelly and David Silver [8] proposed a computer program called MoGo (which uses MCTS and other techniques) which can play the Go at human master level on 9×9 board. MCTS has also been used successfully in other games, such as poker, chess, and shogi.

In summary, Monte Carlo Tree Search is a powerful and versatile algorithm for decision-making in complex domains. With continued development and refinement, it is likely to remain a valuable tool for solving challenging decision-making problems.

Four Phases of MCTS

As illustrated in Fig. 14.2, the general MCTS process could be summarized into the following four phases:

- Selection: Starting from the root node, some predefined child selection rule or policy is recursively applied to descend through the tree to select the most promising child node (e.g., the child node with the highest value or highest Upper Confidence Bound (UCB) score), until the process reaches an expandable node.
- Expansion: The search tree is expanded by adding one or more child nodes to it, according to the available moves (or actions).
- Simulation: Starting from the newly added node(s), a simulation is run according to some rollout policy, which determines the sequence of actions taken from the current state until a terminal state is reached.
- Backup: The results from the simulation (e.g., reward signal) are backed up to the traversed path in the search tree, and the statistics (e.g., average reward and UCB score) are updated.

Here, we discuss some of the common terms in MCTS. A root node is the node that represents the root of the search tree, corresponding to the current game (non-terminal) state. A node is expandable if it represents a non-terminal state and has unvisited (or unexplored) children, meaning there are still moves that have not been tried in the state. A node is fully expanded if all its children have been

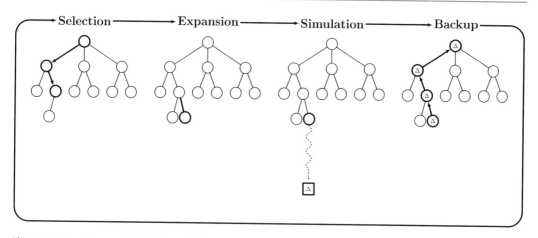

Fig. 14.2 How the general MCTS search algorithm works. The concept was derived from the work by Browne et al. [9]

visited at least once. A leaf node is a node that represents a non-terminal state and does not have any children, meaning no move has been tried yet. The leaf node could be the root node when the MCTS process is just getting started, or it could be some fully unexpanded node in the search tree.

While the tree shown in the selection phase in Fig. 14.2 has multiple nodes at different levels, the tree starts with just a single root node and is built from iteration to iteration in an incremental and asymmetric manner, with each iteration starting from the root of the tree. Figure 14.3 shows a simplified example to illustrate the process. In this simple case, we show a game in which there are only three possible moves for every game state.

When the search process is just getting started (first iteration in the figure), the search tree is completely empty. Given that the tree is completely empty, there is no applicable child node selection phase. A root node is created and added to the tree, and this newly added root node is also the leaf node. A simulation is run to the end of the game from the leaf node, using some rollout policy (such as a uniform random policy), and the reward signal is used to update the statistics of the traversed nodes during the backup phase. For example, if the simulation results in a win with reward signal $r = 1$, then the statistics for the root node would be $N = 1$, $W = 1$, where N represents the total number of visit count, and W represents the total action values.

At the second iteration, the root node is still expandable because it has unexplored moves, so there is no selection phase. According to the standard MCTS procedure, the tree is expanded by creating and adding a new child node (corresponding to a previously unexplored move) to the root node. This newly added child node becomes the current node and is also a leaf node. Again, a simulation is run to the end of the game from this leaf node, and the results from the simulation are backed up all the way up to the root node (as indicated by the heavy weighted edges and nodes in the figure). Let's assume that the simulation results in a loss with reward signal $r = -1$, then the statistics for the newly added child node should be $N = 1$, $W = -1$, and the statistics for the root node become $N = 2$, $W = 0$. Similar processes happen in the third and fourth iterations, but during the backup phase, only the traversed path (indicated by the heavy weighted nodes and edges) will be updated.

Starting with the fifth iteration, all four phases of MCTS are carried out since the root node is now fully expanded (all moves have been tried). The search process selects a (best) child node starting from the root node, according to some tree policy. Let's assume the tree policy selects the middle child node as shown in the figure, and it becomes the current node. Since the current node is expandable, a new child node (leaf node) is created and added to the tree. A simulation is run to the end of the game from

1 iteration   2 iterations   3 iterations   4 iterations

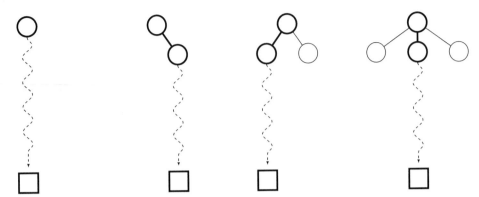

5 iterations   6 iterations   7 iterations   8 iterations

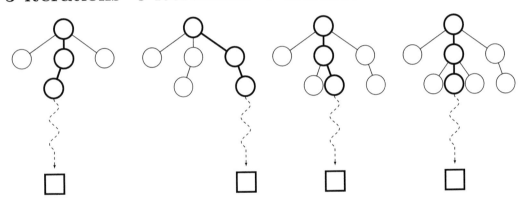

Fig. 14.3 Example of how the general MCTS search algorithm builds the search tree over multiple iterations

this newly added node, and the statistics of the traversed path are updated according to the simulation results once the simulation reaches a terminal state.

As the search continues with more iterations, often hundreds or thousands of iterations, more and more nodes will be added to the search tree, and the statistics of these nodes will be updated according to the simulation results and the traversed path from these different iterations.

After all the search iterations are completed, a child is selected from the root node of the tree according to some predefined rules (such as selecting the most visited child), and the move corresponding to the selected child is played in the actual game. It's important to note that there may be other predefined rules for selecting a move to play, and these can be tailored to specific games or applications.

One of the key strengths of the MCTS algorithm is its ability to select the most promising nodes in the search tree, leading to an asymmetric tree over time.

In MCTS, the algorithm builds a search tree by iteratively selecting nodes to expand and evaluate. The selection of nodes is guided by a tree selection policy, which balances exploration of unexplored regions of the search space with exploitation of promising regions. With the right tree selection policy, MCTS tends to favor more promising nodes without allowing the selection probability of the other nodes to converge to zero, which leads to an asymmetric tree over time.

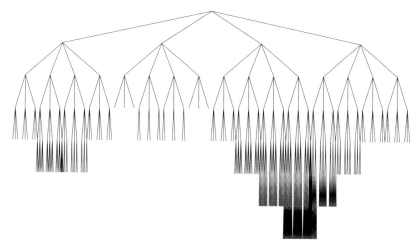

Fig. 14.4 Example of asymmetric tree growth for MCTS, where the algorithm prioritizes exploration of the most promising nodes rather than exhaustively traversing all possible nodes and moves. This selective expansion strategy allows MCTS to efficiently explore large and complex game trees. Concept derived from the work by Browne et al. [9]

This asymmetry is beneficial as it allows the algorithm to focus its search on the most promising and important regions of the search space while still maintaining some exploration (Fig. 14.4). This property makes MCTS highly efficient even when dealing with very large search spaces, such as the game of Go.

The pseudocode for general MCTS is shown in Algorithm 1, adapted from the work by Browne et al. [9]. The search process continues for a fixed computation budget, either in terms of time or number of iterations. Each iteration begins at the root node of the tree, denoted by v_0. The tree policy is applied iteratively to traverse the tree until the most promising leaf node v_l is found. At this point, the leaf node is expanded and simulated to obtain the results Δ. These results are then backed up all the way to the root node. After the computation budget has been reached, the best child is selected from the root node, and the corresponding action is taken in the actual game. Note that the only unknown part of the pseudocode is the function BESTCHILD, which we will discuss in the next section.

Here, v is a node of the search tree, $s(v)$ is the corresponding environment state for note v, $N(v)$ is the visit count for node v, and $W(v)$ is the total values.

Upper Confidence Bounds for Trees

We will now discuss how MCTS selects the best child during the search, which is a crucial component of the tree policy. In MCTS, the search tree is constructed based on how nodes are chosen in the tree, so selecting the best child is critical to the success of MCTS. One naive implementation is to always choose the child node with the highest mean reward, which is the estimated state-action value. However, this approach has a fundamental flaw: it does not explore other actions that may lead to a better overall strategy. This is essentially the exploration-exploitation problem we introduced in Part I of the book, which is an important topic in the context of reinforcement learning.

Before we delve into how MCTS selects the best child, let's first examine another algorithm called Upper Confidence Bounds (UCB). The UCB algorithm was originally developed to solve bandit problems, which is a special case of reinforcement learning tasks in which the agent takes only one action, and the environment immediately transitions into a terminal state, just like the bandit machines

---

Algorithm 1: General MCTS

---

1 **function** MCTSSEARCH (s_0):
2   Create root node v_0 with current state s_0
3   **while** within computation budget **do**
4          $v_l \leftarrow$ TREEPOLICY(v_0)
5          $\Delta \leftarrow$ DEFAULTPOLICY($s(v_l)$)
6   BACKUP(v_l, Δ)
7   **return** a(BESTCHILD(v_0))
8
9 **function** TREEPOLICY (v):
10   **while** $s(v)$ is non-terminal **do**
11   **if** v is expandable **then**
12   **return** EXPAND(v)
13   **else**
14             $v \leftarrow$ BESTCHILD(v)
15   **return** v
16
17 **function** EXPAND (v):
18   Choose $a \in A(s(v))$, where $A(s(v))$ is the set of unexplored actions
19   Add a new child v' to v, where $s(v')$ is the successor state of $s(v)$, and $a(v') = a$
20
21   **return** v'
22
23 **function** BESTCHILD (v):
24   **return** most visited child of v
25
26 **function** DEFAULTPOLICY (s):
27   **while** s is non-terminal **do**
28   Choose $a \in A(s)$ uniformly random
29   Take action a in the simulation environment, and observe successor state s'
30         $s \leftarrow s'$
31   **return** r, reward from terminal state s
32
33 **function** BACKUP (v, Δ):
34   **while** v is not null **do**
35         $N(v) \leftarrow N(v) + 1$
36         $W(v) \leftarrow W(v) + \Delta$
37         $v \leftarrow$ parent of v
38

---

in a casino. In this case, there are no sequences of states or actions in a single episode, just a single state and a single action. As the name suggests, the algorithm computes an Upper Confidence Bound for each of the available actions and tries to select the action that can maximize the UCB score.

The UCB algorithm is a widely used approach for solving the multi-armed bandit problem, which involves selecting actions to maximize a reward over time. The simplest version of UCB is called UCB1, which was proposed by Auer, Cesa-Bianchi, and Fischer in their paper "Finite-time Analysis of the Multiarmed Bandit Problem" [10].

The UCB1 algorithm selects the action that maximizes the UCB1 score, which is computed using Eq. (14.1):

$$UCB1 = \overline{X}_j + \sqrt{\frac{2 \ln N}{n_j}} \tag{14.1}$$

where $\overline{X}_j$ is the average value (or mean state-action value) for action j, and n_j is the number of times action j has been played. N is the total number of times the game (episodes) has been played so far in the context of bandit problems.

The first term of the UCB1 score, $\overline{X}_j$, encourages the agent to select the action with the highest value, which is known as exploitation. The second term, $\sqrt{\frac{2\ln N}{n_j}}$, encourages the agent to explore other actions that may have higher potential reward, as it depends on the number of times an action has been played. As $n_j = 0$, the UCB1 value becomes ∞, so unvisited actions are assigned the largest value.

We can adapt the UCB1 algorithm to select the best child node for Monte Carlo Tree Search (MCTS). This modified algorithm is often called Upper Confidence Bounds for Trees (UCT). The basic idea is to select the child node that maximizes the UCB score, which is computed using Eq. (14.2):

$$UCB = \overline{X}_j + 2C\sqrt{\frac{2\ln N}{n_j}} \tag{14.2}$$

where N is the number of times the current node's parent has been visited, and n_j is the number of times child j has been visited. C is a constant that controls the amount of exploration during the search. If multiple child nodes have the same maximal value, the tie is usually broken by randomly selecting one of those nodes. As with UCB1, when $n_j = 0$, the UCB value becomes ∞. This ensures that all children of a node are visited at least once before any further expansion takes place.

To see how UCT works in practice, let's consider a game where a player moves a pawn on a chessboard. The player can choose from several possible moves on each turn, but doesn't know which move is the best. The goal is to find the sequence of moves that leads to the best outcome (e.g., winning the game).

During the search, the UCT algorithm maintains a tree of game states and possible moves, with each node representing a game state and each edge representing a possible move. The search starts from the root node, which represents the current state of the game. At each step, the algorithm selects a child node to expand based on the UCB score and continues until it reaches a leaf node.

Once a leaf node is reached, a simulation or rollout is executed until the game reaches the terminal state. The simulation result value is then backpropagated up the tree, updating the UCB scores of each node along the path that was taken. This allows the algorithm to gradually improve its estimate of the value of each possible move and to explore promising moves more thoroughly.

The pseudocode for using the UCT algorithm to select the best child is shown in Algorithm 2, where v is a node in the tree, v' is a child of node v, and c is some constant as shown in Eq. (14.2).

Algorithm 2: UCT—best child

---

1 **function** BESTCHILD (v, c):

2   **return** $\displaystyle\arg\max_{v' \in \text{children of } v} \left(\frac{Q(v')}{N(v')} + c\sqrt{\frac{2\ln N(v)}{N(v')}} \right)$

We can then plug in the preceding UCT algorithm to the general MCTS algorithm, which gives us a powerful framework for balancing exploration and exploitation in reinforcement learning problems. By using these algorithms, agents can gradually improve their performance over time by exploring new actions while also exploiting actions that have proven to be effective in the past.

MCTS for Two-Player Zero-Sum Games

The general MCTS algorithm discussed earlier is for single-player games where the agent makes all the moves. However, for two-player zero-sum games like chess or Go, the situation becomes more complex. This is because we need to consider the perspective of both players when backing up the simulation results Δ.

In two-player zero-sum games, the outcome of the game is always a zero-sum result, meaning that for one player to win, the other player must lose. Therefore, we need to calculate the reward for each player separately. For example, in a game of chess, if we use MCTS to plan the next move for the black player, and the simulation results in a win for black, then we should assign a reward of $+1$ to all the edges representing black's turn to move in the traversed path of the search tree. However, for white, the reward should be -1, as this outcome represents a loss for them.

One solution is to switch the sign of Δ for the opponent player to accurately reflect the nature of the game. This is because in two-player zero-sum games, one player's win is always another player's loss. Therefore, using the same Δ for both players would not accurately represent the outcome of the game.

Figure 14.5 shows an example of how MCTS backs up the simulation results based on the toy game Tic-Tac-Toe. In this game, two players take turns playing on a three-by-three grid. One player plays Xs and the other plays Os. The first player to get three marks in a row (horizontally, vertically, or diagonally) wins the game. If the board fills up and neither player has three in a row, then the game is a draw.

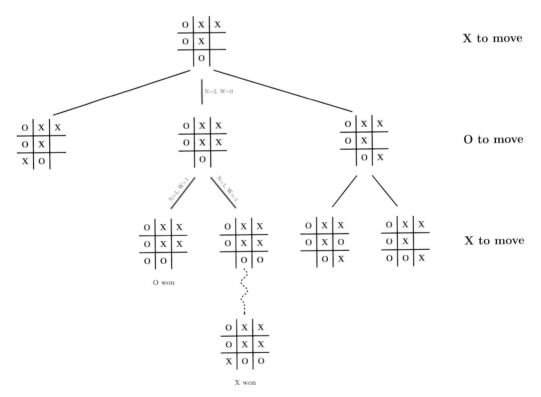

Fig. 14.5 Example of how general MCTS backs up simulation results for a two-player zero-sum game based on the toy game Tic-Tac-Toe

To keep things simple, we focus only on how to back up the results from the node that leads to a terminal state, which is the one labeled with the winner of the game in Fig. 14.5, and we ignore all the intermediate nodes in the search tree.

A simple pseudocode for the backup for two-player zero-sum games is shown in Algorithm 3, where we are switching the sign for the simulation results Δ every time we've used it to update the search tree statistics.

Algorithm 3: MCTS backup for two-player zero-sum games

---

1 **function** BACKUP (v, Δ):
2   **while** v is not null **do**
3          $N(v) \leftarrow N(v) + 1$
4          $W(v) \leftarrow W(v) + \Delta$
5          $\Delta \leftarrow -\Delta$
6          $v \leftarrow$ parent of v

Once we have computed the values of the nodes in the traversed path, it is important to consider how we calculate mean values for two-player zero-sum games. Unlike in single-player games, where the nodes in the search tree all store values for the same player, in two-player games, a high value of a child node may not necessarily represent the best interest of the parent node. This is because the value of a node in a two-player game depends on the perspective of the player whose turn it is.

For example, consider the game of Tic-Tac-Toe. Suppose the current player is X and the search algorithm is evaluating a potential move that leads to a position where X has two in a row and O has one in a row. From X's perspective, this is a good position and should have a high value. However, from O's perspective, this position is bad and should have a low value. Therefore, when computing the mean value of this node, we need to take into account whose turn it is and how the value of the node is computed based on that perspective.

In summary, when using MCTS for two-player zero-sum games, it is crucial to carefully consider how to update the statistics of nodes in the search tree and compute the mean values of their children nodes. This is because the values of each node represent the expected reward that a player can obtain if they select that node and play optimally.

14.3 AlphaZero

In October 2015, DeepMind's AI agent AlphaGo made history by beating the European champion Fan Hui in a five-game Go match [11]. AlphaGo won all five games, becoming the first computer Go program to beat a professional human player on a full-sized board without handicap. In March 2016, another version of AlphaGo, called AlphaGo Lee, defeated the legendary Korean player Lee Sedol with a score of 4-1 in a five-game Go match.

AlphaGo was trained using a combination of deep neural networks, supervised learning, reinforcement learning, and Monte Carlo Tree Search (MCTS) for search and planning. During actual gameplay, AlphaGo also uses MCTS for online planning.

The training process for AlphaGo can be divided into two phases:

- **Phase one**: Trains a policy network P_s and a value network V_s using 30 million positions from the KGS Go Server. This phase uses pure supervised learning, with classification used to train P_s and regression used to train V_s. The policy network P_s outputs the action probability for a given board position, and the value network V_s outputs the evaluation of the board from the current player's perspective. Additionally, a smaller but faster policy network $P_{rollout}$ is trained. This faster policy network has much lower accuracy compared to P_s and is only used in phase two during self-play (as the rollout policy for MCTS simulation). Phase one continues until the networks P_s and V_s can correctly predict the human expert moves in a separate test set to a certain threshold (e.g., 55%).

- **Phase two**: Trains the policy network P and value network V using reinforcement learning over millions of self-play games. Initially, $P = P_s$ and $V = V_s$, and multiple actors run in parallel to generate self-play games, the policy network P is used during the MCTS search when these games are generated. In addition, the faster sample policy network $P_{rollout}$ from phase one is also used to sample actions during the MCTS simulation phase.

In 2017, DeepMind developed a new agent called AlphaGo Zero [12], which is an enhancement to AlphaGo and is much stronger than its predecessor. AlphaGo Zero achieves stronger play in Go than its predecessor, while it is trained using only reinforcement learning and moves from pure self-play games. It also simplifies the training process by eliminating supervised learning and uses a much simpler neural network architecture (a single neural network with shared weights and multiple heads, instead of separate networks).

In late 2017, DeepMind developed a new agent called AlphaZero [13], which is a single system that taught itself from scratch how to master the games of Chess, Shogi, and Go, beating a world-champion computer program in each case.[1] This new AlphaZero agent is a generalization of its predecessor, AlphaGo Zero, which can only play the game of Go.

To summarize, AlphaZero is considered a state of the art for challenging two-player zero-sum games such as Go, Chess, and Shogi. It is trained using only reinforcement learning and moves from pure self-play games, without any human supervision other than the basic rules of the game. During actual gameplay, AlphaZero also uses MCTS for online planning. The training process involves millions of self-play games. The policy network outputs the action probability for a given board position, and the value network outputs the evaluation of the board from the current player's perspective.

In the upcoming sections of this chapter, our attention will be directed toward the intricacies of the AlphaZero agent and its corresponding paper [13]. However, it is important to acknowledge that specific components, such as the neural network structure and the MCTS search algorithm, were originally introduced and extensively examined in the AlphaGo Zero paper [12].

Neural Network Architectures and State Representation

AlphaZero uses a neural network architecture that employs convolutional neural networks (CNNs) to identify patterns in the board positions of games like Go, Chess, and Shogi. CNN has a proven history in tasks like classification and object detection in the domain of computer vision. These patterns are essential for effective decision-making in these games. Professional human players are adept at identifying these patterns during gameplay, and we want our reinforcement learning agent to do the same. For the game of Go, the board is a 19×19 grid. We can represent the board as a $19 \times 19 \times 1$ binary feature plane where 1s represent intersections with stones and 0s represent empty intersections.

---

[1] AlphaZero has demonstrated exceptional mastery in playing games such as Chess, Shogi, and Go at a level comparable to that of a grandmaster. However, it is crucial to note that this impressive capability does not imply that a single AlphaZero agent possesses expertise in all three games. In the case of AlphaZero, each agent is specialized in a particular game and is exclusively trained to excel in that specific domain.

However, the standard representation of the Go board as a $19 \times 19 \times 1$ binary feature plane has limitations. It doesn't distinguish between black and white stones, nor does it provide enough information for decision-making or the Markov property. To overcome these limitations, we use separate binary feature planes for black and white stones, and we stack N history of board positions together.

The final state s_t is a $19 \times 19 \times 17$ image that contains 17 binary feature planes. Eight of these planes X_t represent the current player's stones, with a binary value of 1 if a stone is present and 0 if the intersection is empty or contains an opponent stone. Similarly, the other eight planes Y_t represent the opponent's stones. The additional plane C called the color-to-play indicates which player's turn it is and takes into account the fact that the best move may differ for the two players in the same board position. Additionally, it provides the information on who plays first (default black plays first), which often has some advantages.

To ensure the Markov property holds, we stack up to $N = 8$ previous board positions for each player together, including the current one, creating a 3D image of the board. We concatenate the planes into a single input using the following format:

$$s_t = [X_t, Y_t, X_{t-1}, Y_{t-1}, ..., X_{t-7}, Y_{t-7}, C]$$

AlphaZero uses a single neural network consisting of many residual blocks of convolutional layers and rectifier nonlinear activations (ReLU). Unlike AlphaGo, AlphaZero does not use a separate rollout policy network. We'll discuss the reason for this later when we describe AlphaZero's Monte Carlo Tree Search algorithm. The input features s_t first get processed by a convolution layer of 256 filters of kernel size 3×3, with stride 1, then followed by a batch normalization layer and a rectifier nonlinearity (ReLU) activation. This is followed by 19 (or 39) residual blocks, and the output from the last residual block is then fed into two separate heads for computing the policy and value, respectively.

Each residual block applies the following layers sequentially:

- A convolution of 256 filters of kernel size 3×3, with stride 1
- Batch normalization
- A rectifier nonlinearity (ReLU) activation
- A convolution of 256 filters of kernel size 3×3, with stride 1
- Batch normalization
- A skip connection that adds the input to the block
- A rectifier nonlinearity (ReLU) activation

The policy head applies the following layers sequentially:

- A convolution of two filters of kernel size 1×1, with stride 1
- Batch normalization
- A rectifier nonlinearity (ReLU) activation
- A fully connected linear layer that outputs a vector of size $19 \times 19 + 1 = 362$, which corresponds to the action probabilities for all actions, including the pass move

The value head applies the following layers sequentially:

- A convolution of one filter of kernel size 1×1, with stride 1
- Batch normalization
- A rectifier nonlinearity (ReLU) activation
- A fully connected linear layer with 256 hidden units
- A rectifier nonlinearity (ReLU) activation
- A fully connected linear layer that outputs a scalar value
- A tanh nonlinearity (Tanh) activation that transforms the scalar value in the range $[-1, 1]$

Figure 14.6 shows the AlphaZero neural network architecture for the game of Go, which utilizes 19 residual blocks. A similar architecture is used for other games such as Chess or Shogi. However, the input data for the neural network is slightly different for each game. For Chess, the input data is an $8 \times 8 \times 73$ stack of planes, while for Shogi, it is a $9 \times 9 \times 139$ stack of planes. Additionally, DeepMind employs a slightly different representation mechanism when stacking feature planes to construct the current state s_t for Chess and Shogi.

This architecture allows AlphaZero to process a complete board position as a single input and to identify complex patterns and features that are crucial for effective gameplay.

MCTS Search Algorithm

AlphaZero uses a much simpler MCTS search algorithm. Unlike traditional MCTS, there is no simulation (rollout) during the search. Instead, the neural network evaluates the position, making the faster rollout policy network unnecessary. The overall search process is shown in Fig. 14.7.

As usual, each node s in the search tree represents a specific board configuration or state. Each node contains edges (s, a) for all legal actions $a \in A(s)$, where $A(s)$ is the set of all legal actions in state s. Each edge stores the following statistics:

- $N(s, a)$: The visit count
- $W(s, a)$: The total state-action value
- $Q(s, a)$: The average state-action value, where $Q(s, a) = \dfrac{W(s, a)}{N(s, a)}$
- $P(s, a)$: The prior probability of selecting action a in s, which is obtained from the neural network's output vector p

The original AlphaZero algorithm uses 800 iterations per MCTS search, which is half the number used by the AlphaGo algorithm. It's worth noting that the term "simulation" is sometimes abused when referring to AlphaZero, as it does not use rollouts in the same way that other algorithms might. However, the term is still widely used in the literature.

Selection

In the MCTS algorithm used by AlphaZero, the search starts with the selection phase. At each search iteration, the algorithm begins at the root of the search tree and iteratively selects actions until it reaches a leaf node at time step L. For time steps $0 < t < L$, an action a_t is selected using a variant of the Upper Confidence Bound applied to Trees (UCT) algorithm, along with the set of legal actions at time step t to prevent it violating the basic rules of the game. The UCT algorithm balances exploitation

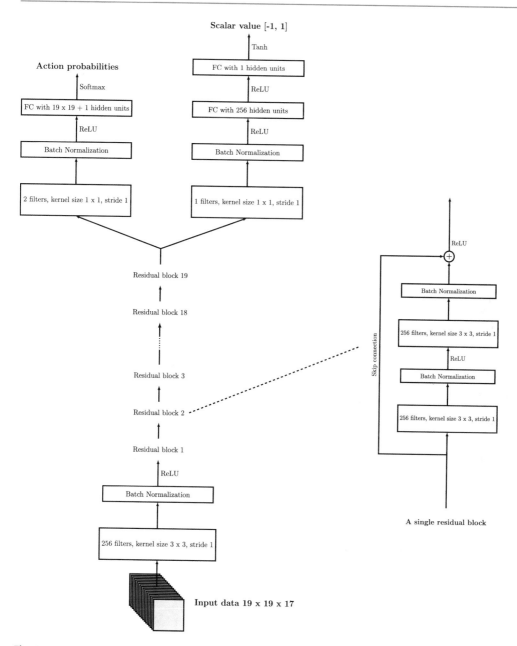

Fig. 14.6 Neural network architecture for the AlphaZero agent for the game of Go

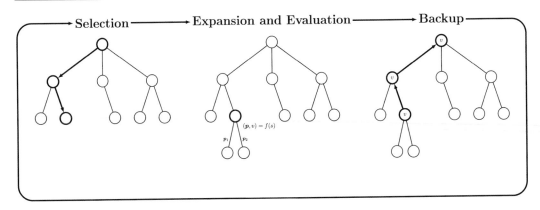

Fig. 14.7 The MCTS search algorithm for AlphaZero

Algorithm 4: MCTS for AlphaZero

---

1 **function** MCTSSEARCH (s_0):
2   Create root node v_0 with current state s_0
3   **while** within computation budget **do**
4           $v_l \leftarrow$ SELECTLEAF(v_0)
5   Evaluate position $s(v_l)$ using neural network: $\boldsymbol{p}, v \leftarrow f(s(v_l))$
6   EXPAND($v, \boldsymbol{p}$)
7   BACKUP(v_l, v)
8   **return** most visited child of v_0
9
10 **function** SELECTLEAF (v):
11   **while** v is fully expanded and $s(v)$ is non-terminal **do**
12          $v \leftarrow \arg\max_a Q(s(v), a) + U(s(v), a)$
13   **return** v
14
15 **function** EXPAND ($v, \boldsymbol{p}$):
16   **for** each action $k = 0, 1, \ldots$ **do**
17   Add child node v_k to v, where:
18          $N(v_k) \leftarrow 0$
19          $W(v_k) \leftarrow 0$
20          $Q(v_k) \leftarrow 0$
21          $P(v_k) \leftarrow \boldsymbol{p}_k$
22
23 **function** BACKUP (v, Δ):
24   **while** v is not null **do**
25          $N(v) \leftarrow N(v) + 1$
26          $W(v) \leftarrow W(v) + \Delta$
27          $Q(v) \leftarrow \frac{W(v)}{N(v)}$
28          $\Delta \leftarrow -\Delta$
29          $v \leftarrow$ parent of v
30

and exploration. Specifically, the algorithm selects the action that maximizes the sum of the average state-action value $Q(s_t, a)$ and an exploration term $U(s_t, a)$:

$$a_t = \arg\max_a Q(s_t, a) + U(s_t, a)$$

Here, $Q(s_t, a)$ is the average value of the state-action pair, and $U(s_t, a)$ is computed using the following equation:

$$U(s, a) = C(s)p(s, a)\sqrt{\frac{N(s)}{(1 + N(s, a))}} \qquad (14.3)$$

where $N(s)$ is the visit count of the parent node, $N(s, a)$ is the visit count of the state-action pair, $p(s, a)$ is the probability of selecting the action, and $C(s)$ is an exploration rate that grows slowly with search time. Specifically, $C(s)$ is given by the following equation:

$$C(s) = \log\left(\frac{1 + N(s) + c_{base}}{c_{base}}\right) + c_{init} \qquad (14.4)$$

It is important to note that the computation method for the exploration rate $C(s)$ described here is not mentioned in the original AlphaGo Zero or AlphaZero papers. This approach is exclusively discussed in the MuZero paper [14], which was published subsequent to the AlphaZero algorithm.

The values assigned to the parameters c_{base} and c_{init} are typically set as 19652 and 1.25, respectively.

Expansion and Evaluation

When the search process reaches a leaf node in the search tree, the game state s_t corresponding to this leaf node is used as the input to the neural network θ to obtain the corresponding output (p, v). Recall that the network outputs a vector p containing action probabilities for all actions, as well as an evaluated value v for the current board, all from the current player's perspective. All children of the leaf node are then expanded simultaneously in a single operation, since we have the action probabilities for all potential actions. The statistics for the edges to this leaf node are initialized as $N(s, a) = 0$, $W(s, a) = 0$, $Q(s, a) = 0$, $P(s, a) = p_a$, where p_a is the probability of selecting action a in state s, derived from the action probabilities in p. In AlphaZero, the evaluation value v replaces the results of random simulation (or rollout) used in traditional MCTS, so there is no simulation phase; thus, we don't need to train a rollout policy network.

In practice, it is often unnecessary to create nodes that represent illegal moves, since an agent should never choose them. This is particularly true in later stages of the game when much of the board is non-empty or when certain moves are prohibited by the game's rules. Limiting the creation of nodes to only those that represent legal moves can significantly reduce memory usage and speed up the expansion process, which can be particularly important when performing large amount of iterations.

Backup

In the final backup phase, the value v, which is coming from the neural network, is backed up for each time step $t \leq L$. During this phase, the visit count $N(s_t, a_t)$ for the state-action pair (s_t, a_t) is updated as $N(s_t, a_t) = N(s_t, a_t) + 1$. The total state-action value $W(s_t, a_t)$ is also updated as

$W(s_t, a_t) = W(s_t, a_t) + v$, where v represents the value that is being backed up. Moreover, the mean state-action value $Q(s_t, a_t)$ is computed as $Q(s_t, a_t) = \frac{W(s_t,a_t)}{N(s_t,a_t)}$.

Since we're dealing with two-player zero-sum games, the value v that is being backed up is occasionally flipped. For example, if the current player is the maximizing player, the backed-up value v will be negated to represent the opponent's utility.

After completion of the three phases, one search iteration is considered complete. The algorithm then proceeds to the next iteration, and this process continues until a limit is reached, such as a time constraint or the maximum number of iterations.

How AlphaZero MCTS Builds Search Tree

Since AlphaZero would expand all children nodes instead of one child, the search tree is built slightly different compared to the general MCTS. A simple example is shown in Fig. 14.8.

To summarize, several modifications have been made to the standard MCTS algorithm for AlphaZero, and we highlight them here:

- After the search reaches a leaf node, there is no rollout. Instead, AlphaZero uses the neural network to evaluate the board position and uses that as an estimated game result to update the statistics in the search tree.
- When expanding a leaf node, all children are expanded in a single operation, rather than the standard MCTS, which expands one child at a time. This means that after node expansion, a leaf node immediately becomes fully expanded.
- AlphaZero uses a slightly different UCT algorithm to select the best child during the selection phase, which incorporates the prior action probabilities $\boldsymbol{p}$ from the output of the neural network.

There are additional techniques that DeepMind adapted to further speed up the MCTS search process, such as using parallel tree search and reusing subtrees. These are advanced features that are nice to have but are much more complex and strongly depend on how we implement the algorithms (e.g., in which programming language). However, these are nice to have but not mandatory in order to make the AlphaZero agent work. Interested readers could find more information on these topics in the original AlphaGo Zero and AlphaZero papers.

Self-Play

To generate high-quality training data to train the neural network, AlphaZero uses a self-play approach with Monte Carlo Tree Search (MCTS).

During each self-play game, for each non-terminal time step $0 \leq t \leq T$, an MCTS search is conducted using the latest neural network parameters, which typically consists of thousands of iterations. After the search is complete, a stronger search policy $\boldsymbol{\pi}_t$ is computed for the current game state s_t. This policy is proportional to the exponential visit count of the children based on the root node, with a temperature parameter τ controlling the level of exploration:

$$\pi(a|s_0) = \frac{N(s_0, a)^{1/\tau}}{\sum_b N(s_0, b)^{1/\tau}} \tag{14.5}$$

where $N(s_0, a)$ represents the visit count of the child node resulting from taking action a in state s_0. The policy $\boldsymbol{\pi}_t$ is stronger than the raw output vector P_t from the neural network and can be viewed as a powerful policy improvement operator. To play the game, a move a is selected based on the policy

1 iteration

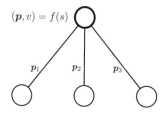

2 iterations

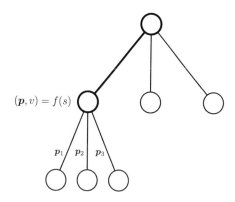

3 iterations

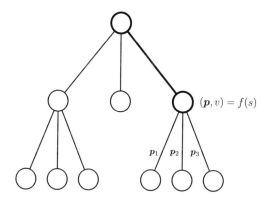

4 iterations

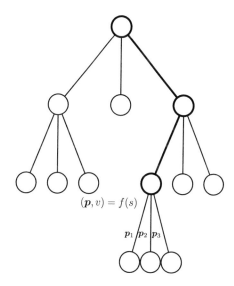

Fig. 14.8 A simple example to illustrate how the search tree is built for AlphaZero's MCTS search algorithm

π_t, and the game transitions to the next state, starting a new MCTS search process. This policy also becomes the target when training the neural network.

A move a_t is then sampled from π_t to play at game state s_t. For the first 30 moves of each self-play game, the temperature in the softmax function (defined as $softmax(x_i) = \frac{\exp(x_i/\tau)}{\sum_j \exp(x_j/\tau)}$) is set to $\tau = 1$. This selects moves proportionally to the visit count of the children in the root node of the MCTS search tree. For the remainder of the game, a temperature close to zero is used ($\tau \rightarrow 0$), which corresponds to always selecting the most visited child of the root node of the MCTS search tree.

Exploration

To encourage further exploration, a small amount of Dirichlet noise is added to the prior probabilities $P(s, a)$ in the root node s_0. Specifically, $P(s, a) = (1 - \epsilon)P_a + \epsilon \eta_a$, where $\eta \sim Dir(0.03)$ and $\epsilon = 0.25$. This is done for each MCTS search.

Resign on Losing Games

During self-play, AlphaZero uses a resign mechanism to avoid wasting computation resources on clearly losing games. The algorithm achieves this by checking the evaluated game results and choosing to resign on those games where the player is clearly losing. This happens after running Monte Carlo Tree Search (MCTS) for 800 iterations. If the mean state-action value of the root node of the MCTS search tree is lesser than a certain threshold, for example, -0.9, the player would choose to resign instead of continuing to play the game.

It is important to note that the automated resign mechanism only happens after a game has reached a certain number of moves, for example, each player has made 10 or 20 moves. It makes no sense to voluntarily resign at the beginning of a game.

However, in order to measure the false positive rate of the resigned games, AlphaZero intentionally disables the resign move for 10% of the self-play games. In such cases, the agent would play the game until reaching the final state. The resign mechanism is a crucial part of the AlphaZero algorithm, as it helps the algorithm focus its computation resources on games where it has a better chance of winning. This, in turn, helps the algorithm learn faster and achieve better results in the long run.

Samples for Training

The goal of generating these self-play games is to collect samples for training the neural network. Specifically, for every self-play game, we collect the transition tuple (s_t, π_t, z_t) for time steps $0 \leq t \leq T$. Here, s_t is the game state at time step t, which is a stacked binary feature plane with the dimensions $19 \times 19 \times 17$ for the game of Go. π_t is the stronger MCTS search policy from time step t, and z_t is the Monte Carlo sample return $G_t = R_t + R_{t+1} + R_{t+2} + \cdots + R_{T-1}$ with no discount. Since all the rewards are zero for non-terminal states, this reduces to $G_t = R_{T-1}$, which is the actual game result, a value of -1 for a loss, 0 for a draw, and +1 for a win.

This mechanism of assigning the final game outcome to every move helps to address the credit assignment issue. So all the decision made by the agent gets assigned the equal amount of credit, depending on the final outcome of the self-play game.

All (s_t, π_t, z_t) are represented from the perspective of the player at time step t. For example, if the self-play game ends up with black winning, then for every time step $t = 0, 1, 2, \ldots, T$, if it's black's turn to move at time step t, we set $z_t = +1$; conversely, if it's white's turn to move at time step t, then $z_t = -1$. The same applies to s_t and π_t.

AlphaZero then stores millions of most recent self-play games into an experience replay buffer, much like the DQN algorithm does. These self-play games are then used to train the neural network. And the self-play actors would periodically switch to the latest neural network to generate more sample games.

Training

The objective of training the AlphaZero is to learn a policy that maximizes the expected return of a particular game, such as Go. Specifically, AlphaZero aims to maximize the expected return of the game from a given state, which is estimated using a value network.

Although AlphaZero uses a neural network to approximate the optimal policy, which is typically what policy-based methods do. However, it is not the same as any of the policy-based reinforcement learning algorithms such as REINFORCE or Actor-Critic we've discussed in the book. For AlphaZero, the objective function to optimize the policy is different from the policy gradient methods. The neural network is trained to maximize the similarity of the neural network's move probabilities p_t to the MCTS search probabilities π_t using a form of supervised learning known as cross-entropy loss. This method is often used in the case of multiclass classification.

The objective for training the policy network for AlphaZero can be written as

$$\mathcal{L}_{policy} = -\sum_a \pi(a|s) \log p(a|s) \tag{14.6}$$

where $p(a|s)$ is the predicted probability of selecting action a in state s by the policy network, and $\pi(a|s)$ is the target probability of selecting action a in state s obtained from the self-play games.

Thus, the policy objective function for AlphaZero can be seen as a form of maximum likelihood estimation. In maximum likelihood estimation, we seek to find the parameters of a statistical model that maximize the likelihood of the observed data. In the case of AlphaZero, the policy network is a statistical model that predicts the probability distribution over actions for a given state. The self-play game provides us with observed actions taken by the network in each state.

Although cross-entropy is often used in supervised learning, we should not categorize the AlphaZero as supervised learning, since the agent does not use any data provided by any form of supervisors (humans, for example). Instead, the neural network is trained using a variant of reinforcement learning called self-play, where the agent learns to improve its performance by playing against itself and adjusting its weights based on the outcomes of those games. This process is similar to how humans learn through trial and error, receiving feedback in the form of rewards or punishments.

During training, AlphaZero learns by playing against itself repeatedly, evaluating the outcomes of each game, and adjusting its neural network weights to improve performance. AlphaZero's neural network is trained using a combination of self-play and Monte Carlo Tree Search to improve its game-playing ability.

MCTS plays a crucial role in AlphaZero by enabling the agent to evaluate and select the best move in each game state. This process is iterative and leads to policy improvement as the system learns from its experiences and gradually improves its game-playing ability.

The neural network for AlphaZero is also trained to minimize the error between the predicted state value v_t and the game outcome z_t from self-play, which is often called the mean-squared error (MSE), as shown in the following equation:

$$\mathcal{L}_{value} = (v - z)^2 \tag{14.7}$$

Since we're using a single neural network with shared weights to represent the policy and value function, the final objective function for AlphaZero is the combined policy objective $\mathcal{L}_{policy}$ and value objective $\mathcal{L}_{value}$, which can be written using matrix forms, as shown in the following equation:

$$l = (z - v)^2 - \pi^\top \log p + c||\theta||^2 \tag{14.8}$$

where z is the true outcome of the game, v is the predicted outcome of the game, π is the policy vector, p is the predicted policy vector, θ is the parameter vector of the neural network, $(p, v) = f_\theta(s)$ represents the output of the neural network given a state s, and c is a hyperparameter that controls the level of L2 weight regularization to prevent the network from overfitting.

Note that in AlphaZero, the MSE loss for the state value is given the same weight as the cross-entropy loss for the policy, which differs from the traditional Actor-Critic algorithm that employs a coefficient to weight the loss of the state value function, in case we approximate the policy and value functions using a single neural network with shared weights.

To train the agent, a batch of 4096 transitions is randomly sampled from the recent 500,000 self-play games. The neural network is then optimized using the stochastic gradient descent (SGD) method. The initial learning rate is set to 0.01, momentum to 0.9, and L2 regularization to 10^{-4}. Moreover, the algorithm decays the learning rate using a staircase method at 400,000, and 600,000 training steps, respectively, with a decay weight of 0.1 during the course of training. The training typically lasts over 700,000 training steps, with a total of 21 million self-play games for Go, 44 million for Chess, and 24 million for Shogi.

Not surprisingly, the AlphaZero was trained using a distributed reinforcement learning architecture, where thousands of self-play actors generate millions of self-play games. And multiple learner agents work in parallel to update the parameters of the neural network.

During training, 5000 tensor processing units (TPUs) were used to generate self-play games. TPU is a special acceleration hardware developed by Google to accelerate the numerical computation, especially for neural networks. This means hundreds or even thousands of machines were used to generate self-play games during training. In addition, 16 second-generation TPUs were used to train the neural network. DeepMind trained separate instances of AlphaZero for Chess, Shogi, and Go, respectively, and the training lasted for 9 hours in Chess, 12 hours in Shogi, and 13 days in Go approximately.

Evaluate Agent

To evaluate the performance of the AlphaZero agent during or after training, there are two main options available.

The first option is to let the AlphaZero agent play against various opponents, such as a (professional) human player or other AI agents, including itself with the weights of the neural network obtained from previous training iterations. And we would rate the performance of the agent using the standard Elo rating during these evaluation games. The AlphaZero agent uses MCTS search to find the best move for each position; this is sometimes called online planning, as the agent is actively using the model to search for the best move during the gameplay. The overall process is similar to how MCTS is applied to generate self-play games, but without using Dirichlet noise for exploration, and it always chooses the best move in every position—the most visited child of the root node of the MCTS search tree; additionally, it's often the case we might let the MCTS search run with more simulations. However, it's crucial to ensure that these evaluation games are not part of the training data if we're evaluating the performance of the agent during training.

The second option is to use the trained neural network to predict moves and game outcomes from historical (professional) human gameplay. This involves building an evaluation dataset using historical games played by (professional) human players. The AlphaZero agent's neural network is then used to predict the moves and actual game outcomes.

In practice, both methods are combined together to evaluate the agent, especially during training as is the case for the original AlphaZero agent. This dual evaluation mechanism can give us more confidence on the actual strength of the agent. This is because if we only evaluate the agent by setting it up to play against some opponent, for example, some online player or the previous model, it's hard to assert the opponent's true strength. And this will be a problem since the Elo rating assumes we have good estimate about the rating for both players to produce some accurate ratings. For instance, if the agent plays 100 games all against some amateur player that's weaker than the agent, then the agent might get a very high rating after winning all these games; thus, it might give us a false impression that the agent is doing pretty well, even though it's not the case.

On the other hand, if we use historical human games to evaluate the agent's performance, the quality of the historical (professional) human game also will impact the results of evaluation. For example, if we only have games that are played by amateur human players, then the prediction accuracy might be very high; this might also give us a false impression that the agent has achieved a strong play. On the other hand, if the (professional) human games are played by high-ranking professional players, then the agent might not be able to accurately predict the human moves and game outcomes, especially at the beginning of the training session. This might cause us to have doubt about the performance of the agent. It's important we use a variety of games from different levels and players to build the evaluation dataset, so we can get a more accurate estimate of the performance of the agent.

14.4 Training AlphaZero on a 9 × 9 Go Board

It is important to recognize that training an AlphaZero agent is a time-consuming endeavor, taking weeks or even months on a 19 × 19 Go board, even with powerful GPUs. DeepMind's original paper mentioned that they trained the AlphaZero agent using thousands of servers and TPUs. Despite such significant computational resources, it still took 72 hours to train the 20-block version of AlphaZero and 40 days for the 40-block version. Unfortunately, we lack access to a comparable level of computational resources and budget to conduct similar experiments.

Considering that our objective is not to create the strongest agent capable of defeating the world champion, we have made certain adjustments to our approach. These adjustments include using scaled-down settings, such as a smaller neural network, fewer self-play actors, and fewer games played, to train the AlphaZero agent. Specifically, we are focusing on Go with a 9 × 9 board size, which requires less computational resources and therefore less time to train the agent. After 50 hours of training on moderate hardware, our agent has achieved a strong level of play. Our best model can sometimes defeat an amateur 1d level opponent.[2]

Distributed Training

We employed a distributed RL architecture for our training process, utilizing approximately 120 actors, 1 learner, and 1 additional process responsible for evaluating the agent's performance. All of these components were executed on a single server equipped with 128 CPUs and 8 GPUs; each GPU has 24 GB of vRAM. The program utilized only the multiprocessing module from Python, without relying on any third-party tools or libraries for distributed computing.

---

[2] For the game of Go, players' skill levels in the game have been categorized using kyu and dan ranks. Kyu ranks, ranging from 30 k to 1 k, are considered ranks for students or beginners. On the other hand, dan ranks, spanning from 1d to 7d, represent advanced ranks for amateur players. Generally, beginners who have recently learned the rules of the game start around the 30th kyu rank. As they progress and improve, they move numerically downward through the kyu grades. Once players surpass the 1st kyu rank, they are awarded the 1st dan rank. From this point onward, their progress is measured numerically as they ascend through the dan ranks. Additionally, there are special professional dan ranks (1p–9p) specifically for professional players.

Neural Network Architecture and State Representation

Our neural network architecture and state representation were closely aligned with the approach described in the AlphaGo Zero and AlphaZero papers. To construct the state, we utilized the last eight board configurations, resulting in a final state represented as a $9 \times 9 \times 17$ image. While we largely adhered to the AlphaZero neural network architecture, we made some modifications by employing fewer blocks and convolutional filters, since a 9×9 board is much simpler than a 19×19 board.

In our initial experiment, we began with a version comprising 20 blocks, but with only 64 convolutional filters for the convolutional layers and 64 hidden units for the value head. This particular neural network consisted of 1,434,493 parameters. However, after 24 hours of training, we observed that the agent exhibited a high frequency of pass moves, which abruptly ended games within approximately 20 steps. This behavior was observed in both self-play actors and the evaluation process. We speculated that the 20-block version might be excessively powerful for the 9×9 problem, despite our attempt to scale it down by reducing the number of convolutional filters and hidden units. Consequently, we made the decision to halt the training session and explore alternative approaches.

Following the unsuccessful initial experiment, we tried different variants of neural network architectures. These included a 10-block version with 64 filters and hidden units, as well as a 15-block version with the same number of filters and hidden units. However, none of these configurations yielded satisfactory results after 24 hours of training. Eventually, we settled on an architecture consisting of 12 blocks, which included 1 initial convolutional block and 11 residual blocks, with 64 convolutional filters and 64 hidden units for the value head. We decided to train this revised neural network for over 40 hours to assess its progress. The 12-block version had a total of 842,621 parameters and showed improved performance compared to our previous attempts. However, when tested against strong opponents like other computer engines, it failed to consistently score wins.

In our final experiment, we decided to use a more powerful neural network architecture, which consists of 11 blocks (1 initial convolutional block and 10 residual blocks). In this configuration, we increased the number of convolutional filters, with each block utilizing 128 filters for the convolutional layers and 128 hidden units for the value head. This neural network had a total of 2,998,461 parameters. The results of our final experiment revealed that this more powerful neural network architecture achieved significantly stronger play than the 12-block version. As a result, we will focus our further discussions on the model with the 11-block version.

Self-Play

Our approach closely follows the methodology described in the AlphaGo Zero and AlphaZero papers. However, due to the relative simplicity of the 9×9 problem and its smaller action space, we limit the number of simulations to 200 per Monte Carlo Tree Search (MCTS) during each game. To distribute the workload for position evaluation during MCTS, we evenly utilize the available GPUs on the server.

To optimize computational resources, we have set the maximum game length to 162 moves ($9 \times 9 \times 2$). Once a game reaches this length, no further moves are allowed, and we compute estimated scores for both players to determine the winner. Similar to the AlphaGo Zero paper, we employ a resignation mechanism. We initialize the resignation threshold at -0.85, dynamically adjusting it to maintain a resignation false positive rate below 5%. To measure the false positive rate, we disable resign moves for 10% of the self-play games and let the agent play until natural termination (both players have passed in two consecutive moves or reaching the maximum game length). To avoid premature resignations, we only allow the agent to consider resigning after a certain number of steps

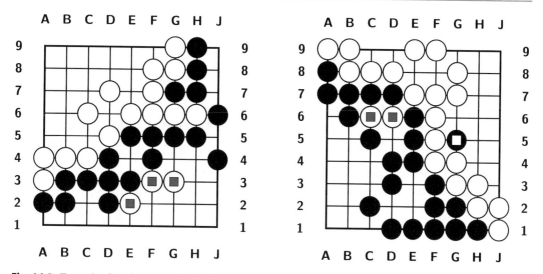

Fig. 14.9 Example of dead stones at the final state of a 9 × 9 Go game. In the left chart, the white stones at E2, F3, and G3 are considered dead because they have been completely surrounded by the opponent's stones, regardless of whether they have been captured. The right chart presents another example where dead stones are present from both players. The white stones at C6 and D6 are dead, as well as the black stone at G5

into the game, such as 30 or 40 steps. Additionally, we disable resignation for the first 50,000 self-play games, as the neural network's predictions are less accurate in the early stages.

After the game concludes, we score the games using a simplified version of the Tromp-Taylor method, incorporating a komi of 7.5 for the white player. It is important to note that these scores are estimates, as the Tromp-Taylor rules require players to capture the stones they believe to be dead before calculating the final scores. Failure to capture these stones results in scoring them as if they were alive. However, accurately detecting such dead stones at the end of a game is a challenging task for computer programs. For instance, in Fig. 14.9, dead stones in a Go game are marked with a small square and should be removed from the board before calculating the final score. Nonetheless, accurately detecting these dead stones is difficult due to the absence of explicit logic. This is why additional techniques such as simulation or neural network prediction are often employed to estimate scores.

MCTS Search

We have implemented the MCTS (Monte Carlo Tree Search) algorithm described in the AlphaGo Zero and AlphaZero papers, using identical configuration and hyperparameters. Our implementation involves 200 simulations per search, which proves sufficient for a 9 × 9 board. The exploration parameters, denoted as c_{base} and c_{init}, are set to 19652 and 1.25, respectively. We introduce Dirichlet noise to the root node at each step, with $\epsilon = 0.25$ and $\eta = 0.03$, as specified in the AlphaZero paper. During self-play games, we adjust the temperature in Eq. (14.5) to 1.0 for the initial 16 steps and 0.1 for subsequent steps. To enhance the efficiency of the search process, we utilize parallel MCTS search with eight parallel leaves, incorporating virtual loss as outlined in [15]. Furthermore, we reuse the subtree for the remaining steps within the same game.

However, it's important to note that in evaluation games, where the current model plays against the previous model, we don't add Dirichlet noise to the root node, and we do not reuse the subtree.

This decision is based on the fact that the two players are distinct, and thus they are likely to produce different predictions when presented with the same state.

Training

We strictly follow the standard procedure and mechanisms outlined in the AlphaGo Zero and AlphaZero papers to train our agent. Our training approach is based on pure self-play games collected from dedicated self-play actors, with no involvement of human-generated games or other compute engines or AI programs to bootstrap the learning process.

Given that we have 120 actors, we employ a single learner agent for the neural network update, which is a reasonable choice. To optimize our training, we utilize the SGD optimizer from PyTorch, setting the initial learning rate to 0.01, momentum to 0.9, and applying an L2 regularization of 0.0001. The learning rate follows a staircase pattern, progressively decreasing to 0.001 and 0.0001 at 200,000 and 400,000 training steps, respectively.

During each training step, the program randomly samples a batch of 1024 examples from the most recent 250,000 self-play games. To enhance the training process, we incorporate data augmentation techniques such as random rotation and flipping, which are the approach used in training the original AlphaGo agent. Additionally, we generate a checkpoint every 1000 training steps.

As 9×9 Go presents a smaller problem space compared to 19×19, we adjust our approach accordingly. For each checkpoint, we play a total of 5000 self-play games instead of 25,000 and limit the training data to the most recent 250,000 self-play games instead of 500,000. We create the initial checkpoint after the actors have played a total of 20,000 games, ensuring a diverse range of game states are explored before we start the learning process (e.g., update the neural network). During the entire training session, a total of one million self-play games were used for training the neural network.

Evaluation

During the training process, we assess the agent's performance at each new checkpoint. We employ the latest model to play against the previous checkpoint's model, similar to self-play. However, there are some distinctions: the two players are not identical (black represents the latest model, while white represents the previous model), and we do not require the agents to engage in additional exploration (no noise is added to the root node of the search tree). After every Monte Carlo Tree Search (MCTS), we always select the best move (the child node with the highest visit count). The Elo ratings are computed using a simplified implementation. Specifically, we adopt the equation from the original AlphaGo Zero paper to estimate the probability that player a will defeat player b. This equation is represented as $P(a \text{ defeats } b) = \dfrac{1}{1 + \exp\left(c_{elo}(e(b) - e(a))\right)}$, where the standard constant c_{elo} is equal to $1/400$. Additionally, we utilize a variant of the k factor depending on the player's current ratings. If the current ratings are below 2100, the k factor is set to 32. For ratings between 2100 and 2400, the k factor is 24, and for ratings above 2400, the k factor is 16.

We also evaluate the model's accuracy in predicting human gameplay using an evaluation dataset consisting of a total of 10,000 games and 620,000 positions. One challenge we encountered when evaluating 9×9 Go is the limited availability of games played by professional human players (professional dan ranks) on such a small board. In fact, our evaluation dataset comprises less than 3% of games played by Japanese professional Go players ranging from 1p to 9p. As a result, we had to collect the remaining games from the Internet, predominantly from the CGOS server. It is worth mentioning that while we refer to the evaluation dataset as human play games, it is highly likely that a substantial portion of the games were played by computer engines or AI programs rather than actual humans.

To ensure the high quality of the evaluation dataset, we only include games in which both players have an Elo rating of at least 2100. This is because an Elo rating of 2100 is approximately equivalent to the level of an amateur 1 dan [16]. Furthermore, although we filter games based on Elo ratings, it is still possible that the initial ratings were not accurate. Additionally, to ensure a diverse dataset, we limit the inclusion of games from the same player to a maximum of 200. Moreover, we only include unique games in the evaluation dataset. Two games are considered duplicates if they were played by the same two players, lasted the same number of steps, and ended with the same outcome (i.e., the same winner and the same points for the winner).

Results

We now present the training results of our experiment on a 9 × 9 Go board. As mentioned earlier, we utilized a neural network architecture consisting of 11 blocks for training. The training process took place on a single server equipped with 128 CPUs and 8 GPUs. The agent was trained for 200,000 steps, which required approximately 50 hours.

Figure 14.10 shows the training losses of the neural network. Both the cross-entropy loss for the policy and the MSE loss for the value show convergence as the training progresses. This is expected behavior, as for the AlphaZero algorithm, the training objective is solely focused on minimizing these losses, as the policy is improved through the MCTS search operator. Notably, the cross-entropy loss converges at a much faster rate, indicating that accurately predicting the game outcome is a considerably more challenging task.

Figures 14.11 and 14.12 illustrate the model's prediction accuracy based on an evaluation dataset containing 10,000 games and 620,000 positions. The creation of this evaluation dataset was previously described. Notably, after 100,000 training steps, the model achieves a 40% prediction rate for moves within the evaluation dataset. Moreover, we observe an enhancement in the model's ability to accurately predict human game outcomes.

Figure 14.13 shows the Elo ratings of the agent during training. These ratings were computed from evaluation games, where the latest model played as the black player against the model from the previous checkpoint as the white player. The results indicate that the agent's performance peaked at around 150,000 training steps. As training progressed, the performance began to diverge and did not

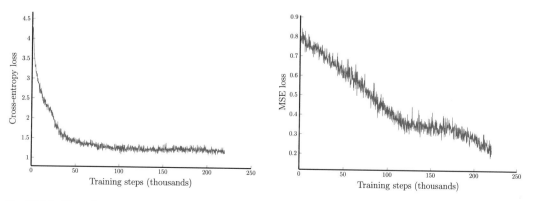

Fig. 14.10 The policy cross-entropy loss (left) and value MSE loss (right) during training of the AlphaZero agent on a 9 × 9 Go board

Fig. 14.11 Accuracy of the neural network's prediction on human moves. It measures the percentage of top 1 prediction accuracy, which is the ones with highest probability matching the actual human moves

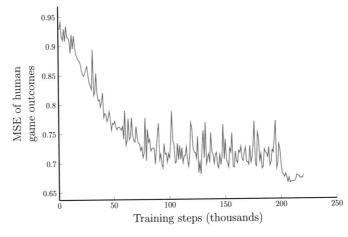

Fig. 14.12 MSE of the neural network's prediction on human game outcomes. It shows unscaled error between the value from neural network's prediction and the actual game outcome

recover. We suspect that the primary reason for this issue is the large learning rate. In our experiments, the initial learning rate was set to 0.01 and scheduled to decrease in a staircase manner to 0.001 and 0.0001 at training steps 200,000 and 400,000, respectively. However, since the agent likely reached a certain level of proficiency after 100,000 training steps, it may be more appropriate to reduce the learning rate earlier, such as decreasing it to 0.001 and 0.0001 at training steps 100,000 and 200,000, respectively.

To assess the true strength of our agent, we conducted a series of matches against a computer program known as CrazyStone [17]. Utilizing deep neural networks for move prediction, CrazyStone proved to be a formidable opponent. Our most advanced models (11-block version from training steps 154,000) displayed remarkable performance, winning 16 out of 20 games against the amateur 1 dan level opponent of CrazyStone. This outcome undeniably attests to the significant level of strength achieved by our agent, despite its comparatively low Elo ratings in the self-play evaluation games. However, it should be noted that our agent did not consistently secure victories when pitted against

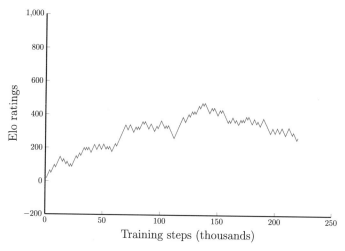

Fig. 14.13 Performance of the AlphaZero agent on a 9 × 9 Go board. Elo ratings were computed from evaluation games between the current (latest) model and the model from the previous checkpoint

the 2 dan level CrazyStone opponent. We find this outcome satisfactory, given the agent's relatively short training duration and limited exposure to a small number of games. It is important to highlight that we have not conducted extensive fine-tuning on the agent's training methods or hyperparameters.

In Fig. 14.14, we present the final state of the first 12 games played against the CrazyStone amateur 1 dan level opponent. Our agent played as white, while CrazyStone played as black. The scores were determined using the Chinese rules, with a komi of 7.5 for white.

14.5   Training AlphaZero on a 13 × 13 Gomoku Board

For readers who find Go too complex, either because they struggle with understanding its rules (which are indeed quite complex) or lack the computational resources to experiment with it, there is an alternative: training AlphaZero on simpler games like Gomoku. Gomoku, also known as Five in a Row, is a straightforward board game that shares some similarities with Go. In Gomoku, players take turns placing their stones on a 15 × 15 board, without strict restrictions on stone placement like in Go. The objective is simple: the first player to get five stones of the same color in a row, horizontally, vertically, or diagonally, wins the game. If the board is filled with stones and no player achieves a five-stone row, the game is a draw. Unlike Go, Gomoku does not involve pass or resign moves. The combination of similarities to Go and simpler rules makes Gomoku an ideal problem for testing and validating the AlphaZero agent. To put it simply, in the context of reinforcement learning, if Go were equivalent to Atari games, then Gomoku would be akin to the cart pole problem.

Gomoku can also be played on smaller boards, such as 9 × 9 and 13 × 13. However, on smaller boards, like 9 × 9, the game often ends in a draw if both players consistently make optimal moves. This is not the case on larger boards, where the first player (black) has an advantage and can always win if they make the best moves throughout the game. Therefore, it is not surprising to see games on larger boards typically concluding in around 20–30 moves when the black player is highly skilled, unless the game is played with additional specific rules. For instance, some variations of the game state that having more than five connected stones does not result in an automatic win, and certain positions may be restricted during the opening moves.

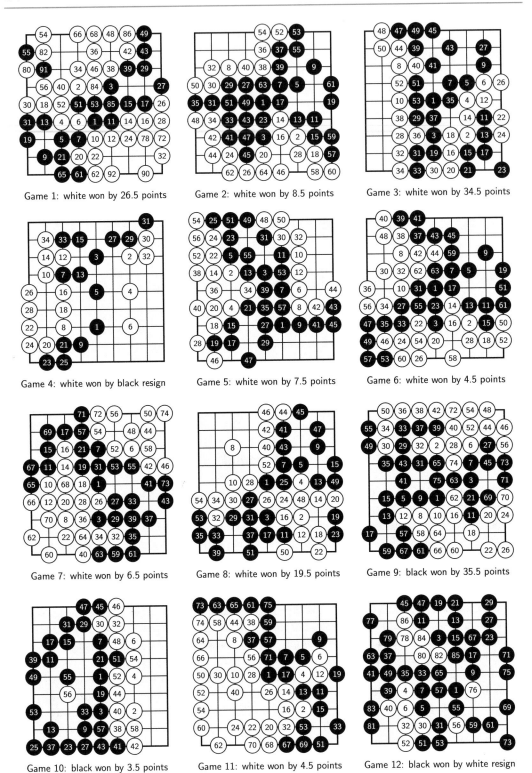

Fig. 14.14 In these games, our agent (playing as white) is against CrazyStone (playing as black). We used the 11-block model version and carefully selected the best model produced after 154,000 training steps. To provide a suitable challenge, CrazyStone was set at an amateur 1 dan level. The games were scored using the Chinese rules, with a komi of 7.5 points to white

There's an interesting fact about Gomoku: the standard AlphaZero agent performs quite well in normal cases. However, there is a particular scenario where the opponent attempts to win by connecting five stones at the edge of the board, especially during opening moves. Surprisingly, we have discovered that the agent fails to block the opponent in such cases, despite its overall strong gameplay, as shown in Fig. 14.15. This shortsighted behavior can be attributed to the agent's lack of exposure to such games during training. After all, it is highly unlikely for the self-play agents to engage in these specific moves.

To address this issue, we have opted for a slightly modified neural network architecture tailored to the unique nature of the Gomoku game, as mentioned earlier. In our experiments, we have found that employing a padding of three, rather than the usual one, for the initial convolutional block effectively resolves this edge problem in Gomoku.

In our experiment, we trained an AlphaZero agent to play freestyle Gomoku on a 13 × 13 board. Freestyle Gomoku means no additional strict rules; a string of same colored stones equal to or greater than five is considered a win.

We followed a similar configuration as outlined in the previous section, where we discussed training the AlphaZero agent on a 9 × 9 Go board. However, we made the following changes:

- Our neural network consists of 14 blocks, including 1 initial convolutional block and 13 residual blocks. We used 40 convolutional filters for all convolutional layers and 80 hidden units for the value head.
- To address edge case problems in Gomoku, we used a padding of three instead of one for the initial convolutional block in the neural network.
- Due to the larger action space on a 13 × 13 board, we increased the number of simulations per Monte Carlo Tree Search (MCTS) to 400 for both self-play and evaluation games.
- We trained the neural network using only the most recent 150,000 self-play games, with a batch size of 256. And the first checkpoint was created after 5000 self-play games.

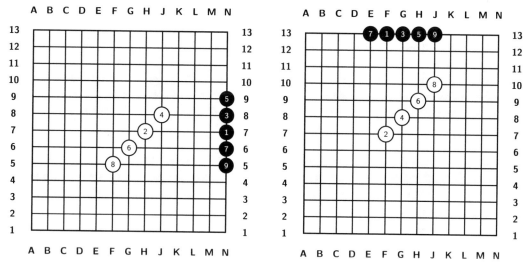

Fig. 14.15 Example of an edge case problem for a standard AlphaZero agent on a 13 × 13 Gomoku board, where the opponent (black) is trying to quickly win the game by connecting five stones on the edge of the board. In such cases, the agent (white) often fails to block the opponent. This issue persists even when the agent is trained for extended periods and exposed to a greater number of self-play games during training

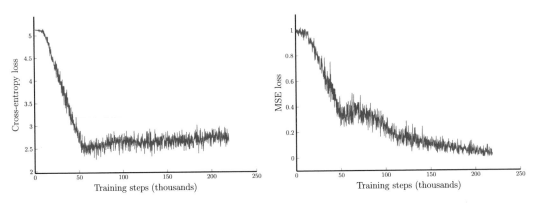

Fig. 14.16 The policy cross-entropy loss (left) and value MSE loss (right) during training of the AlphaZero agent on a 13 × 13 Gomoku board

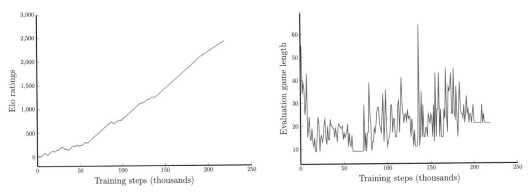

Fig. 14.17 Performance of the AlphaZero agent on a 13 × 13 Gomoku board. Left: Elo ratings were computed from evaluation games between the current (latest) model and the model from the previous checkpoint. Right: Length of the evaluation games

- Since Gomoku does not have a resign move, we did not include a resignation mechanism in the self-play games.
- Instead of using human play games to evaluate the neural network's prediction accuracy, we relied solely on Elo ratings computed from evaluation games between the latest model and the model from the previous checkpoint. This decision was made due to the limited availability of quality Gomoku game records.

Apart from the mentioned changes, the settings remained the same. The training was performed on a single server with 128 CPUs and 8 GPUs. We collected 5000 games per checkpoint and generated a checkpoint every 1000 training steps. The training process involved one million self-play games and took 16 hours to complete 200,000 training steps.

Figure 14.16 shows the training losses of the neural network. As training progresses, it becomes evident that both the policy and value losses exhibit a consistent decrease. This behavior closely resembles that observed during training on a 9 × 9 Go board.

Figure 14.17 shows the Elo ratings of the agent computed during evaluation games, where the latest model plays against the model from the last checkpoint. The agent consistently achieves remarkably high Elo ratings, confirming our previous explanation. In Gomoku, particularly on larger board sizes, the black player holds a significant advantage as it can secure a win by consistently playing the best move. Alongside the Elo ratings, we also provide valuable insights into the length of the evaluation games, which serves as an additional metric for evaluating the agent's performance. For Gomoku, the duration of the evaluation games becomes a crucial factor to consider. Typically, if the agent lacks the ability to effectively block its opponent, the game concludes within approximately ten moves, especially during the initial training phase. However, as the agent's learning progresses, it acquires the skill to successfully block the opponent's moves, which often means longer and more engaging games.

Nevertheless, it is important to note that the results significantly differ on smaller board sizes, such as 9×9. In such cases, the Elo ratings tend to plateau after a certain period of training. Additionally, the length of the evaluation games often extends up to $9 \times 9 = 81$ moves. This phenomenon occurs because, on a smaller board, the first mover (black) is constrained by limited space and cannot execute intricate moves. Consequently, the game often concludes with a fully occupied board before either player can secure a victory.

Figure 14.18 illustrates the training progress of the agent on a 13×13 Gomoku board, showcasing its performance in evaluation games. Remarkably, the agent demonstrates significant learning within a short span of 18,000 training steps, successfully connecting five stones horizontally to secure a victory. By approximately 28,000 training steps, the agent exhibits the ability to block the opponent's moves in simple scenarios. Around 68,000 training steps, a notable breakthrough occurs as the agent discovers an effective strategy of winning the game by connecting stones diagonally. Furthermore, it swiftly learns to counter the opponent's attempts at achieving a diagonal victory. The agent's prowess continues to improve, evident around the 148,000 training step mark, where it consistently displays strong play. In fact, the agent becomes virtually unbeatable, demonstrating optimal moves that render white's best efforts futile. Even with further training, the agent consistently exhibits this exceptional behavior.

14.6 Summary

In the final chapter of this book, we delved into the renowned AlphaZero algorithm, which has achieved remarkable success in reinforcement learning (RL) by defeating world champions in complex and challenging strategic games.

We began by explaining why model-free reinforcement learning falls short in solving such difficult problems and why a model-based approach is necessary for effective planning. Subsequently, we explored the limitations of traditional search algorithms when confronted with games like Go, which have enormous search trees.

A significant portion of the chapter was dedicated to elucidating the Monte Carlo Tree Search (MCTS) algorithm, which lies at the core of the AlphaZero agent. We provided an overview of the general MCTS algorithm and described its different phases during a search session. Furthermore, we demonstrated how the MCTS algorithm can be applied to two-player zero-sum games.

Having established a solid understanding of the MCTS algorithm, we proceeded to explain the AlphaZero algorithm. Our focus remained exclusively on AlphaZero due to its generality, power, and relatively straightforward implementation compared to its successors. We elucidated the state representation for the game of Go and outlined the architecture of the neural network employed in AlphaZero.

Next, we introduced the modified MCTS algorithm used by AlphaZero. Unlike the general MCTS algorithm, AlphaZero does not simulate the game until the end during rollout; instead, it employs the neural network's predictions as estimated game outcomes. This approach significantly speeds up the process compared to full simulation.

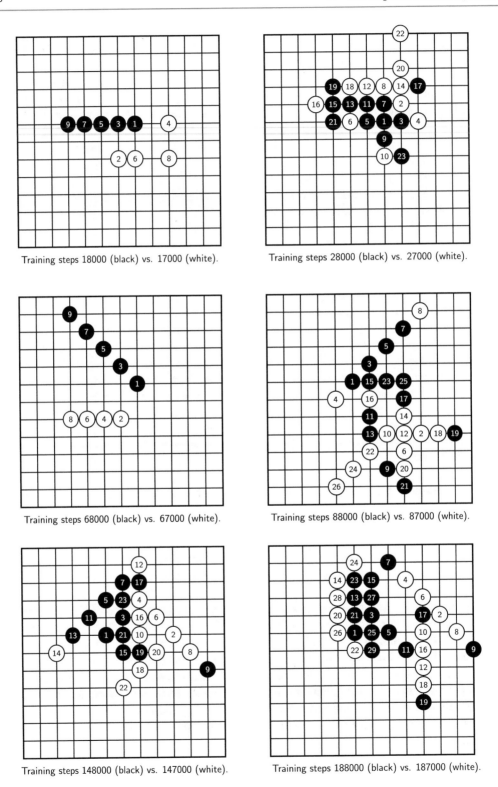

Training steps 18000 (black) vs. 17000 (white).

Training steps 28000 (black) vs. 27000 (white).

Training steps 68000 (black) vs. 67000 (white).

Training steps 88000 (black) vs. 87000 (white).

Training steps 148000 (black) vs. 147000 (white).

Training steps 188000 (black) vs. 187000 (white).

Fig. 14.18 Evaluation games to show the different stages of the agent during training on a 13×13 Gomoku board, where we use the latest model (black) to play against the model (white) from the last checkpoint. This provides evidence of the agent's acquisition of diverse winning strategies over the course of the training sessions. Ultimately, the agent reaches a strong level of play, demonstrated by its ability to make moves that render white incapable of winning, despite white's best efforts to block the opponent's progress

In conclusion, we presented the results of our experiments training the AlphaZero agent on a 9 × 9 Go board, which requires fewer computational resources and a smaller budget compared to the traditional 19 × 19 Go board. Remarkably, our trained agent achieved an amateur 1 dan level of play despite only 50 hours of training, one million games, and no fine-tuning of the training process or hyperparameters.

For individuals who may not have access to the necessary resources or those who may not be interested in Go, we have additionally conducted a comparable experiment utilizing a 13 × 13 Gomoku board. Although Gomoku is a less complex game compared to Go, this experiment unveils interesting insights into the agent's performance. It sheds light on the difficulties encountered by conventional neural network architectures when dealing with edge cases in Gomoku while also showcasing how a simple modification to the neural network architectures can rectify this issue.

By conducting extensive experiments and analyses, we have drawn our final conclusions. However, we humbly acknowledge that our understanding may not be complete, and there may be areas we have not fully explored. Like an iceberg, the depths of RL are vast, and this book represents only a fraction of what lies beneath the surface.

For example, as we near the completion of this book, an exciting development has emerged from DeepMind's recent publication. They have unveiled a new RL agent called AlphaDev [18], which is a modified version of the AlphaZero agent discussed in this chapter. According to their claims, the AlphaDev agent has made significant breakthroughs in discovering faster sorting algorithms. This announcement has generated tremendous enthusiasm, as it holds the promise of transforming our understanding of computation and coding. Furthermore, it provides compelling evidence that RL and the AlphaZero approach can be successfully employed to tackle real-world challenges.

As we reach the end of this book, we would like to extend our deepest gratitude for your unwavering interest and express our hope that this comprehensive exploration of reinforcement learning (RL) fundamentals, mathematics, and algorithms has proven to be both enlightening and enjoyable for you.

Throughout our journey, we humbly acknowledge that we may have faced challenges and made mistakes along the way. We are committed to continuous improvement and value your feedback to help us refine and enhance the content. Please don't hesitate to reach out with any concerns or corrections.

We encourage you to continue your exploration of the ever-expanding field of RL. May your continued journey lead you to new insights and groundbreaking discoveries as you delve deeper into this thrilling realm of research. Above all, we sincerely hope that this book has provided you with a solid foundation and has served as an inspiration for your future endeavors.

References

[1] CWI. 66th NHK Cup. https://homepages.cwi.nl/~aeb/go/games/games/NHK/66/index.html, 2018.
[2] John Tromp and Gunnar Farnebäck. Combinatorics of go. In H. Jaap van den Herik, Paolo Ciancarini, and H. H. L. M. (Jeroen) Donkers, editors, *Computers and Games*, pages 84–99, Berlin, Heidelberg, 2007. Springer Berlin Heidelberg.
[3] John von Neumann and Oskar Morgenstern. *Theory of Games and Economic Behavior*. Princeton University Press, 1944.
[4] IBM. Deep blue. www.research.ibm.com/artificial-intelligence/projects/deep-blue/, 1996.
[5] Markus Enzenberger. The integration of a priori knowledge into a go playing neural network. 1996.
[6] Richard S. Sutton. Learning to predict by the methods of temporal differences. *Machine Learning*, 3(1):9–44, Aug 1988.
[7] Sylvain Gelly and Yizao Wang. Exploration exploitation in Go: UCT for Monte-Carlo Go. 12 2006.
[8] Sylvain Gelly and David Silver. Achieving master level play in 9 × 9 computer Go. *AAAI*, 8:1537–1540, 2008.

[9] Cameron B. Browne, Edward Powley, Daniel Whitehouse, Simon M. Lucas, Peter I. Cowling, Philipp Rohlf-
     shagen, Stephen Tavener, Diego Perez, Spyridon Samothrakis, and Simon Colton. A survey of Monte Carlo tree
     search methods. *IEEE Transactions on Computational Intelligence and AI in Games*, 4(1):1–43, 2012.

[10] Peter Auer, Nicolò Cesa-Bianchi, and Paul Fischer. Finite-time analysis of the multiarmed bandit problem.
     Machine Learning, 47(2):235–256, May 2002.

[11] David Silver, Aja Huang, Chris J. Maddison, Arthur Guez, Laurent Sifre, George van den Driessche, Julian
     Schrittwieser, Ioannis Antonoglou, Veda Panneershelvam, Marc Lanctot, Sander Dieleman, Dominik Grewe,
     John Nham, Nal Kalchbrenner, Ilya Sutskever, Timothy Lillicrap, Madeleine Leach, Koray Kavukcuoglu, Thore
     Graepel, and Demis Hassabis. Mastering the game of Go with deep neural networks and tree search. *Nature*,
     529(7587):484–489, Jan 2016.

[12] David Silver, Julian Schrittwieser, Karen Simonyan, Ioannis Antonoglou, Aja Huang, Arthur Guez, Thomas
     Hubert, Lucas Baker, Matthew Lai, Adrian Bolton, Yutian Chen, Timothy Lillicrap, Fan Hui, Laurent Sifre,
     George van den Driessche, Thore Graepel, and Demis Hassabis. Mastering the game of Go without human
     knowledge. *Nature*, 550(7676):354–359, Oct 2017.

[13] David Silver, Thomas Hubert, Julian Schrittwieser, Ioannis Antonoglou, Matthew Lai, Arthur Guez, Marc Lanctot,
     Laurent Sifre, Dharshan Kumaran, Thore Graepel, Timothy Lillicrap, Karen Simonyan, and Demis Hassabis.
     Mastering chess and shogi by self-play with a general reinforcement learning algorithm, 2017.

[14] Julian Schrittwieser, Ioannis Antonoglou, Thomas Hubert, Karen Simonyan, Laurent Sifre, Simon Schmitt, Arthur
     Guez, Edward Lockhart, Demis Hassabis, Thore Graepel, Timothy Lillicrap, and David Silver. Mastering atari,
     Go, chess and shogi by planning with a learned model. *Nature*, 588(7839):604–609, Dec 2020.

[15] Guillaume M. J. B. Chaslot, Mark H. M. Winands, and H. Jaap van den Herik. Parallel monte-carlo tree search. In
     H. Jaap van den Herik, Xinhe Xu, Zongmin Ma, and Mark H. M. Winands, editors, *Computers and Games*, pages
     60–71, Berlin, Heidelberg, 2008. Springer Berlin Heidelberg.

[16] FICGS. Wiki - English Go ranks and ratings. www.ficgs.com/wiki_en-go-ranks-and-ratings.html.

[17] Remi Coulom. Crazy stone. www.remi-coulom.fr/CrazyStone/.

[18] Daniel J. Mankowitz, Andrea Michi, Anton Zhernov, Marco Gelmi, Marco Selvi, Cosmin Paduraru, Edouard
     Leurent, Shariq Iqbal, Jean-Baptiste Lespiau, Alex Ahern, Thomas Köppe, Kevin Millikin, Stephen Gaffney,
     Sophie Elster, Jackson Broshear, Chris Gamble, Kieran Milan, Robert Tung, Minjae Hwang, Taylan Cemgil,
     Mohammadamin Barekatain, Yujia Li, Amol Mandhane, Thomas Hubert, Julian Schrittwieser, Demis Hassabis,
     Pushmeet Kohli, Martin Riedmiller, Oriol Vinyals, and David Silver. Faster sorting algorithms discovered using
     deep reinforcement learning. *Nature*, 618(7964):257–263, Jun 2023.

Index

A

Action, 10
Activation functions, 133–136
Actor-Critic
  Ant locomotion task, 203
  Atari visual complex game Pong, 195
  baseline function, 189, 190
  entropy loss to encourage exploration, 195
  Humanoid locomotion task, 203
  neural network architecture, policy network, 186
  neural networks, 190–192
  on-policy methods, 206
  policy gradients, 189, 190
  vs. REINFORCE with baseline, 192, 193
  solving reinforcement learning problems, 190
  state-action value function, 190
  value-based and policy-based methods, 191, 196
Adaptive KL penalty, 216–219
Advantage function, 169, 208
  dueling network architecture, 170, 171
  neural networks, 169
  and reinforcement learning, 169
Agent, 9
Agent-environment boundary, 10
Agent-environment loop
  action, 10
  agent, 9
  environment, 7–8
  model, 10–11
  policy, 10
  reward, 9
  state, 8
Agent's policy, 207
AlphaDev, 279
AlphaGo, 4
  MCTS search algorithm, 260
  neural network architecture, 256, 259
  training process, 255
AlphaGo Zero, 5
AlphaZero algorithm
  architecture, 258
  capture-based games, 246
  dynamics function and reward function, 247
  expansion and evaluation, 261

function, 265
game of Go, 245–246
Go board, 257
on a 9 × 9 Go Board, 267
  CrazyStone, 272–274
  distributed RL architecture, 267
  Elo ratings, 271–273, 277
  evaluation, 270–271
  MCTS search, 269–270
  neural network architecture and state
    representation, 268
  policy cross-entropy loss, 271
  self-play, 268–269
  training, 270
on 13 × 13 Gomoku Board, 273
  advantage, 273
  Elo ratings, 276
  neural network architecture, 275
  self-play games, 275
  training, 276–278
IBM's Deep Blue, 247
MCTS search, 260, 263
neural network, 256, 259, 265
performance of, 276
policy network, 265
reinforcement learning, 245
simulation and planning process, 246
training, 255, 264
Alternative Bellman equations, value
    functions
  Bellman equation, 42
  dynamics function P, 42
  environment state, 41
Ant, 199
Arcade learning environment (ALE), 5
Artificial intelligence (AI)
  breakthroughs, 3
  in games
    Atari 2600, 3–4
    Go, 4–6
Atari 2600, 3–4
Atari games
  DQN (*see* DQN, Atari games)
Automatic differentiation, 141

Printed in the United States
by Baker & Taylor Publisher Services